High Performance

Audio Power Amplifiers

for music performance and reproduction

Newnes
An imprint of Butterworth-Heinemann
Linacre House, Jordan Hill, Oxford OX2 8DP
225 Wildwood Avenue, Woburn, MA 01801–2041
A division of Reed Educational and Professional Publishing Ltd

Ɋ   A member of the Reed Elsevier plc group

OXFORD   BOSTON   JOHANNESBURG
MELBOURNE   NEW DELHI   SINGAPORE

First published 1996
Reprinted with revisions 1997, 1998

British Library Cataloguing in Publication Data
A catalogue record for this book is available from the British Library

ISBN 0 7506 2629 1

Typeset by P.K.McBride, Southampton
Printed and bound by Antony Rowe Ltd, Eastbourne

High Performance
Audio Power Amplifiers

for music performance and reproduction

Ben Duncan, A.M.I.O.A., A.M.A.E.S., M.C.C.S

*international consultant in live show, recording & domestic
audio electronics and electro-acoustics.*

 Newnes

Errata

- p. 7, Fig. 1.1, line 10 should read: 'And how sensitivity around 3.5 kHz becomes increasingly acute at high levels.'
  MAF = Minimum Audible Field.

- p. 141, Fig 4.53, labels 'TR 1, 2, 3' are missing. TR3 at top. TR1 lowermost.

- pp. 156, 158, 160 each mention of 'shunt' and 'series' to be transposed and vice-versa.

- p. 242, 8 lines from bottom, '485 v' to read '457 v'.

- p. 264, line 2 should read: 'Power in each successive octave band doubles (+3 dB/octave up-slope)'.

- p. 264, line 8 should read: 'Analogous to pink or reddish light, the energy and power per octave is equal in pink noise'.

- p. 268, line 3, '300 million' should read '100 million'.

- p. 296, line 7, delete 'minimum'. Line 14, change 'linear' to 'logarithmic'.

Foreword

Ben Duncan is one of those rare individuals whose love and enthusiasm for a subject transcends all the usual limits on perception and progress. In fact, without the few people of true independent spirit, progress in the world would be swamped by the xylocaine of vested interest, narrow attitude and corporate monoculture.

Amongst my early experiences of Ben Duncan's thinking, many years ago, were his contention that electronic components have qualitative audio properties and his recommendation that we listen to the sound of capacitors of various dielectrics. The outcome was the exclusive use of polypropylene capacitors in all Turbosound's passive hi-pass networks. This is not only illustrative of the depth to which the man goes, but also his extensive seen and unseen influence on the whole audio community. He is an holistic thinker and I believe there are very few things in the Universe that he has not, at one time or another, considered having an effect on audio quality. Does he keep his flights of fancy and strokes of brilliance to himself? Not one bit of it. He communicates compulsively and in large quantities as anyone who has followed the general audio press for the last dozen or so years will tell you.

A memorable early experience of power amplifiers was with the then relatively new transistor variety powering a P.A. I had built for the Pink Fairies, that was at the original Glastonbury in 1971. After the sixth failure of an HH TPA100, for no apparent reason, I was running out of working stock. On sitting down to consider the hopeless situation it became worse when I found the live soldering iron. My next immediate thoughts were about a change of career. Anyhow, the point of this sad little tale is that in those days power amplifiers were absolutely horrible things because despite the fact that they had somewhat puny voltage swings they were, nevertheless, always blowing up at the slightest opportunity and particularly in the hour before show time. These days things have progressed a long way and sound system operators bask in the luxury of equipment that is almost indestructible and capable of audio quality usually associated with esoteric hi-fi as well as delivering arc welding levels of power.

I am extremely grateful to Ben that he has undertaken the Herculean task of collating all the relevant facts on, and to do with, power amplifiers ranging from the in depth assessment of household mains to determinations as to whether it actually sounds any good. The breadth of the book enables an average human to purchase or design power amplifiers knowing that all relevant information is at their disposal and as such this book should be considered a positive contribution to the sum total of mankind. I hope it has a similar effect on his bank balance.

Tony Andrews, Hoyle, Surrey

March 1996

Contents

Preface

There has never been a book like this one, in its interleaving of electronics and audio, engineering ideality, and musical and practical reality. There haven't actually been many books dedicated to audio amplifiers, period. On any level.

Beginning with the electronics, amplifiers for driving loudspeakers are actually rather hard. So hard, that after 75 years, there is not a lot of convergence – not compared to say washing machines, which are similarly old. In spite of a century of consumerism, Music remains on a higher, primal level that interfaces with levels of human perception that precede and can outstep the logical. There have been many gifted minds at work in amplifierland, but they haven't had even half the answers. Many have come unstuck, or lost the plot completely, confused by mathematical catastrophes in audio's higher dimensions.

Audio power amplifiers are unsung key tools in the immense growth of human mass-attuned consciousness during the 20th century. Imagine amplifiers were dis-invented. Without speaker-driving-devices, there wouldn't be hi-fi systems, radios or (wild applause) TVs. There wouldn't be PA systems, and there wouldn't be any festivals bigger than village garden fetes, or at best 2000 seater amphitheatres. There wouldn't be cinemas and recording studios, and no solid-bodied or electronic instruments. The huge emotional and psychic amplification, through music (its *own* capabilities also *vastly* expanded by electronic amplification, recording and processing) and the sound-tracked cinema (and video offspring) and their mass broadcast and affordability, that has made the 20th century vibrant like none before it, would be naught. The human world without good amplifiers – or any audio amplifiers – *would be far less linked to spiritual and emotional heights* – and probably not a lot quieter.

'*High Performance*' means that the book does not cover equipment where makers knowingly make significant 'corner cuttings' that degrade audio quality, reliability and utility, particularly so called 'consumer' and much so called 'M.I' equipment. This seems a more natural dividing line than any of the more common ones, like pro.*vs*.domestic. Everyone who is *serious* about music, wants much the same things, however much it needs adapting to suit their particular environment. What 'High Performance' *does not* mean is any particular price or other label of exclusivity. The amplifiers covered in this book could cost (at 1996 prices) £135 or £13,500. They could deliver 25w or 2500w, be used in the home, in the studio, a stadium, or in a field, so long as their aim or suitability is to permit the faithful reproduction of some kind of music and all its nuance.

Across this book, you will discover that the contents' focus purposefully veers from a wide pan across the most global, broad-minded considerations including amplifiers employing valves (tubes) and/or 'zero'-feedback, wherein generalities are enough, through to a narrow concentration on the majority of modern transistor ('solid state') power amplifiers with global NFB, where the detail is magnified. Consideration of esoteric types has been exponentially tapered to avoid the book expanding to infinite volumes, while dove-tailing with the burgeoning number of new and reprinted books about alternative valve amplifier technology.

More than any other you will see, this book fills-in and connects-up 101 missing details about audio power amplifiers.

Ben Duncan, Oxbows, Co. of Lincoln, England

January 1996

To Amy and Jake,
and to the many gifted musicians,
singers and producers who have
inspired my work

Acknowledgments

I am indebted to the following makers, designers, theoreticians, teachers, mathematicians, technologists, sound engineers, musicians, friends and colleagues – many of them 'masters of sound' – for their input either during the production of this book, or at some earlier juncture; and where asterisked (*), for their kind assistance in proof-reading, criticising or enhancing particular sections. They should not however be held liable for any errors or omissions! Ξ indicates valued assistance with the supply of pictures.

In the UK:

* Andy Salmon, MS&L.
Bill Bartlett.
Bill Huston, Aanvil Audio.
* Bruce Hofer, Audio Precision.
Ξ Charlie Soppelsa, Rauch Precision.
Chris Hales, C-Audio.
* Chris Marshman, YEC.
Ξ Danny Cooklin, Turbosound.
David Dykes.
* Dave Newson.
* David Cole, Turbosound.
David Heaton, Audio Synthesis.
Ξ David Neale, BSS Audio.
Prof. Malcolm Omar Hawksford,
  University of Essex, Dept. of
  Electronics.
* Duncan Werner, Music technology
  Course leader, & colleagues,
  University of Derby, Dept. of
  Electronics.
Eddie Cooper, Audio Precision.
* Gary Ashton, Fuzion.
Graham Lust.
Harry Day, Reddingwood Electronics.
Heather Lane, AES .
Ξ Ian McCarthy, MC² Audio.
Prof Jack Dinsdale
Ξ Jerry Mead, Mead & Co.
John Hurd.
* John Newsham, Funktion One.
* Lawrence Dickie.
Dr. Keith Holland, ISVR.

Keith Persin, Profusion.
Ken Dibble, Ken Dibble Associates.
* Matt Dobson, Coastal Acoustics.
* Mark Dodd, Celestion.
* Martin Colloms, Colloms
  Electroacoustics.
Martin Rushent.
Michael Gerzon.
* Neil Grant, Harris-Grant Associates.
Norma Lewis, senior assistant, BDR.
Norman Palmer, De Aston.
Paul Freer, Lynden Audio.
Paul Holden, ATMC.
Paul Jarvis.
Paul McCallum, Wembley Loudspeakers.
Paul Reaney.
Peter Baxandall.
Peter Brotzman, Britannia Row Prods.
Phil Newell.
Phil Rimmer.
* Richard Vivian, Turbosound.
* Richard Dudley, B&W Loudspeakers.
* Russ Andrews, RATA.
Stan Gould, BSS Audio.
Stephen Woolley, Fender Electronics.
Ξ Steve Harris, Hi-Fi News.
Steve Smith, Sound Department.
* Terry Clarke, MC² Audio.
* Tim Isaac, ATC.
* Toby Hunt, Funktion One Research.
* Tony Andrews, Funktion One.
Vince Hawtin, fanatic.

In the USA:
Adam Savitt-Maitland.
Bill Steele, Spectrum Software.
Bob Carver.
Ξ Brian Gary Wachner, BGW.
Dan Parks, NSC.
Ed Dell, Audio Amateur Publs.
Harvey Rosenberg, NYAL.
Joe Buxton, Analog Devices.
John Atkinson, Stereophile.
John Szymanski.

Ξ Patrick Quilter, QSC.
Roger Cox, Fender.
Skip Taylor & Larry Hand, Peavey.
Ξ Tim Chapman, Crest Audio.
* Walt Jung, Analog Devices.

Overseas
Colin Park, ARX Systems, Australia.
Conrad Eriksen, Norway.
Tommy Jenving, Sweden.

Acknowledgment of other picture sources and production services:

A.Foster & Sons, British Standards Institution, Canford Audio, Citronic, Crown Inc., Electronics World, Hi-Fi News & Record Review, Lincolnshire County Library Service, MAJ Electronics, National Physical Laboratory, Peter Gilyard-Beer, Pro Mon Co, SB, SoundTech, Stereophile magazine, Studio Sound & Broadcast Engineering, Spectrum Software.

Front cover

Upper picture shows Brittania Row Productions' amplifier racks at work backstage during Pink Floyd's 1994 World Tour. Each contained four BSS Audio EPC-780, rated (per rack) at 10kW, with drive split between four frequency bands, to power Turbosound *Floodlight* and *Flashlight* horn-loaded speaker systems.

Lower picture shows MC^2 Audio model MC-650, which has microprocessor controlled auxiliary and protective functions, and it widely used in recording studio control rooms, as well as for PA systems.

Rear cover

Pass Labs' Aleph 5 is a modern, high-end domestic amplifier, working in single-ended Class A, with an absolute minimum signal path, comprising only two MOSFET gain stages.

Method of Capitalisation

The names of scientific units are capitalised broadly in accordance with the SI (Système International d'unités) convention, but there are exceptions when English Grammar, typographic values and visual communication take precedence.

Units that are named after people are proper nouns and as such, should be capitalised: Amp(ère), dB (deci-Bell), Farad (Faraday), Henry, Hertz, Joule, Ohm, Volt (Volta), Watt. The SI convention (due to French origins) contrarily requires abbreviated forms to be capitalised. However, for typographic and visual purposes, the symbols 'V' and 'W' are usually set lower case when associated with numbers on their own: 3v, 100w, as the capitalised forms V or W are otherwise too dominant.

Ampère is mostly spelt out in full – as well as being capitalised – in this book, to avoid confusion with amplifiers, since the two words – which will be encountered frequently – may both be abbreviated to *amp* and *amps*.

System of presentation

To write in depth *and* breadth about audio power amplifiers, many special terms are required, from disparate disciplines. Many of these are *italicised* when initially introduced, and most are translated or defined in the Glossary. If an unfamiliar word appears, look there. To ease reader's immediate comprehension, occasional terms have been briefly clarified or explained parenthetically (in brackets).

References appear at chapter ends, ideally numbered in the order in which they first appear in the chapter. 'Further Reading' then lists apposite literature that is not specifically cited, as it is either incidental or else so germane that it would be referred to throughout that chapter.

Differences in technical terminology and practices outside of Britain, particularly those in the USA, are acknowledged (e.g. in brackets) where possible. Throughout this book, levels in dB referred to maximum output or clip reference point (r) will be cited as '–dBvr' if voltage (v), also abbreviated dBr, where v is understood. Less often dBwr will be used if a power delivery level is being referred to full power (w) delivery.

High performance audio power amplifiers is a long-winded description if repeated too often. Yet it alone is what is being focused on, a point not to be forgotten when the subject of the many sentences to follow is sometimes abridged, down to *audio power amplifiers* or just plain *audio amplifiers* or *power amplifier*, throughout this book.

In places, familiarity with the capabilities of the PC (Personal Computer) is assumed throughout, as it is today the *de facto* workhorse in the world of all serious engineering, and much else. The creation and/or processing of nearly every squiggle of ink in this book was created with a trio of them *Hercule*, *Hilary* (after Sir Edmund) and *Adelos*.

References and Further Reading

Books are distinguished from journals by having no months associated with the year, and if available, the ISBN is cited.

Where journals span two months, the first month is cited.

Where journals have floating publication dates, issues are referred to by nominal quarters, e.g. Q2 means 2nd issue in that year.

Anonymous, faceless publications are placed under 'Nameless'.

Abbreviations employed

AN	Application Note
EPD	Electronic Product Design (UK)
EPR*	Electronic Power Review (UK)
ETI	Electronics Today International (UK, Austr.)

EW, EW+WW Electronics World, formerly *Electronics & Wireless World*, formerly
           Wireless World, formerly *The Marconigraph*, (UK).
HFN, HFN/RR Hi-Fi News & Record Review (UK).
H&SR*   Home & Studio Recording (UK).
ISBN   International Standard Book No.
JAES   Journal of the Audio Engineering Society.
LCCCN   Library of Congress Cataloguing card No. (USA)
L&SI   Lighting & Sound International (UK).
Q   Annual Quarter (1-4, or more).
S&VC   Sound & Video Contractor (USA).
TAA   The Audio Amateur (USA).
WW   Wireless World (UK). See EW, above.
Journals believed to be no longer published.

Publications

Back issues of, or photocopies from, most journals cited, can be obtained by contacting
the publishers.

Journals

Audio, 1633 Broadway, NY 10019, USA.
Audio Engineering Society, AES, Room 2520, 60 East 42nd St, NY 10165 2520, USA. Or
local branches worldwide.
Electronic Industries Association, EIA, 2001 Eye St, N.W., Washington, DC 20006.
Elektor - see TAA, below.
Electronics Today International, ETI, Nexus House, Boundary Way, Hemel Hempstead,
Herts, HP2 7ST, UK.
Electronics World (EW, formerly WW), Quadrant, Sutton, Surrey, SM2 5AS, UK.
Glass Audio - see TAA, below.
Hi-Fi News & Record Review (HFN), Link House, Dingwall Ave, Croydon, Surrey, CR9
2TA, UK.
Institute of Electrical Engineers (IEE), Savoy Place, London, WC2, UK.
Institute of Acoustics (IOA), PO Box 320, St.Albans, Herts, AL1 1PZ, UK.
Lighting & Sound International (LSI), 7, Highlight House, St.Leonards Rd, Eastbourne,
East Sussex, BN21 3UH, UK.
Sound & Video Contractor (S&VC), 9800 Metcalf, Overland Park, KS, 6621-22215, USA
Speaker Builder - see TAA, below.
Stereophile, 208 Delgado, Sante Fe, NM 87501, USA.
Studio Sound & Broadcast Engineering, 8th Floor, Ludgate House, 245 Blackfriars Rd,
London, SE1 9UR, UK.
The Audio Amateur (TAA), PO Box 576, Peterborough, New Hampshire, NH 03458 0576,
USA. (Same address for SB, GA, Elektor).

Newsletters – concerning software used to create the graphs in this book:

Audio Precision (test equipment), Audio Precision, PO Box 2209, Beaverton, Oregon, 97075-
3070, USA.
Spectrum News (MicroCAP simulation software), 1021 South Wolfe Road, Sunnyvale, CA
94086, USA.

Introduction and fundamentals

"Why do rhythms and melodies, which are composed of sound, resemble the feelings; while this is not the case for tastes, colours or smells ?"

Aristotle

1.0. What are audio power amplifiers for ?

In sound systems, power amplifiers are the bridge between the loudspeakers and the rest of *any* sound system. In everyday parlance, 'Audio Power Amplifier' gets abbreviated to 'amplifier' or 'amp'. But *all* audio power amplifiers (other than those that drive vinyl disc cutter-heads) are really *'loudspeaker drivers'*. The definition is global if earpieces and headphones are included.

Sometimes, amps are combined with the speakers, forming 'powered speakers'; or they may be packaged with the preceding equipment functions, e.g. as in domestic 'integrated' hi-fi amplifier + preamplifier, or a band's 'mixer-amp'.

1.1 What is the problem ?

A sometime bass player, and foremost international writer on, and reviewer of, audio quality explains *"If you read electrical engineering textbooks, you're left with the impression that the audio amplifier is a well-understood, lowly sort of beast, compared with radio-frequency circuits. ... All I want is an amplifier that performs its simple task in an accurate, musically honest manner. I can't think, however, of an amplifier which can do this without crapping out at high levels, or obscuring low-level detail, or flattening the soundstage ... , or changing the (tonal) balance of the speaker ... , or adding metallic sheen, or loosing control of the speaker's bass so it booms, or gripping it so tight that music looses its natural bloom, or doing something - whatever it is - that destroys the music's sense of pace."* [1]

1.2 What is audio ?

Throughout this book, *audio*, music or *programme* (*program* in US) are all short-hand for 'music *re*production, or production or live amplification'. Other than the sounds of acoustic instruments and the outputs of electronic ones, *music programme* may include speech, gasps, birdsong, street sounds, sonar beeps, toad clucks, or any other sounds whatsoever that are employed for musical purposes.

'Sound' is synonym for *audio* in modern usage, at least in the context of sound system' and 'sound reproduction'.

1.3 What's special about audio ?

The amplifiers in this book are about reproducing music but they are equally applicable to the amplification or reproduction of speech, where the highest qualities and nuances of the living voice are of importance. These include religious and spiritual ceremonies, plays, poetry and chant.

What's special about all of these – all really variants of music – is that they involve sounds that make direct contact with powerful, pre-verbal centres of the human mind, affecting conscious states, ultimately in the higher direction of ecstasy [2] [3]. Music is not just '*the art that the other arts aspire to*', but it is in there with the highest, most transcendent mind events that can be experienced by human beings.

The human eye and ear are both amazing for the range of levels, or dynamic range, over which they can resolve differences and operate without damage. The range in both cases is up to at least 1000 million times (160dB). The eye can perceive a single photon. The ear can perceive the result of air moving over a distance equal to the radius of a hydrogen atom, the smallest building block of all matter in the cosmos.

Of the two, listening is humankind's most wideband sense, spanning 10 octaves. Whereas *everything we see* is compressed into just *one* octave of light!

The reproduction of music is a multi-dimensional event. Music involves instantaneous changes in Sound Pressure Level (SPL). The listeners' instantaneous perception of musical values depends on what has gone before. The changes are driven by two things *that have no physical reality* – the way music is structured in *time* and in *pitch*. As these two are not causally related, and the sound already has a 'where' (humankind's everyday 3D co-ordinates), at least 5 dimensions must be involved [4,5,6]. As we live in lower, four-dimensional space-time (3D = 3 dimensions of space, and 1T = one dimension of time), the human brain is not able to 'see' the whole picture, only segments at a time. This broadly explains why the traditional scientific method has been so badly dented by its attempts to prescribe the optimum means of music reproduction.

Some two hundred years ago, William Blake, the English visionary polymath wrote:

"For man has closed himself up,
 'till he sees all things
 thro' the chinks of his cavern."

William Blake, 1757 to 1827

Another reason, is our increasing mis-apprehension of *quality*. This symptom of 20th century living is considered again in chapter 9.

1.4 The ramifications of quality on audio

It is important for all sound system users to be aware that music's subtler qualities and intended communication may be restricted or even prevented when an amplifier damages or twists the signal which represents (*is an analog of*) the music. To 'damage' and to 'twist' are forms of distortion. There are many names for the different ways in which this can happen. Some ways are blatant, others subtle.

When music is distorted, it not only looses it subtler essence; it can also hurt physically. Undistorted music, even at extremely high peak SPLs, as high as $140dBC_{SPL}$, is not painful to engaged listeners and will not immediately harm healthy ears. The majority of hearing damage is mainly caused by, or greatly exacerbated by, industrial and urban noise [7]. Explosive and percussive sounds can have instantaneous levels that are 20dB (10x) above what ordinary SPL meters and acoustic spectrum analysers capture. The likelihood of any short or long-term hearing impairment is greatly exacerbated by distorted sound systems. Inadequately rated or designed power amplifiers are just one contributor to this.

1.5 Some different aims of sound reproduction

"Experience which is not valued is not experienced .... Value is at the very front of the empirical procession"

Robert M. Pirsig, Lila [8]

Historically, since the birth of sound recording in the late 19th century, the idealist aim of quality recording, and the quest of Hi-Fi ('High Fidelity') sound reproduction equipment has been to capture, then reproduce at any later time, the captured sound with as much accuracy as possible. Intention thus defined, perfection has been attained only when the reproduction has more accuracy than the sharpest human perception [9]. This approach is still widely mis-named (as if narrow-mindedly) '*concert hall realism*'. A better, more global (if clumsy) description of what is sought, is '*full sonic capture of a music event*'. Such ideals always beg the question 'which seat ?', since the 'reality' of all musical events depends on where the participant is. As in any other perfectionist 'event recording', the 4 dimensional manifold (w,ht,d+t) of human perceivable reality is probed. There is no singularity here.

Some people prefer and justify the use of considerably inaccurate replay systems (including power amplifiers) on the grounds that source material is mostly very considerably inaccurate. Unable to perceive pleasure as something beyond time and linear measure, they cannot see that spending £1000's to get immense pleasure from music that might total just 1% of a collection, or *just* 680 minutes (say) could possibly be justified. The closely guarded secret is that ecstatic states are *timeless*.

Others may naively hope to, and the fine artisans do achieve, some kind of degree of cancellation of the inaccuracies (e.g. partnering a 'slow', dull amplifier with a 'fast', speaker with 'brittle' treble). Others, loving alcohol and rich food too much, perhaps, bask in the creation (often with valve equipment) of a euphonic, edgefree or ethereal sound that never existed in the recording session. Or their stance may simply reflect that for them, music is a second-division interest, into which they cannot afford to invest any further.

Since the mid 60s, along with rock'n'roll music (in its diverse forms), an alternative, openly hedonistic definition of what some recording producers and users of sound reproducing equipment are seeking, has developed: that music is sought and created by humans to generate or aggregate ecstatic and blissed states, and the purpose of sound reproduction is to make such higher states more available [2,3].

1.6 About people and their hearing

A significant number of people (possibly 0.1% of the population – which makes several million worldwide) have unexpectedly sensitive hearing. Compared to the average figures reeled off in acoustics and electronics text books, the perceptive ability of some individuals with music and chant, (*not* necessarily speech) extends up to ten times further out in audible frequency, pitch, harmonic content, signal delay, level or phase – amongst others.

One example is that some people are as sensitive to 4Hz, a frequency often described as 'subsonic' (inaudible except through bone conduction), or strictly 'infrasonic', at one fifth of the frequency where ear hearing in most people ceases. Another is pitch discrimination. It has been noticed that some musically trained listeners can detect the difference between tones only 0.1% apart, up to at least 10kHz. This implies the human ear + brain combination is sometimes capable of resolving timing differences of around 100nS or a tenth of a millionth of a second.

Such abilities are natural in some people, and learnt in others. Differences that are identifiable to some of these 'golden eared' listeners (and that can even cause involuntary physical reactions such as a feeling of sea-sickness) can correlate with differences in basic, conventional audio measurements of only 1 part in 100,000 or even less. Other differences may be just 1/30th of the immediately preceding signal, yet it has taken 10 or 20 years of discussion before they are measured [10, 11]; commonplace, simplistic measurement techniques with unrealistic test signals cannot 'see' them at all.

Sensitive listeners, once they have self-confidence, will notice differences in music recordings they love and know well, when sound equipment is changed. After dismissing well established, simple reasons for sonic differences (such as a slight level mismatch causing the louder unit to sound brighter, a facet of loudness perception), the fact remains that what is judged to be sonically better nearly always reveals details and ambiences that were not previously audible. Moreover, nearly every audio amplifier is perceived by skilled listeners to have a sound signature of its own.

Yet such basic technology as Hi-Fi is (presumably) assumed by the general public to have reached 99% of what is possible. Instead, the experience of the high-end should be warning us that what the public hear is probably barely 5 to 10% of what is possible, and the best systems are still pushing at the 50% barrier and all differ in their particular 'bestness'. An amplifier – as our servant, and the speaker's master, is not supposed to add or subtract from the performance. One of the worlds' grandmasters of high-end amplifier design, Nelson Pass, reminds us that, past a point, there is no 'best' amplifier "... just as there is no best painting or best wine" .

1.7 Limits of 'objectivity'. Why listen ?

The traditional 'brute force' approach to overcoming individual sound signatures is to make conventional performance-indicator measurements very good so differences don't matter, which presumes that what is being measured says all about what is heard. Loudly written specifications claiming 'zero distortion' or 'ultra-low distortion' has over the years tricked many millions of users into thinking the sound *had* to be good. This approach, known as 'hearing with the eyes' is still taken to extremes in obsessively technical, 'hard line objectivist' circles.

Much conventional measurement is like claiming a house to be utterly perfect because its sides are at 90.0000°, i.e. it has highly accurate orthogonality – while other relevant aspects that are not being measured (in this case, say the audio equivalent of a house's damp-proofing) may be catastrophically bad. Many measurements are made because they are easy to make because test equipment exists for them, and because such equipment is commonplace, easy to use and easy to buy. But the original relevance of the tests has mostly been forgotten and is rarely questioned. Even the more modern and sophisticated measurements are made to look small by such a complex signal as music. As a measure of this, there are still no real-time, error-computing spectrum analysers able to span audio's 0-200kHz with a 160dB dynamic range. British loudspeaker designer Mark Dodd sums the situation up in the equation:

(Music + recording + playback electronics + ear + brain) = complex problem

Since the mid 70s, amplifier users who are confident in their aural judgment have increasingly learnt to trust their ears with little or no recourse to specifications and measurements, except to check and pass them for basic standards. Outside of electronics and the purely technical, *the subjective approach to decision making is the norm on every level.*

An amplifier is not supposed to add or subtract from the performance. Yet *if* conventional measurements *are* all, then the audibility of some variations of the parts in amplifiers indicates that some human brain-ears are reacting to (as mentioned earlier) differences well below 1% (1 part in 100) and in some cases, 1 part in 100,000 or even less. To make sense of this without invoking the 4 spacial dimension, consider the ink in a book. The ink makes the 500 pages into 1 million words of knowledge. But it only makes the book weigh one gram heavier. If the book can only be analysed by weighing on scales, then the quality difference between it and another same-sized 500 page book on a completely different topic will be hard to measure. Humans who can 'read' may be detecting differences in ink weight of 1 part in 10 million. The order is implicate! [12].

Overall, to be suitable for its intended purpose, even the cheapest audio equipment has to designed with a more global and detailed attention to engineering detail, than any equivalently power-rated 'industrial' amplifier.

1.8 Why are power amplifiers needed for audio ?

For the most part, the processing of audio signals can be performed with only minuscule power, either input or dissipated. Analog signals pass through the majority of the overall signal path at average levels in the order of 100mV to 1 volt. Load impedances may be as high as 100kΩ but even if as low as 5kΩ, only 120μW (a hundred and twenty *micro*watts; or about a tenth of a thousandth of one watt) would be dissipated. At this rate, it would take about eight million hours or hundreds of years of playing, for the load to absorb or use one unit (1kWh) of electricity !

Most loudspeakers used to reproduce audio are highly inefficient. Typical efficiencies of common direct radiating speakers are 1% to 0.05%. By comparison, the efficiency of an internal combustion engine (considered highly inefficient by ecologists) is between 2500% and 50,000% greater. A medium sized car uses about 70kW to move 4 people or hundreds of pounds of goods, and its own weight – altogether at least half a tonne, at speeds of say 70mph.

In some sound systems, to move just the weight of air molecules to reproduce a bass drum, as much as 7kW of electrical 'fuel' can be burned in bursts. And yet a loudspeaker only needs to convey 1 *acoustic* watt to the air to recreate music at the highest practical sound levels in a domestic space, i.e. about 120dB$_{SPL}$. And a tenth of this level (0.1 acoustic watts) will still suit most of the loudest passages in the less extreme forms of music. If speaker efficiency is taken as 0.1%, and $^1/_{10}$th of an acoustic watt is enough, then an *electrical* input power of 1000 times this is needed, i.e. 100 watts.

The highest SPLs in music can be considerably greater than 0.1 acoustic watt. Loudspeaker drive units exist that can handle short term *electrical* power bursts (the norm in much music) of 5000 watts (5kW) or more. With 2% efficiency, today's most capable drivers can generate 100 acoustic watts each. With horn loading, efficiency can be raised to 10% or more, allowing one drive unit to produce 500 acoustic watts for large scale PA. This allows fewer sound sources to be used, improving quality.

When comparing SPL figures it is helpful to remember that at medium SPLs (sound levels) and mid frequencies, a *tenfold* increase in watts offers only an approximate doubling in loudness to the ear. But at the lower bass frequencies and at higher SPLs, *considerably smaller* changes in wattage, say just x3 to x5, have the same doubling effect.

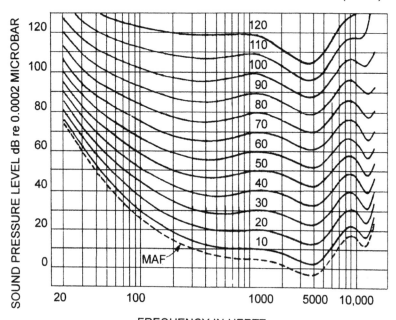

Figure 1.1

Robinson & Dadson *free-field equal-loudness contours, showing how average human hearing sensitivity to pure tones, varies with frequency and level, when facing the sound. Made at the UK's* National Physical Laboratory *in the 1950s, these are superior in accuracy to the older, more famous curves, made in the 1930s by Fletcher & Munson, at Bell Labs, USA. But they are still only approximate for music and for individual ears. On the left is the SPL (sound level) at your ears. The levels on each curve are in* Phon, *a level unit which like each curve, follows the average ear's sensitivity. Notice how the 120dB dynamic range of the midrange is squeezed down to about 60dB (a 1000-fold 'space' reduction) at a low bass frequency (20Hz). And how sensitivity around 3.5kHz because increasingly acute at high levels. 'MAF' is the average threshold of perception in a very quiet space. But it is not the end of sound; some people and many animals can hear 20 or more dB below the MAF level.* Courtesy National Physical Laboratory

1.9 Music fundamentals

Music has a number of key qualities. In the beginning, there are (amongst other things) particular tones. Some tones (fundamentals) are harmonically (numerically) related to others having higher frequency. These belong to a tonal subset called harmonics. Together, fundamentals and harmonics, and their phase relations, along with the *envelope* (the 'shape' of the sound developed by the averaged amplitude) create a *timbre*.

Tones which are not harmonically related to anything may be discordant. If they are harmonically related, but adversely (usually odd harmonics above the 5th) they are dissonant.

When a tone changes in intensity, its 'frequency' (as perceived by the ear) changes. Pitch is (crudely) the musician's ears' own version of frequency and level co-ordinates.

Sounds (often from percussion) that have no dominant, identifiable tones are *a*tonal. They are akin to noise bursts.

Continuous sound, both tonal and atonal, get boring after a while. The tonal waveform changes over a short period (its period is usually measured in milliseconds) but each subsequent cycle is identical to the first. Continuous atonal sound is considerably more interesting, or relaxing as a waterfall can be, say.

Music's higher vital component, its dynamic, differentiates time. Tonal and atonal sounds fade and increase, stop and start, and change in pitch and frequency in diverse patterns, creating wave-patterns (as seen on an oscilloscope) that rarely repeat exactly, and would appear madly chaotic to an alien creature without ears. The overall amplitude (size, loudness) pattern is called *the envelope*.

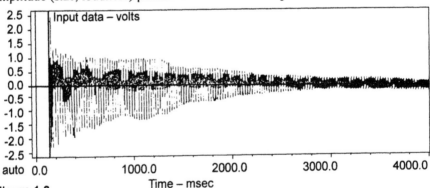

Figure 1.2

In this 'amplitude-time domain' recording off a Fender Precision bass guitar's E-string, the waveform (read from left to right) is seen decaying over 4000mS (milliseconds) or 4 seconds. The envelope is the outside shape. Hence a 'decaying envelope'. The darker parts are where the waveform density is greatest, due to higher frequency inflexions. Notice the initial transient spike, up the top left side, is positive going. Otherwise the envelope is asymmetric - as the signal has a higher negative amplitude, at least for the first two seconds. Reproduced from Stereophile magazine, with permission

Timing retains the music's meaning, concerning the precise schedule of the beginning (attack) and build up of each tonal and atonal building block, and its sustain (levelling off), decay and release.

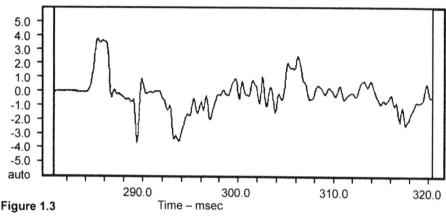

Figure 1.3   Time – msec

Fender Precision bass guitar's E-string, with the amplitude-time window magnified, out to 40mS (0.04 seconds). Shows overall asymmetry. The transient is the positive dwell point (at 288ms), followed by negative spiking (at about 289 mS). Reproduced from Stereophile magazine, with permission.

In most recording locations, sound is reflected off nearby surfaces, causing multiple early reflections or 'reverb'(eration). The added complexity is heard as richness. See '*Dynamics*', above, and on page 12.

Complex distortions, both gross and subtle, caused by mics, speakers, electronics and cables can cause deviations in what the musically adept and experienced ear expects. Tonal qualities can be unduly emphasised or retracted, timing thrown out of sync, subtle dynamic contrasts and 'edges' blurred, and spacial qualities bizarrely warped or flattened.

1.10 Adjectives that describe sound

Despite the fact that music drives a large part of all human art throughout history, and predates all technology, and despite the fact that everyday English speaking calls upon tens of thousands of words, the right words are oddly sparse when it comes to describing how 'sound' sounds. Even the long established technical vocabulary of music composition is small in comparison to other fields.

In a technically educated person's vocabulary of some 10,000 to 50,000 words, it's hard to find even 100 that are widely used for audio performance description. On the following pages are 96 terms, set in a musical context.

1.10.1 Tonal qualities

Adjectives which describe narrow tonal qualities or irregularities are listed in the following table, in order of frequency.

The dividing lines between bass, mid and treble are always arbitrary.

' + ' indicates the adjective is used to describe an excess in the frequency area. '++' means 'even more than'. ' − ' indicates a deficit. *No symbol* means the word is not simply a tonal indicator, or is used mainly for frequency area identification.

Words on cascaded lines are synonyms.

⇧ Highest Treble

Airy	see '*Broader tonal perspectives*'.
Sheen	Very high treble, above 16kHz, often absent or really hiss.
+Tizzy	Excess around 12-16kHz, usually overemphasizing cymbals' high harmonics.
+Hard,	
+Metallic,	
+Brittle	Excess of high, metallic-sound, usually around 8-16kHz.
+Bright,	
+Brilliant,	
++Glassy	Excess around 4-8kHz.
Sibilance	5 to 7kHz.

Treble, down into the High Midrange

+Crisp	Peak about 3 to 4kHz.
Presence	Centered on 2kHz.
+Nasal	1kHz emphasis.
+Honky	Like a Cockney saying 'oi', or like poor or improperly used mid/hf horn speakers. Around 600 to 800Hz.
+Chesty	Excess in the 200 to 400Hz area, particularly with pure male vocals.
+Boxy	As if the singer is inside a cardboard box. Aggravated by cube shaped monitors. Around 250-450Hz.
+Barky	
+Woody	A characteristic mid-bass resonance in some larger speakers.

Low Midrange into Bass

Boomy	see *dynamics*.
Punchy	Around 120–160Hz, a high definition area.
+Balls,	
+Ballsy,	
+Gutsy	Low bass that is visceral, i.e. can be felt.
Boof-Boof	Around 80–90Hz, soft bass area.
+ Chunky	80–90Hz 'sample' bass with added harmonic definition.
– Gutless	Absence of low bass.

⇩ Lowest Bass

1.10.2 Broader tonal descriptors

Airy	Smooth, apparently effortless high treble, seeming to extend further than the music. Suggests ultrasonic capability.
Closed in	Treble lacking above 10kHz, almost the opposite of airy.
Dark	Sound that tilts down from the bass upwards.
Dull	General lack of treble.
Enclosed	Dull, coloured, airless.
Lean	Slight, gentle reduction below 500Hz, or very clean, transparent bass.
Muffled	Where high frequencies are reducing rapidly, above 2kHz.
Open	See *Airy*.
Recessed	General lack of midrange
Rich	A downtilt in level *above* 300Hz. Also, a slight excess of reverb.
Thin	Overall lack of bass.
Tune playing	A consistent quality (especially power) throughout the bass range. Opposite of 'One note bass'.

1.10.3 General sonic adjectives and nouns

Aggressiveness	Preponderance of mid-high energy (3-6kHz), often phasey and distorted.
Ambiance	Not a sonic adjective, but may be portrayed in recordings. To do with mood and feeling. Elusive Neptune–Venus stuff.
Ambience	Usually subtle, low level non-musical background sounds captured in a recording, that are usually only subliminally appreciated, but add to the sense of the occasion.
Analytical	When sound equipment seems to reveal too much of the stitchwork in music. Sometimes used when a system has distortions that unduly emphasise detail or 'edges'.
Articulation	When you can hear the inner detail of complex sounds, particularly those in the main vocal range (300Hz-3kHz).
Clinical	Suggests sound that is clean, bright, sharp, detailed, but may be mildly pejorative, as cleanliness in one area shows up dirt elsewhere. Also suggests emotional qualities are held back.
Detail	– see *Space*
Dry	Sound tending to lack reverberation.
Euphonic	Erring on the side of being pleasing at the expense of accuracy.
Fuzzy	A spiky yet soft texture caused by high distortion and compression.
Glare	Distorted mid treble. Also tonal imbalance or forwardness.
Grainy	Excess texture. A kind of distortion usually in the high midrange.
Gritty	Like grainy, but harsher and coarser.
Grunge	Like gritty, but more muffled. (Actual *Grunge music* is closer to being *gritty*)
Hardness	Fatiguing 'wood block' type of midrange emphasis.
Harsh	Dissonant and/or Discordant. Unpleasant.

Loose	Badly damped bass
Lush	See *Rich* under Dynamics.
One note bass	Poor damping of major resonance(s) in low bass.
Phasey	Symptomatic of a frequency response that undulates like a comb. Co-exists with a manic, zig-zag phase response, literally 'phasing' our hearing system.
Transparent	When you feel you're hearing just the music, not the replay equipment. A sense of 'nothing in the way'. Being able to hear back to the recording venue. In Martin Collom's words "Vital aims !"
Woolly	- see *Loose*.

1.10.4 Dynamics

The next group deals with how accurately the music at many levels and frequencies is output over time. The comparison can be loosely likened to lowering a small plaster model of a huge mountain range over a copy, to check the fit. Representing 50 miles of complex 3D surface (Time, Energy and Frequency), it has fine detailing to the nearest inch of real mountainside, symbolic of the ear's ability to resolve millionth-sized differences buried deep in the main mass of sound.

Boomy	Poor bass damping. A bad loudspeaker/amplifier combination.
Congested	See *Smeared* and *Thickened*, which are facets of the same effect.
Dynamic contrast	Subtle changes in level or pitch embedded amongst much larger changes.
Dynamic range	In audio engineering, the amplitude performance envelope of a sound system. In music parlance, the programme's intensity range. Audiophiles call this 'Dynamic contrast' to distinguish.
Fast	Incisiveness of attack, particularly of bass fundamentals, but as bass doesn't 'move fast' by definition, most likely a reflection of rapid damping, proper synchronisation between the fundamental and harmonics (see Chapter 7), and correct reproduction of all associated harmonics.
Incisive	Conveying the 'slicing' sound of close miked snares, like a sonic machete knife. Indicative of good attack synchronisation, like '*Fast*' and '*Slam*'.
Lifeless	Superficially perfect, anodyne reproduction conveying nil emotion or interest. Commonly caused by forcing equipment or system to manifest a perfect measured frequency response without regard for factors affecting space or dynamics.
Micro-dynamics	Lifelike energy (transients) in small, low-level sounds.
Muddy	Especially applied to bass; see '*Smeared*'.
Pace	Ability to make music seem to unravel at the pace (or BPM) it was recorded at, rather than slower. See '*Fast*'.
Punchy	Similar to *Slam*, but can have a pejorative element of '*One note bass*'.

Rich(ness)	Lots of coherent reverb. More usually applied to program rather than equipment.
Rounded	Loss of attack transients, due to poor damping, poor hf response, or slew limiting.
Rhythm	Ability to put across the infectious 'vibe' inherent in a live show, that makes people want to dance or move in rhythm.
Slam	Convincing, correctly synchronised attack for a fundamental in the 125Hz area.
Slow	Rhythm *seems* slower.
Smeared	Caused by excess incoherent reverb, too much harmonic and intermodulation distortion and/or timing errors.
Solid	Well damped bass.
Squashed	May be caused by hard limiting. Seeming absence of most dynamic contrasts.
Thickened	Can be caused by compression or soft limiting, or more subtly by any path component, from mics to resistors. Reduced dynamic contrasts.
Transient	Abrupt, short lived events in music. Skilled ears can resolve differences in attack slopes and harmonic synchronisation down to tens of microseconds.

1.10.5 Space

Dimensional qualities are embedded in recorded music. Even a mono soundfield can yield spacial information. Good stereo can create 3D images, but as a low grade hologram, you can only experience it from one side, not be in it. It is nonetheless capable of offering a deep experience, and is still revealing its holographic recovery capabilities, over 60 years after its simultaneous invention in the UK and USA. Users of higher dimensional encoding systems (eg. Binaural, *Ambisonics*, *Holophonics* or equipment like the *Soundfield* Mic or the *Azimuth Co-ordinator*) can create full sonic holograms, ie. soundfields (or dynamic sculptures) you can walk around and get inside.

Detail	More spacial version of *Dynamic*
Contrast	Also the diametric of *Muffled*.
Etched	Finely detailed.
Focus	Sharpness of detail. May vary across the soundfield, in all 3 dimensions.
Image, Imaging	Ability to portray width, depth, and sometimes height.
Layering	Sounds having a precise depth in a soundfield, with the implication of many depths or infinite gradation.
Hologram	When coherent and correctly focused light or sound enables higher dimensions to unfold.
Pinpoint	When the image is very stable and finely etched, like some metal sculpture.

Smeared	When an otherwise sharp image seems to be portrayed through butter-smeared glass. See also under Dynamics. Also Timing.
Sound stage	The space between and around two speakers in which sound in stereo (Greek for 'solid') appéars to emanate from.
Timing	Time is a another kind of spacial dimension. The timing between sound components at different frequencies coming from one or more instruments may be unlike the original sound. Compare bass to mid, bass to treble, etc. Delays of milliseconds or less can be audible.

1.10.6 Botheration or Abomination

When something sounds atrociously bad, e.g.

"Like a box of rifles"	An unbearably crashy, thuddy sound.
"Like sandpaper:"	An unnervingly scrapy, scratchy sound.
"Like a train crash"	A frighteningly loud metallic aggression attack.
"Blanketed"	A heavily dulled, possibly slow, tiring sound.

1.11 Nature and range of music (alias programme)

(i) Asymmetry

All sounds comprise alternate compressions and rarefactions of the air (or other physical medium), ie. pressure cycles back and forth. For a completely pure tone, the net change in air pressure is nil (assuming an average of whole numbers of cycles). For most musical sounds however, the waveshape that describes the pressure moment-by-moment is for periods 'more in one direction than the other', skewed or lopsided. This is called asymmetry. See Figures 1.2 and 1.3. In a well sealed room containing music, it would have the effect of very slightly varying the atmospheric pressure over periods of several seconds.

When music programme is converted into an analogous electrical waveform, asymmetry causes DC voltage shifts, that can upset operation. This is most likely in amplifiers of certain topologies, and that have only been tested with conventional test signals, which are all perfectly symmetrical.

(ii) Transients

All kinds of music in its raw, live state often has unpredictable bursts which can be 6, 10 or 20dB higher than the average pertaining up to now, or for some time. Fast *edges*, lasting less time than an SPL meter needs to respond, can be 20 to 40dB above the average short term levels. The ear barely registers these, but notices when they are missing, and notices for sure something nasty is happening when edges cause the signal chain to hiccup for a much longer period.

In the recording process, and in PA systems, transients are 'rescaled' with *compressor-limiters*. These 'squash' rather than excise the effect. As a result, transients in common recordings are at most often only 15dB above the average level.

The *average* difference between the average (or mean or rms) level, and the highest *peak* levels of programme, has a name. It is called P.M.R, alias Peak-to-Mean Ratio see section 2.4.4. This term subtly distinguishes it from *Crest Factor*, which has a similar definition but to be applied only to regular, uniform waveforms. PMR is initially determined by the genre of music. Some approximate examples are:

Orchestral works	18 to 30dB
Rock	9 to 30dB
'Muzak'	3 to 9dB

(iii) Variability

Music is not *pre*dictable. There is no algorithm (as yet) and computers are still incapable of recognising signals which are music and those which are noise. Within the parameters 0, +145dB$_{SPL}$, 10 – 100,000Hz and zero to *n* thousand seconds, almost anything can happen.

1.12 Bass and subsonic* content
* acousticians prefer 'infrasonic'

Bass (lf, LF, low frequency) energy in music is the content with frequencies below 300 Hz (Hertz). Taking 20Hz as the lowest limit, the region we characterise as bass encompasses half the octaves that music spans [13]. Not forgetting our logarithmic perception of frequency, this means that every Hertz difference in frequency counts for more, the lower the frequency [14]. The *Robinson and Dadson* equal loudness contours (Figure 1.1) are an essential starting point to understanding how sounds at bass frequencies differ from midrange sounds, and particularly how their useful dynamic range is compressed naturally by the ear.

Below 150Hz, bass becomes increasingly visceral amd tactile. At the high (110dB+) SPLs at which heavy metal or hard rock is performed, kick-drum frequencies centered around 120Hz can be felt in the solar plexus. These are literally 'hard' and many listeners may find them offensive or at least unpleasant.

Below 100Hz, bass softens. Reggae, Funk and House music make full use of the 'pleasure' region, centered on 80Hz. PA system SPLs in this range have many times been observed peaking at up to 135 to 145dB, i.e. 15 to 20dB beyond the supposed 'pain threshold' without harm or even physical discomfort. Indeed, many listeners describe the experience as cathartic, healing and giving rise to raised consciousness. These lowermost audible frequencies are also the ones most strongly experienced when hearing underwater – an environment like that in which all human life began.

Below 40Hz, bass (from a more limited acoustic, but still infinite electronic repertoire of instruments) becomes more tactile again, and by 16Hz (or some lower frequency depending upon physiology), it is no longer audible through the ear, but solely through bone conduction. This alters the way we hear, since sound is transmitted much faster through solids (the earth, the floor, the feet, the skeleton) than through the air. It is therefore possible for sub-bass signals to be 'felt' *ahead* of the higher, audible components. This effect may be noticed in thunderstorms.

In most music, there is little explicit, large signal 'sub-sonic' musical content. The lowest, large amplitude fundamentals are made by electronic means; or acoustically by pipe organs, giant gongs, and long horns. Other instruments may make sub-harmonics (eg. the 3rd *sub*-harmonic of 41Hz, the fundamental frequency of the lowest note (E) on a bass guitar, is approximately 14Hz), but only in small amounts.

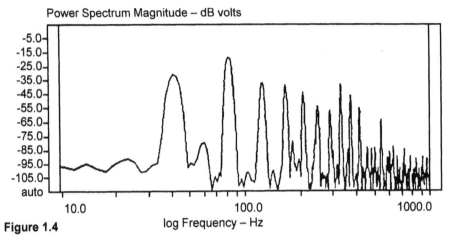

Figure 1.4

Fender Precision bass guitar's E-string, showing amplitude-frequency, i.e. spectral content, during the initial decay period after the Figure 1.2 transient. *Reproduced from Stereophile magazine, with permission.*

At frequencies below the fundamentals of musical instruments, say below 15Hz, much sub-bass is present in the real world even if it is just the rumble of the stage boards, of underground trains and the air conditioning. If these sounds are wholly excluded from a recording, it can seem disconnected from its reality – at least on systems capable of reproducing low bass at high levels giving it audibility. Also, the abrupt filtering needed to remove subsonic content delays wanted, legitimate bass signals at slightly higher frequencies, with respect to the midrange.

1.13 HF dynamics and ultrasonic content

High frequencies (hf, HF, treble) begin at around 5kHz. Higher still, peoples' hearing rapidly becomes insensitive at their *upper limit*. Depending on your genetic makeup, diet, health, age, and cumulative exposure to *non*-musical percussive sounds (particularly hammering and gunshots), the upper conscious limit at quite high SPLs (100dB) typically varies between 12 and 20kHz. In spite of nearly 20 years exposure to high intensity music, up to 145dB$_{SPL}$, the author's hearing presently rolls-off at about 17.5kHz.

Energy Vs. Power

Above 3 to 5kHz, the average *power* levels of nearly all kinds of music, *integrated over a minute or longer,* are less than the levels at lower frequencies, and reduce further with ascending frequency [15]. As a rule of thumb, the power in the mid to low bass regions is at least ten times (20dB) more than the power at 10kHz.

A great body of music revolves around relatively abrupt change. Sounds stop and start. Rhythmic sounds may have durations under 1 second, and the hf components of these can have momentary (but repeated) levels *as high as the largest bass* signals. In this way, music can have low hf power, but high hf *energy*.

Music's HF dynamics are most apparent in PA and recording monitoring, and on the more lovingly mastered digital recordings. In most vinyl recordings, the hf dynamics have been compressed, to avoid badly designed pickup arms jumping out of the groove.

For PA and recording, the ability to capture hf dynamics up to 20kHz has been aided by capacitor microphones (which can have a far smoother, higher-extending, and less compressed response at hf than most dynamic kinds), and on stages, by active mic splitters. These buffer (strengthen) the signals emerging from each microphone, preventing subsequent losses in long cables.

Ultrasonic pleasure

While human conscious hearing stops around 20kHz, higher, *ultrasonic* frequencies in music, up to at least 80kHz, can be perceived by the brain. This much has been long established by listening tests carried out by veteran recording mix-console designer Rupert Neve and producer and monitoring system researcher Philip Newell, amongst others. When frequencies in programme that are above 20kHz are filtered out, sensitive listeners notice a lack of vitality. More recently, it has been demonstrated 'objectively', in the sense that specific neural activity and chemical production has been measured [16], that the subliminal perception of the ultrasonic sounds associated with music enhances pleasure.

System limits

In conventional digital recordings thus far, being under 18 bits, frequencies above 20kHz have to be wholly absent (or at least heavily filtered) for anti-aliasing purposes. By contrast, moving coil cartridges are well known for retrieving ultrasonic content up to 200kHz or more, off vinyl disc recordings. Some moving coil and capacitor mics are similarly responsive above 20kHz.

Perspective

Putting ultrasonic frequencies into perspective, their range is not that remarkable in our hearing's logarithmic terms, with 200kHz only reaching three and-a-half octaves above the average conscious hearing limit of about 17kHz. 200kHz is also the frequency for BBC Radio 4 in the middle of the longwave band used for public radio broadcasting in Britain and Europe. Unwanted reception is a problem for audio power amplifier design that we will meet in Chapter 3.

References

1   Atkinson, John, *As we see it – I wish*, Stereophile, Feb 1993.
2   Rosenberg, Harvey, *The Search for Musical Ecstasy, Book 1 In the home*, 1993, Image Marketing Group, USA, ISBN 1-884250-01-7.
3   Duncan, Ben, *Black Box column, Investigating a rave*, Hi-Fi News, September 1995.
4   Heyser, Richard, *A view through different windows*, Audio, Feb 1979.
5   Heyser, Richard, *Geometry of sound perception*, AES preprint 1009, 51st convention, May 1975.
6   Heyser, Richard, *Catastrophe theory and its effect on audio, parts 1 to 3*, March, April & May 1979.
7   Duncan, Ben, *Earlash*, Studio Sound, June 1995.
8   Pirsig, Robert.M, *Lila – an inquiry into morals*, Black Swan/Bantam Press, 1991. ISBN 0-552-99504-5.
9   Black, Richard, *A 'subjectivist' writes*, HFN/RR, Dec 1988.
10   Hawksford, Malcolm, *The Essex echo*, Hi-Fi News, Aug 1985; Aug & Oct 1986; & Feb 1987.
11   Duncan, Ben, *Loudspeaker cable differences*, Proc.IOA Vol.17, part 7, 1995.
12   Bohm, David, *Wholeness and the Implicate Order*, Routledge & Kegan Paul, London, 1980.
13   Duncan, Ben, *The spirit of bass*, EW+WW, Feb 1993.
14   Colloms, Martin, *Basso Profundo – bass perception and low frequency reproduction*, Stereophile, Dec 1991.
15   Stuart, J.R, *An approach to audio amplifier design, part 1*, Wireless World, August 1973.
16   Ohashi, T, E.Nishina, N.Kawai, Y.Fuwamoto & M.Imai, *High frequency sound above the audible range affects brain activity and sound perception*, 91st AES preprint 3207 and JAES Dec 1991.

Further reading

17   Atkinson, John, and Will Hammond, *Music, fractals and listening tests*, Stereophile, Nov 1990.
18   Colloms, Martin, *Pace, Rhythm and dynamics*, Stereophile, Nov 1992.
19   English, Jack, *The sonic bridge – understanding audio jargon*, Parts 1 & 2, Stereophile, May & June 1993.
20   Harley, Robert, *Looking through a glass clearly*, Stereophile, March 1992.
21   Holt, J.Gordon, *Sounds like ?*, Stereophile, July, August & Sept 1993.
22   Schroeder, Manfred.R, *Self similarity and fractals in science and art*, J.AES, Vol.37, Oct 1989.
23   Zuckerkandl, *Sound and Symbol – Music & the External World*, Princetown University Press, 1969.

Overview of Global Requirements

2.1 Common formats for power amps

Audio power amplifiers are most commonly encountered in one of six formats, which exist to meet real requirements. In order of generally increasing complexity, these are:

1 A **monoblock** or *single* channel amplifier. Users are mostly audiophiles who require physical independence as well as implicit electrical isolation (*cf*.3); or else musicians needing clean, 'mono' instrument amplification (Figure 2.1).

2a A **stereo** or two channel unit. This is the almost universal configuration. In domestic, recording studio 'nearfield' and home studio monitoring use, the application is stereo. For professional studios, and for PA, the two channels may be handling different frequency bands, or the same bands for other speakers, but usually it is the same 'stereo channel', as amplifiers are normally behind, over or underneath the L, R or centre speaker cabs they are driving.

2b Dual monoblock – as 2 but the two channels are electrically separated and isolated from each other – the intention being so they can handle vastly different signals without risk of mutual interference. However, being in proximity in a single enclosure and possibly employing a common mains cable, together with having unbalanced inputs, inevitably allows some form of crosstalk through voltage-drop superimposition; and magnetic and/or electrostatic coupling and interaction, between wiring.

3 Multi-channel – most often 3,4 or 6 channels. Originally for professional touring use, for compactness, eventually working within the constraints of the 19" wide 'rack-mount' casing system, the *de-facto* amplifier casing standard for pro audio gear worldwide. Three and six channel (Figure 2.2) mono and stereo 'Tri-amp' units have been made so the three frequency bands needed to drive many actively-configured PA speakers, can come from a single amplifier box. Multichannel power amps are also applicable to home cinema and home or other installed Ambisonic (higher-dimensional) systems.

Figure 2.1

The *Otis* Power Station, *a monoblock all-purpose musician's hi-fi, monitoring, PA and instrument amplifier, half-rack 1u. The soft clip LED indicator is discussed in chapter 3. One unit resides in a Japanese museum of classic Anglo-American rock'n'roll technology.* Copyright and design by Jerry Mead, 1988.

Figure 2.2

The Turbosound TMA-23, team designed in 1982, was an early realisation of a high power, rack-mounting stereo tri-amp for quick-connection to the first full-range, touring PA speaker boxes. Note Lemo input connector (centre) and two EP6 speaker outlets, one in use. From the author's collection.

4 Integrated power-amp + preamp. Not to be confused with monolithic integrated circuits (ICs), this is the familiar, conventional, budget domestic Hi-Fi 'amp'. The control functions are built in, saving the cost of a separate pre-amplifier in another box. But sensitive circuitry (such as high gain disc and tape inputs) may not sit comfortably alongside the stronger AC magnetic fields commonly radiated by power amplifiers' transformers and supply and output wiring. Careful design is needed to reap cost savings without ending up with irreducible hum and degraded sound quality. In practice, most integrated amplifiers are built *because of* a tight budget, and so amplifier performance is traded off in any event. But some high grade examples exist and the trend is increasing at the time of writing.

5 'Powered'. The power amp(s) is/are built into the speaker cabinet, to form a 'Powered' or 'Active cabinet' (Figure 2.3). This approach has been slow to catch on. It has seen some niche use in the past 20 years in smaller installations, and in the home, usually in conjunction with an 'on-board' active crossover. Having one or more amplifiers potentially within inches of the loudspeaker parts they are driving has the clear advantage that the losses, errors and weight in speaker cables are brought down towards the minimum. This is most helpful in large systems where speaker

Figure 2.3

The rear of a well known active loudspeaker, used both in studios and domestic systems, showing the exterior of the integral amplifier, design by Tim Isaac. Note XLR input and 'bull bars', useful to 'tiewrap' the cables to. Power amplifiers within active speakers can benefit greatly from the well-defined load. (Courtesy ATC Ltd.)

cables are most often at their longest. Since an amplifier in a speaker cab does not need its own casing, there can be savings in cost, and the total system weight (of amps + speakers) can also be reduced. In the home, the need to live with the conventional amplifier's bulky metal box is avoided.

One downside, at least for touring, is that even if there is an overall weight reduction, the speaker cabinets assume added weight, which may cause flying (hanging) restrictions. There's the need to runs mains cables *as well as* signal cables to each speaker cabinet. This is more of a nuisance in large systems. For touring sound, health and safety legislation is also unwelcoming to powered cabs, particularly when *flown*, on several counts. Also, if flown, maintenance can be onerous and adjustment impossible without remote control.

Although beyond the remit of this book, it is worth noting that musician's 'combo' amplifiers are an older, simpler and far more widespread variant of the powered cab.

2.2 Loudspeakers

Knowing a little about loudspeakers, the load that is audio amplifiers' *raison d'etre*, is a pre-requisite to understanding amplifiers. In the following sections – indeed most of the rest of this chapter those features of loudspeakers that most define or affect the design and specification of power amplifiers are introduced. In subsequent chapters, some aspects of speakers' behaviour are covered in greater depth.

2.2.1 Loudspeaker drive-unit basics

There are six main types of speaker drive-units or *drivers* used for quality audio reproduction. Another name for a driver is a *transducer*, a reminder that they *transduce* electric energy into acoustic energy, *via* mechanical energy.

Cone drivers

The most universal, everyday form of the 'moving coil' or 'electro-dynamic' type of drive-unit (Figure 2.4) has a moving cone, with a neatly wound coil of wire (the 'voice coil') attached to its rear. The coil has to be connected to and driven by an amplifier. The coil sits in a powerful magnetic field, and can move back and forth without rubbing against anything. When driven, a signal-varying, counteractive magnetic field is set up, causing the coil and the attached cone to vibrate in sympathy with (as an analogue of) the driving signal. The principles are akin to an electric motor, except that the vibration is linear ('in and out'), rather than rotational.

Moving coil drive-units can be made in many ways. Most have to be mounted in some kind of enclosure before they can be used. Drive unit size (strictly, the piston diameter) broadly defines frequency range. Most cone drivers range from 1" (25mm) up to 24" (0.6m) in diameter, for use at high treble down to low bass. There are at least 15,000 different types of cone materials, textures and weights available. Most are made of paper pulp, but plastics, metals, composite materials, and laminated combinations are also used. Every one sounds different, and measures differently.

Figure 2.4

A moving coil drive-unit, with key parts identified. (Courtesy Funktion One Research)

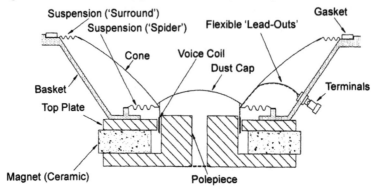

There are as many permutations again for the voice coil's diameter, height and wire gauge; the type of dust cap, magnet, the chassis, and the flexible jointing, called the *surround* (at the front), and centering device at the rear, the suspension or *spider*. With all the moving parts, ruggedness and stiffness is pitted against the need for agility, hence levity. This is the main reason why the radiating part is cone shaped. This shape can stiffen the most limp paper against the axial force applied to it by movements of the voice coil.

Compression drivers

The second most common type of driver, at least in professional sound, is the compression driver (Figure 2.5). This is simply a specialised form of moving coil drive-unit. The depth of the cone is replaced by a much shallower, and usually opposite-facing and dome shaped radiating surface, called the *diaphragm*. The voice-coil is attached peripherally between the edge of the dome and the suspension. This type is made for some midrange but mainly hf speakers which are *horn loaded* . All bass-bins and most midrange horns employ specially adapted but ordinary-looking cone drivers; these alone can handle the larger excursions required. A compression driver cannot handle more than very small excursions. To avoid large excursions and po-tential ripping of the diaphragm, a compression driver must *never* be driven with program having frequencies below its rated range, and not driven without being attached to a suitable horn. Usually, the diaphragm is pressed out of a plastic film, or a phenolic resin-impregnated cloth or other composite, or from very light, but stiff metal, usually aluminium, else titanium or beryllium. Size ranges from about 6" (150mm) for midrange, down to 1" (25mm) for high treble and above.

Soft and hard dome drivers

The equal-second most common type of speaker drive-unit is familiar enough: It has an almost hemispherical diaphragm shaped like some compression drivers' but the dome is forward (like a fried egg) and nearly always working into free air. This is used on its own, instead of a small diameter cone, as a tweeter (hf drive unit). The material can be any of those used in cone or compression drivers.

Figure 2.5

Two types of hf compression driver made by Emilar, widely used in PA systems from the mid 70s.

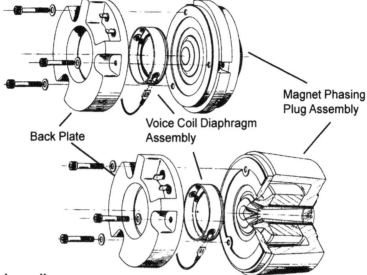

Magnet Phasing
Plug Assembly

Voice Coil Diaphragm
Assembly

Back Plate

Common voice coil

The three types of drive units discussed so far all have similar voice coils. They may range widely in weight, diameter (from 0.75"/19mm up to 6"/150mm) and power handling (commonly from 3 watts to 1000w), but they will all mostly have a DC resistance of five to ten ohms, and a nominal (AC, 400Hz) impedance of 8, 15 or 16 ohms.

The ribbon driver

The ribbon speaker is a fourth kind of electro-dynamic drive-unit. Instead of a voice coil attached to the radiating part, the amplifier signal is connected across a length of flat (planar) conductor foil or 'ribbon', which is again placed in a magnetic field like a voice coil, but also radiates sound like a cone, diaphragm or dome. Compared to ordinary voice coils, this arrangement can be lighter and certainly presents a much purer ('resistive') impedance to the amplifier. The classic ribbon had a very low DC resistance, and was transformer coupled. Modern ribbon speakers have longer strips, amounting to 3 or 5 ohms of near pure resistance, benign to most audio amplifiers connected to it. When 'built big' as a *panel loudspeaker*, a ribbon drive-unit forms a wide-range loudspeaker in its own right, i.e. no cabinet required. There is little breakup in the ribbons' surface to mar the sonic quality. And, unlike other drive units, absence of a cabinet means the sound source radiates as a dipole, i.e. from both sides. This can be important to the amplifier, in so far as room inter-action can change the impedance seen, by *reflection*. Small ribbon drive-units are used as tweeters. They may be horn loaded to magnify their rather low output. Sonic quality can be very high, although naturally favouring the reproduction of stringed instruments.

The electrostatic source

The Electrostatic Loudspeaker (ESL) employs the inverse or dual principle of the electro-dynamic or 'motor' types of drive-units that we have just looked at. The movement is provided by electrostatic (electric field) force, rather than magnetic attraction and repulsion. The vibrating part is a thin, critically stretched sheet called the diaphragm. The fixed part, after the capacitor it mimics, is called a plate. Electrostatic drivers are commonly made in the form of panels, like ribbon speakers. A power source (usually from the AC mains) provides the high EHT DC voltage of over 1000 volts, that is needed to *polarise* the plates. A high signal voltage swing is also required. This, together with isolation from the EHT is attained by interposing a transformer. In practically-sized and costed electrostatic speakers, the transformer and the diaphragm have a surprisingly limited capacity for handling high levels at low frequencies. In primitive designs, overdrive in the bass can cause the diaphragm to short against the opposite plate. In modern ESLs, the diaphragm is insulated. A well know electrostatic employs an aggressive *crowbar* circuit for protection. If the ESL is subjected to potentially damaging high levels at low enough frequencies, this shorts the speaker's electrical input, possibly blowing up the amplifier, or at least blowing a fuse or shutting down the music. Under most other conditions, the ESL appears as an almost purely capacitative load, with resistive damping across it.

The piezo driver

The two fundamental types of drive unit 'motor' looked at so far all date back (in principle) to the early years of this century, or even to the beginnings of the modern harnessing of electricity, two hundred or more years ago.

The *piezo* drive-units' principle is the dual of the familiar household act of creating large voltages by squeezing crystals. Although piezo-electricity precedes human-kind, as it can occur naturally, it has only been widely harnessed in the past 50 or so years, first in crystal mics and pickups, and more recently in fuel-less 'push button' gas fire lighting. The dual, or reverse process, that of making a crystal vibrate by applying electricity to it, was first harnessed by Motorola, who have been producing hf drive-units employing this principle since at least 1977. This type of drive unit looks capacitive, rather like an electrostatic, but has a higher DC resistance – so it can draw no long term power. Despite potentially useful high hf performance, since there is still a limited range of piezo drive units, most being fitted to integral, out-dated horn designs, piezo tweeters are not used much in high performance systems, but they are occasionally used in PA systems, and may be found optimally applied in refined custom speaker systems.

The Motorola piezo element cannot be 'burned out' by too much 'power' as it presents a high impedance. But it is rated at about 25v rms, and excess voltage will quickly destroy the crystal. The crystal can even be harmed by room heating. For use with amplifiers having headroom above 25v rms, operating 2 or 3 in series is suggested. This should not degrade damping as it would with a low impedance speaker.

Inductive coupling

Eli Boaz at Goodmans, part of the *TGI Group* (comprising speaker manufacturers Tannoy, Goodmans and Martin Audio in the UK) spearheaded the development of 2-way drive-units where the hf driver is inductively coupled (*Inductive Coupling Technology* or ICT). It comprises a radiator with a conductive collar (Figure 2.6). Placed *within* the bass/midrange voice coil, it acts as a single turn transformer, picking up magnetic field most efficiently at hf. This arrangement is limited to use at HF, but there is no need for a crossover, and it is highly rugged – a tweeter that cannot readily burn out. And without having the capacitative loading region of a conventional tweeter (and the load dip of any passive crossover), an ICT driver's load impedance at HF is benign.

Figure 2.6

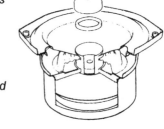

The exploded hf dome above this Tannoy drive-unit has no ohmic connections, and cannot be burnt out. It employs inductive coupling technology (ICT), the first completely new type of drive-unit to enter mass production for many years. Each new drive-unit type has its loading peculiarities and this adds a new layer of variables to the considerations of amplifier users and designers alike. Courtesy Tannoy Ltd.

2.2.2 Loudspeaker sensitivity *vs.* efficiency

Loudspeaker drive units have to be 'packaged' to be usable in the real world. *Together, enclosures and drive-units define the efficiency* of the resultant loudspeaker. Efficiency (or its derivative, sensitivity) then decides the scale of amplifier power needed. With different high performance loudspeaker types, efficiency varies over an unusually wide range of at least a hundredfold, from about 20% down to 0.2%.

Efficiency is not often cited, but can be inferred from the vertical and horizontal polar radiation patterns, the impedance plot, and the sensitivity. Sensitivity is the derivative of efficiency that makers use to specify 'how much SPL for a given excitation'. In part sensitivity is universally specified because it's easier to measure. It is given as an SPL with a given input (nearly always 1 watt) at a given distance at close range (1 metre normally). So the spec is the one that reads:

'Sensitivity 96dB @ 1w @ 1m'.

96dB is a high sensitivity for most domestic speakers but low for professional types. The sensitivity is but a broad measure of efficiency differences, since two factors are missing.

One is how the sound energy is spread in space. If it is all focused forwards, sensitivity (dB SPL @ 1w @ 1m) is raised as the sound 'density' at the measuring position increases. At low frequencies, rated sensitivity commonly falls as the sound radiation becomes more nearly spherical, while efficiency is unaffected.

Factor two is the impedance. Where mainly resistive, efficiency is about the norm, as computed by integrating the SPL over all the solid angles. But around the (or a) resonant frequency where the impedance changes rapidly from capacitive to inductive, efficiency is high, as little energy is dissipated.

With these 4 dimensions of variables (3D space + 1D impedance), converting efficiency into sensitivity figures and vice-versa is not straightforward. But as a rough idea, an 86dB@1w@1m rated domestic speaker is about 0.5% efficient. While with a two sided (planar) speaker, the efficiency might be the same 0.5% but sensitivity would ideally halve towards 80dB.

2.2.3 Loudspeaker enclosure types and efficiencies

Horn-loading is by far the most efficient technique. It is between ten and over a hundred times more efficient than any others. 'Efficiency' means it gives the most acoustic intensity for a given power input, from the amplifier. Only when a horn (or 'flare') is coupled to a transducer with a low output (e.g. a ribbon driver) is the overall efficiency *not* 'streets ahead' of all the other driver+enclosure combinations.

The most efficient drivers are the familiar electro-dynamically-driven cone, dome and compression types, particularly those with an optimum balance between the strength of magnetic and electric coupling, the levity of the moving parts and the *compliance* of the suspension. In the midrange, some ESLs can be as efficient as the cone-driver, both in the context of a refined domestic speaker.

The least efficient enclosures are:

(i)   **none** (this holds true at low frequencies only),

(ii) the **sealed box** (SB) or 'infinite baffle' (IB), and

(iii) the **transmission line** (TL) – used to extend bass response.

Of these, the latter two are important, practical forms that have to be lived with. They can in any event be made relatively quite efficient by making the enclosure *big*. To some extent, *Colloms' law* holds here:

"*Loudness (per watt or volt) is inversely proportional to bandwidth and smoothness*".

This is fine until we come to consider the refined horn-speakers which do not attempt 50% efficiency, and where a minimum of three types are needed to cover the audio band. While at least ten times more efficient, there is little or no bandwidth narrowing over ordinary speakers.

Compression- and Piezo-drivers are those usually coupled to horns (flares), and may need no other boxing or at least not any specific enclosure, as their rear chamber is usually already sealed. Sound radiation is then mainly defined by the horn, subject to mounting.

Ribbon drive units may be also horn mounted; or if 'Planar' types, then along with ESLs, they may be simply mounted in a frame that has little effect on the sound radiation which is *dipolic*, ie. two sided, like a harp's.

The other two types of drive units – the cone and soft-dome are usually mounted in closed ('infinite baffle' or 'sealed box') enclosures, or in the case of cone drivers alone, in ported ('Thiele-small', 'vented' or 'reflex') enclosures.

Cone drivers are also used 'coupled to' horns – either midrange, or bass ('bins'). In practice, as the rear of the cone's basket mounting frame is open, and the fragile magnet is also unprotected, cone drivers in bins and horns are almost always mounted inside the overall enclosure.

Horns, transmission lines, sealed and vented boxes, and other loading types may form complete loudspeakers in free permutation. Of these, only combinations of horns or of sealed boxes can cover the *full* audio frequency and dynamic range within their own family, i.e. without involving each other or the other types.

2.2.4 Loudspeaker configurations: a résumé

Few or no *single* loudspeaker drive units offer overall, high performance audio reproduction. The nearest contender is an ESL or ribbon type panel, which can work as the sole drive-unit for the kinds of music that have no loud, low bass content. But to listen without restriction and risk of damage to every other kind of music, two, three or more drive units must be used to cover low, medium and high frequencies (which from 10Hz to 20kHz span a wavelength range of some 2000 fold!), and over the 120dB+ dynamic range required for high performance sound reproduction.

Matching levels

Often, the sensitivities (loudness) of the individual units that are optimum for each frequency band, differ. Commonly, the tweeter is more sensitive than the driver covering bass/mid frequencies. If efficiency is unimportant, the mis-matched sensitivities (which would otherwise cause a uneven, 'toppy' frequency response) may be overcome by adding a series 'padding' resistor in line with the hf drive unit (tweeter).

Parallel connection

Alternatively, in touring PA and wherever else efficiency matters, or wherever high SPL capability is sought, an overall flat response may be attained by using two, parallel connected bass/mid drive-units. Applicability is *always subject to coherence in the acoustic result*, hence suitable mutual positioning of the paralleled drivers, so they work together. If 'ordinary' drivers (i.e. electro-dynamic types), a 15 or 16Ω rating will be likely chosen, so the resultant load is about 8Ω, rather than 4Ω if the paralleled drivers were each the usual 8Ω.

Again *always subject to coherence in the acoustic result* (and thus suitable mutual positioning, generally closer than a quarter of the shortest wavelength), drive units or complete speakers can be paralleled across either a given amplifier; or when this runs out of drive-capability, across amplifiers *ad infinitum* that are driven with an identical signal, and have either identical, or acoustically justifiable different gains.

Despite the above, the fewer drive-units or speakers reproducing a given programme in a given frequency range, the better the sonic results. *As is so often the case in high performance sound reproduction, least is best – if it is usable.*

Why crossovers?

When two or more drive units are used to cover the audio range, it is usually *very* important that they only receive programme over their respective, intended frequency ranges. Programme at other frequencies must be omitted, as it will usually degrade a drive-unit's sonics, by exciting resonances and sub-harmonics. At moderate to high drive levels, physical damage is also likely. The 'frequency division' or 'frequency conscious routing' that ensures different drivers receive their intended range, and not other frequencies, is the *crossover* (no relation to crossover *distortion*).

The crossover may also perform phase shifting or signal delay in one or more bands. Principally this is to synchronise (UREI use the phrase 'time align') the signals delivered by the drive-units, e.g. because the sound from some has slightly further to travel.

A variety of crossover types exist, the more sophisticated designs aiming to have the bands meld neatly in the crossover regions, without boosting or cutting any frequencies.

Crossover point (XOF)

The crossover frequency or 'point' is chosen by speaker designers on the basis of:

(i) **power handling.** Drops dramatically if XOF too low for the higher driver. Caused by the exponential increase in excursion as LF limits reached.

(ii) **distortion.** For the same reason, rises rapidly for the higher driver below a comfortable XOF.

iii) **frequency response**, both on- and off-axis. If either of these changes abruptly, near the proposed XOF, the XOF had better be moved. But with some drive-unit combinations, there is no ideal XOF.

Passive crossovers

In the majority of domestic speaker systems, but particularly where low cost or simplicity are paramount, the crossover is passive (unpowered) and operates at 'high level', being placed within, and supplied as part of the speaker cabinet (Figure 2.7).

Figure 2.7

Passively crossed over cabinet.

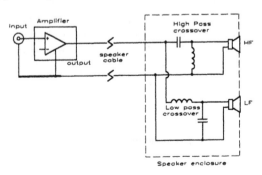

In this form, the crossover comprises physically large and heavy, high voltage-rated capacitors, high current-rated inductors (coils), and high power-rated resistors.

Passive low level

It is possible (but not very common) to have a passive crossover operating at line level, installed before the signal enters the power amplifier, or otherwise before the power stage (Figure 2.8). The crossover parts can then be smaller and lighter as the voltage and current ratings (that in part determine size) can then be 30 to over 100 times lower. This is a fine arrangement for all-integrated active cabinets.

Figure 2.8

Low-level passive crossover.

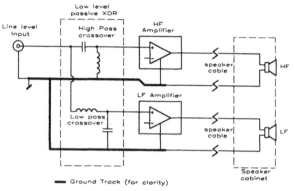

Otherwise, in ordinary 'mix and match' sound systems, one problem arising is that the crossover is physically divorced from the speaker, so careful connection is needed. Another, less daunting, is that any existing, passive high level crossover component values cannot be simply transferred, since they will have been 'tweaked' to best suit the vagaries of the drive-units' impedances.

Active crossovers

The active crossover (first suggested by Norman Crowhurst in the 1950s) takes the preceding concept a stage further. Frequency division is accomplished *actively*. This means using active devices – and a DC power source – to provide filtering in a highly predictable manner. For example, the filters are able to work in an ideal environment, having well defined and resistive loading. This, and the filter function are defined potentially very precisely by active electronics, usually employing high NFB (Figure 2.9).

The main disadvantage of active crossover systems is cost, not just of the active crossover, but of the added amplification and cabling. In DIY domestic set-ups there is also the *bulk* of equipment (if using say three stereo amplifiers, placed centrally; or six monoblock amps and two mono crossovers, half to be placed by each speaker) and their cabling. Such inconvenience is irrelevant in concert sound systems, and even in recording studios. It is also absent in active powered enclosures. Rather the reverse – l integrated active cabs are 'Plug in and go' systems. For everyone else, more care needed when connecting-up an active system, taking care that the drive units are going to be fed their appropriate band. In professional sys-

Figure 2.9

Classic 3 way active crossover.

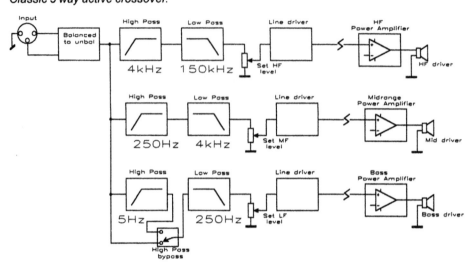

tems, multipin speaker and line connectors are commonly used so the two or more frequency bands' connections are always correctly routed.

Notable advantages [1] compared to the common, high level passive crossover are:

(i) reduced 'congestion' and similar intermodulation distortion symptoms as each power amplifier handles only a section of the audio range.

(ii) differences in driver sensitivities (considered under 'matching levels', above) are ironed out and without compromise, except the requirement for, or use of amplifier headroom, by simply adjusting gain controls.

(iii) higher dynamic headroom *by diversity*, as the programme peaks occurring in the respective out-of-band frequency ranges do not steal any headroom. Also, brief clipping (overdrive) of the bass band (etc) has little effect on clarity when the other bands are not driven into clipping at the same time.

(iv) the amplifiers are connected directly to their respective drive-unit(s). There are no reactive components (i.e. the passive crossover) to steal current, but the drive-unit's impedance dips may still demand significant current headroom from the amplifier.

(v) the ease and capability of creating highly conjugate, highly matched and closely toleranced crossover functions.

Of these, the low-level *passive* crossover potentially has all the same advantages – except possibly this last item.

Active manifestations

Active crossovers commonly take one of three forms. In pro-audio 'the crossover' is commonly used packaged in its own box, like other signal processors, as a *standalone*.

As a power amplifier presents a (literally) well placed opportunity to share a box and a power supply, a few power amp makers offer a pluggable option or 'octal socket', where a crossover module or card can be retrofitted. Often, the performance of the card's filtering (e.g. the frequency and slope rate) may be decided upon by the user. This is often a low-budget option limited to installation work, but while it can save money, it need not be shoddy. If carried out with as much care as any crossover is due, turning an amplifier into a 'speaker-driving filter' has the advantage of minimising superfluous hardware and signal path complexity.

Otherwise, the active crossover along with the power amplifiers, is placed in the loudspeaker enclosure. This creates an 'active enclosure', 'active speaker' or 'Tri-amp cab' (if 3 way; else 'Bi-amped-', 'Quad-amped-cab', etc) that has the advantage of hiding the complexity, and offers a *fait accompli*, but takes away the flexibility that touring PA users often need.

Active systems are widely used in pro-audio, for installed and touring PA systems, and studio control room monitors. In nearly all cases, stand alone active crossovers, discrete power amplifiers, and individual enclosures are brought together. In high-end hi-fi, discrete active systems are comparatively rare, except in DIY circles. Active speakers are becoming somewhat more common, at least in the UK and Europe (see Appendix 2).

Bi-wiring

Bi-wiring is a 'part-way house' to having a low level active (or passive) crossover. A separate speaker connecting wire is provided for each drive unit, or for each frequency band (Figure 2.10). This lessens interaction and intermodulation that is otherwise caused by communal speaker cabling. We will meet this later, under the banner 'good noding'.

Figure 2.10

Bi-wiring improves sonic quality by avoiding superimposition voltage drops over the greater length of the output-stage to speaker connection. Otherwise LF signal currents upset the HF's driver signal's purity - and even vice-versa.

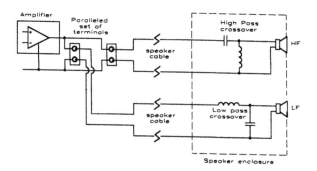

Other networks

Whether the crossover is passive and high level; or passive and low level; or active, other components may be associated with each drive unit:

i) **DC protection capacitors** are connected in series with drive units to block steady current flow should a DC voltage appear across the box's, the speaker's or the driver's input terminals. They are *not* required for the hf (and mf) drivers with passive crossovers, where the high and 'bandpass' filters already include the required series capacitor – as part of the crossover.

In some designs, more complex, active crowbar circuitry is used, in order to obviate the need for a series capacitor.

ii) **Zobel** and other 'conjugate matching' networks. Comprising networks of capacitors and resistors, and less often inductors, these act to smooth out the impedance variations of the drive units, either singly or altogether, and as seen by the preceding crossover and also amplifier.

iii) **Music overdrive protection** – in some designs a lightbulb, usually a rugged 12v type, is connected in series with hf drive units which in practice require protection most of all. At worst, the lightbulb will be quicker, easier and cheaper to change than the hf driver or diaphragm. The effect the lightbulb has on sonic quality can be small and sonically benign if the lamp is not visibly glowing during normal loud passages.

'Auto resetting' 'thermal trip devices', alias Ptc (Positive Temperature Coefficient) thermistors are also used. These are usually in the form of a cement coated disc. At room temperature, they exhibit a low resistance. When *eventually* tripped by excess current, the hot resistance rapidly increases to about 100 fold, and the protected driver's power dissipation drops 10,000 fold or *pro-rata*. The effects on sonic quality of series ptc thermistors are as yet questionable.

Other protection

Loudspeakers have also been protected by add-on boxes, containing historic power-reading circuitry for example, which crudely opens a relay in line with the speaker or line level signal if the drive-unit is seen heading towards a burnout. The use of fuses within power amplifiers for speaker drive-unit protection is considered in chapter 5.

2.3 The interrelation of components

2.3.1 What loudspeakers look like to the amplifier

There is a tacit presumption that most amplifiers can comfortably drive any 'reasonable' loudspeaker. Beyond whatever is 'reasonable', discomfort may occur - to all parties. To the power amplifier which is nothing but a loudspeaker driver, the most salient information about any loudspeaker it is expected to drive, and the stress that may engender, is that loudspeaker's *impedance*.

Impedances

A speaker's nominal impedance is commonly (and over-simplistically) described by a single round figure, usually 15, 8 or 4 ohms for the majority of moving-coil drive units. With ribbon drive units, or whenever several drive units are paralleled to increase handling or coverage, lower impedances of 3, 2 or 1 ohms or even less are the norm. With electrostatic and piezo (hf) drive-unit types, the load impedance can be higher, but are also more or predominantly *capacitive* (like a capacitor) across the audio range. This can be far more taxing to the amplifier.

Low vs. high impedances

At this juncture it is helpful for those unfamiliar with electronics jargon to grasp a counter-intuitive fact: that the *lower* impedance, the *heavier* the (current) loading on the amplifier. To remember this and that 4 ohms is *harder* to drive than 16 ohms, think of hill slopes: a 1-in-4 hill is far harder to drive or climb up, than a 1-in-16, ie. an impedance in ohms is the reciprocal of the relative loading. Remember also:

A low impedance demands more current, and less signal voltage is needed for a given current.

A high impedance requires more signal voltage, to be driven with a given current.

Figure 2.11

The impedance of a 15" drive unit mounted on a nominal baffle. In some cabinet designs, there could be two or more resonant peaks. Note the labelling of the resistive, capacitive and inductive impedance zones.

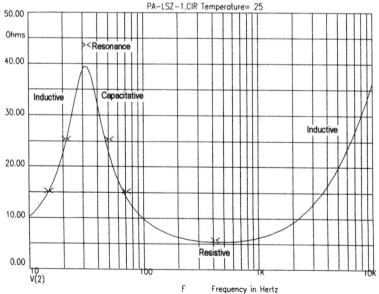

Variation vs. frequency

Loudspeakers' impedances nearly always vary over the frequency range of use. Figure 2.11 shows how a nominal 8 ohm, 15" bass drive-unit typically varies from

5.5 ohms at 450Hz, peaking up to about 40 ohms or so, at the mechanical resonant frequency, which typically lies between 20 to 120 Hz for a bass driver. Here it is 31Hz. Together, the drive-unit and the speaker enclosure largely determine this. Impedance also rises to a maximum at (and beyond) the highest usable frequency.

At and about the resonant frequency, the impedance variation at the loudspeaker's terminals is due to the reflection of mechanical energy storage, and damping, back to the electrical domain. Figure 2.12 shows that the loading is capacitative on the right side of the resonant peak, where impedance is falling with increasing frequency, while the impedance that slopes upwards with increasing frequency, on the left side of the resonant peak, is inductive. Figure 2.12 shows this in another domain. When the phase (lower graph, left scale) is positive, the impedance is capacitative; when negative, it is inductive. When towards the centre, it is resistive. Dead centre is pure resistance.

Figure 2.12

The impedance of the previous figure (upper graph), shown alongside the phase map (lower graph), clearly shows the relationship between pure resistance, and inductive and capacitative phase - at least in terms of voltage. In some instances, a plot of current-phase might be more appropriate.

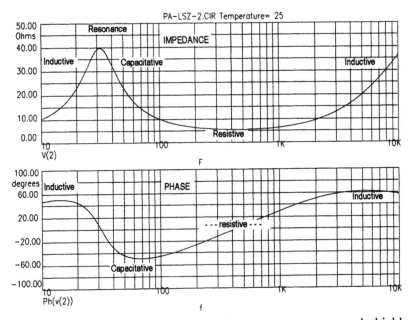

The resonant frequency area(s) of any bass speaker system are commonly highly stressful to amplifiers, and with many BJT amplifiers, when contact is prolonged by a low-enough frequency and perhaps an insistently-enough pounded note, it has frequently been fatal. Other amplifiers have been known to simply burst into uncontrollable oscillation.

At or above the highest usable frequency, the impedance rise is again inductive. It represents the effects of the voice coil inductance. Eddy currents, as such, or manifest as *skin effect* and *proximity effect,* may also contribute to the inductivity. Inductive effects are the cause of *back EMFs* ('kick back' voltages) that unless damped can upset sound quality, and can even destroy an unsound amplifier design.

Passive crossover effects

Most high performance speakers for domestic and small studio use contain passive (unpowered) crossovers. Such enclosures are driven from a single amplifier. Passive crossover networks are 'in line with' the drive units' impedances. The combination is complex; it may increase or decrease peak current demand, hence loading. As an example of the latter, extra parts may be added to create a *conjugate* crossover, which makes the overall loading look resistive, but also absorbs power.

Static vs. transient

Conventionally, impedance values are taken after applying a steady and repetitive test signal (e.g. continuous sine wave) and allowing a few moments for the recovered signal amplitude that represents impedance, to settle. In the short term, impedance can be considerably lower. *At worst* [2,3] *it is possible for an ordinary dynamic (moving coil) type of loudspeaker to demand current as is it had 1/6th of its nominal impedance.* In other words, an 8 ohm speaker can sometimes look like 1.4 ohms. This will not happen all the time or even very often, nor for very long at a time – but for high quality sound reproduction, and not forgetting that music involves repetition, the possibility *must be allowed for.*

Acoustic contribution

The impedance (load) characteristics of drive units can be affected by cabinet air leaks, also by reflections in the room, hence the positioning of the enclosure. Horn loaded drive-units are usually the most sensitive to this.

The upshot is that most loudspeaker loads are a wide variable, not just between different models and types, but depending on programme dynamics and excitation frequencies.

2.3.2 What speakers are looking for

The fact that most loudspeakers do not employ conjugate impedance compensation, and so they have impedance curves that vary 'all over' with frequency, means that for high performance, the amplifier kind the speaker *needs* to see is a 'voltage source'.

Why voltage?

The signal *voltage* must be almost unaffected (ideally far below say 1% change) whether the speaker is connected or not; and likewise, regardless of whether it's drawing 50 milliamps or 50 amperes. That means a 'stiff power source', alias a *low impedance* or 'high-current-capable' source.

If the source impedance isn't low (enough), then as the speaker's impedance varies with frequency, the change will be superimposed on its own frequency response as a tonal aberration [4, 5]. A source impedance that is almost as high as speakers' own minimum impedances is a major failing with power amplifiers having low or nil global negative feedback, and the outcome is a tonal anomaly as gross as 5dB [4]. This may not be all bad, but it will certainly be arbitrary.

Energy control

In part, a *voltage source* is required to drive speakers, because loudspeakers store, as well as convert, energy. The fundamental resonance is the place (in the frequency domain) where this is most true. Some of the stored energy 'kicks back', and needs to be quickly damped (dissipated) to avoid transient distortion or 'smearing'. The same reactive effects may also demand surprisingly high peak currents from the amplifier at other times, when driven by music signals.

In both cases, the answer is a high current sourcing *and sinking* capability. Both of these features are implied but neither are guaranteed by a low source impedance. The overall requirement is a 'current-capable-enough voltage-source'. So far, most power amplifiers throughout history have aimed to be this, but some have come closer than others.

Damping factor?

The majority of high performance amplifiers are solid state and employ global (over-all) negative feedback, not least for the unit-to-unit consistency it offers over the wild (eg. +/–50%) tolerances of semiconductor parts. One effect of high global NFB (in conventional topologies) is to make the output source impedance (Z_o) very low, potentially 100 times lower than the speaker impedance at the amplifier's output terminals. For example, if the amplifier's output impedance is 40 *milli*ohms, then the nominal damping factor with an 8 ohm speaker will be 200, ie. 40 milliohms (0.04) is 1/200th of 8Ω. This 'damping factor' is essential for accurate control of most speakers.

Yet describing an amplifier's ability to damp a loudspeaker with a single number (called 'damping factor') is doubtful [6]. This is true even in active systems where there is no passive crossover with their own energy storage effects, complicating especially dynamic behaviour.

Figure 2.13 again takes a sine-swept impedance of an 8 ohm, 15" driver in a nominal box to show how 'static' speaker damping varies. Impedance is 70 ohms at resonance but 5.6 ohms at 450Hz. Now, at the bottom, is plotted the output impedance of a power amplifier which has high negative feedback, and thus the source impedance looking up (or into) it is very low (6 milliohms at 100Hz), though increasing monotonically above 1kHz. The traditional, simplistic 'damping factor' takes this ideal impedance at a nominal point (say 100Hz), then describes attenuation against an 8 ohm resistor. This gives a damping factor of about 3 orders, ie. 1000, but up to 10,000 at 30Hz. Now look at the middle curve: This is what the amplifier's damping

Figure 2.13

Views of the damping surface in 2D. Lower plot shows the very low steady-state output source impedance of a typical transistor amplifier with high NFB. The middle plot shows how this degrades after passing down a few metres of reasonably rated cable, and a series capacitor (which might be the simplest crossover, or for fault protection). The upper plot repeats the impedance vs. frequency behaviour of the 15" bass driver. The effective damping factor is the smaller and highly variable difference between the upper and middle plots, not the difference between the highest impedance on the upper plot and the lowest on the lower plot, used by amplifier makers!

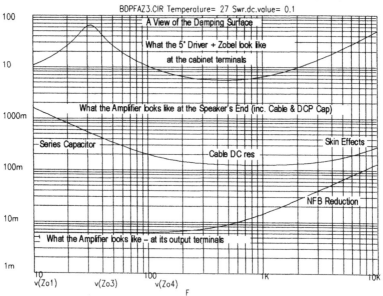

ability is degraded to, after is has traversed a given speaker cable and passed through an ideal 10,000µF series capacitor, as commonly fitted in many professional cabinets for belt'n'braces DC fault protection. The rise at 1kHz is due to cable resistance, while cable inductance and the series capacitance cause the high and low-end rises respectively above 100 milliohms.

We can easily read off static damping against frequency: At 30Hz, it's about x100. At mid frequencies, about x50, and again, about 100 at 10kHz. However, instantaneous 'dynamic' impedance may dip four times lower, while the DC resistance portion of the speaker impedance increases after hard drive, recovering over tens to thousands of milliseconds, depending on whether the drive-unit is a tweeter or a 24" shaker.

Even with high NFB, an amplifier's output impedance will be higher with fewer output transistors, less global feedback, junction heating (if the transistors doing the muscle work are MOS-FETs) and more resistive or inductive (longer/thinner) cabling. Reducing the series DC protection capacitor value so it becomes a passive crossover filter will considerably increase source impedance – even in the pass-band. The ESR (losses) of any series capacitors and inductors will also increase source impedance, with small, but complex, nested variations with drive, temperature, use patterns and aging. The outcome is that the three curves – and the difference between the upper two that is the map of damping factor – writhe unpredictably.

Full reality is still more complex, as all loudspeakers comprise a number of complex energy storage/release/exchange sections, some interacting with the room space, and each with the others. The conclusion is that damping factor has more dimensions than one number can convey.

Design interaction

While high performance loudspeakers are being designed and optimised, and certainly before they are finalised for production, in-depth listening is a pre-requisite. This means that amplifiers are required to drive them while they are being tested and optimised in the design process. Many drive-unit and speaker manufacturers are limited (or limit themselves) to using just one amplifier make to test their designs and production. The situation is rarely publicised.

It follows that many loudspeakers are inevitably looking for that one kind of amplifier that was used when they were 'voiced' and 'tweaked' by their designer(s). With no less potential for habitual patterns, listeners are looking for the amp that interacts with a loudspeaker in a particular way. The combined behaviour or 'chemistry' is complex, and can be unpredictable and frustratingly unrelated to the type or class of amplifier.

Here is just one reason why the ideal high performance power amplifier/speaker combination can be determined *only by trying them together*, and why quite disparate amplifier designs and topologies may shine equally through a particular speaker.

Knowledgable 'high-end' loudspeaker designers and manufacturing companies employ a well chosen group of different power amplifiers. One or two will be the best sounding 'references'; some will be widely-used models, and not particularly good performers; others will be niche models, discovered accidentally over the years, that expose loudspeaker problems.

2.3.3 What passive crossovers look like to amplifiers

For conventional full range loudspeakers with passive crossovers, the crossover components stand between the amplifier and the speaker. Unless the speaker's drive unit(s) is/are blown, or have been disconnected or removed, then the crossover won't usually be 'seen electrically' (by the amplifier) on its own. Figure 2.14 shows the impedance *that would be seen* by the power amplifier *if* the loudspeakers were 8 ohms resistors, and how it drops from about 10.5 ohms to just over 5 ohms, at the crossover point. As both drivers are being driven at this point, it's what you might expect. Figure 2.15 shows how much the picture changes when the resistors are replaced by real speakers: now there are two dips.

In the crossover's pass-band where there should be no (attenuative) action on the signal voltage at all, the crossover should be 'transparent'. In practice, one pass

Figure 2.14

If speakers were simply resistors, the load on the amplifier might appear as here, with the apparent 11 ohm load simply halving to 5.5 ohms at the one point where the two drivers are both drawing substantial current, the crossover point. Note that the crossover dip may as well be a resonance, and as such adds to the amplifier's load stress.

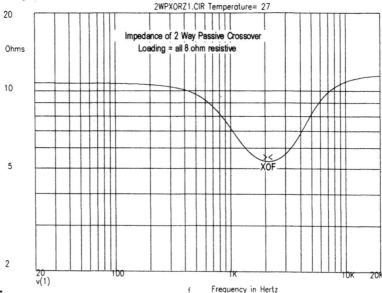

Figure 2.15

Here, the resistors are replaced by drive units having the impedance characteristics shown in the lower graph. The upper graph shows how the impedance seen by the amplifier has changed - notably two dips where there was one.

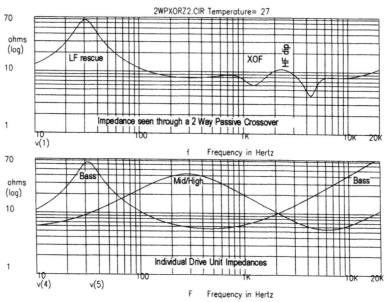

band is another driver's stop-band. Even a steady state test signal will experience the added reactive loading at most frequencies. In turn, the crossover is liable to add to the peak current demanded by the drive-units. With music signals, a passive crossover stores energy and can 'kick back' like a speaker, potentially adding to the speaker's demands. Overall, for the amplifier's own good, passive crossovers benefit no less than the drive unit, from being coupled to an amplifier with very low source impedance, with ample current sourcing *and sinking* capability.

2.4 Behaviour of power amps as voltage sources

In the preceding section 'Damping factor?', it was mentioned that most high performance audio amplifiers employ some global (overall) negative feedback and that this confers a very low output source impedance, usually well below 100mΩ (<<0.1Ω). This condition is known as a 'near pure voltage source', indicating the amplifier will *try* to double the power it delivers, each time the load's resistance is halved, and pro-rata for any other 'reasonable' load.

In practice, power into 4 ohms is typically between 195% of and 165% of the power into 8Ω (short of 200%), and power delivery into 2 ohms (if it doesn't trip-out or blow the amplifier) is even further short of doubling again, say 165% to 135%. For familiar economic reasons, the amplifier's power supply and power device, heat dispersal, and other *cost-linear-related* ratings, ultimately limit the safe or possible current delivery in every power amplifier. Exactly where this limit lies, how quickly it kicks-in, what the speakers are, and the kind of music, can have a large bearing on audio quality.

That amplifiers approximate voltage sources has been true since output 'transformerless' transistor amplifiers (invented in the mid/late 50's) arrived *en masse* with silicon transistors, around 1965. By contrast, nearly all valve amplifiers employ output transformers, have much lower levels of global NFB, and deliver their maximum power into the impedance that the transformer is set to; power delivery falls away gently as the speaker load deviates in *either* direction.

2.4.1 Drive-unit power ratings after EIA/AES

The power ratings of loudspeaker drivers used to be 'as long as a piece of string', but today, most reputable makers now specify drive-units' power handling according to the AES standard [7]. The AES test is an adoption of the standard pioneered by EIA (Electrical Industries Association, US), [8].

Testing employs pink noise, shaped by well defined clipping and bandpass filtering, so the resulting power spectrum is representative of globally averaged music program. This test signal amounts to a random, multi-sinewave stimulus, occurring almost equally at all audio frequencies over time, with plenty of waveform clipping (Figure 2.16). It replaces the older 'continuous rms' (strictly an average) power rating taken at a spot frequency.

Figure 2.16

Typical version of 'AES' pink noise, used for speaker rating. 3 plots each of peak (upper) and rms (lower) values, across the audio band. The level rises slightly across the band, to keep the PMR (the average different between the upper and lower curves) to about 10dB. The spikes in the peak response at LF are caused by the test system trying to settle on the random noise signal.

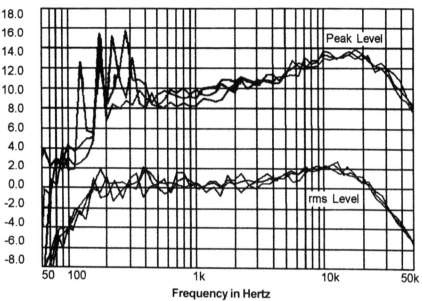

Frequency in Hertz

The driver's AES power rating is derived from power input needed to raise the temperature of the voice coil to the maximum safe limit. For a small tweeter, voice coil temperature can reach the limits in a few seconds, while a large bass-driver's coil can take several minutes. *But in the long-run, the maximum safe power input is decreased by the accumulated heating of the magnet structure* – even with HF drive-units, let alone giant LF drive-units, the magnet can take up to hours to reach thermal equilibrium. Afterwards, a drive-unit's performance parameters are retested, and if they've changed more than +/10%, the AES test has been failed.

2.4.2 Output power capability requirements

Audio amplifier power delivery capability (in watts) has a relationship to cost that is roughly linear, hence £ or $ per watt. But amplifier output capability has a highly non-linear, slack relationship with perceived loudness. Thus the particular power rating can matter surprisingly little (unless you are adept at thinking geometrically or logarithmically), so long as it is in the right area (numerically) and plays loud enough in practice.

The *order of* power required depends on a host of factors. These include:

i) the **sensitivity** (or efficiency) of the speaker system. Sensitivity is usually cited for 1 watt of drive, at 1 metre distance. The 1 watt is alternatively (and preferably) specified as 2.38v rms (the voltage for $1w \Rightarrow 8\Omega$) into *any* impedance. Sound level drops off at between 3dB and 6dB *per each doubling* in this distance, depending on speaker directivity and the acoustic conditions.

ii) the **maximum SPL** required. For some kinds of music and situation, 105dB will do; for others, $140dB_{SPL}$ will be required.

iii) the **room size**. How many metres distance are the above SPLs required at? If the room is at all big, then unless the listening position is near to the speakers relative to the room size, room acoustic data, such as reverberation time, will be involved.

iv) the **headroom** required. Like a car with a larger than normal engine, an over-sized amplifier won't easily be stressed, and sonics can benefit accordingly. Pro audio systems routinely employ amplifiers rated at five to even ten times the long-term 'AES/EIA' rating of the speakers they are used with.

For domestic hi-fi, home cinema, and recording studio control rooms, there are usually few listeners, often 1 or 2 and perhaps 10 at most. Here, room size is the key factor. In small rooms, listeners are typically only 2 to 4m distant. In large rooms, distance ranges 4m to 12m at most. Domestic speaker sensitivities average 87dB @ 1w @ 1m but range from below 80 up to 97dB @ 1w @ 1m. Maximum SPLs are usually satisfactory if between 100 to 120dB (C-wtd, peak) at 1m. There are already a range of variables, without mentioning the many other factors.

In practice, power requirements are never calculated from first principles by users. Instead, rules of thumb and experience give figures which are near enough most of the time.

For **domestic Hi-Fi**:
- with ordinary speakers and commonly acceptable sound levels – 30 watts to 300w. Go higher for bigger rooms and/or higher levels.
- with horn-loaded speakers, 0.3w to 10 watts, and upwards to 1kW if the neighbours can handle it, else as above.

For **studio nearfield** monitoring 50 to 500 watts overall; or 30 to 300w per band if active.

For **main studio** monitoring 250w up to 2kW per frequency band.

For **live sound** and 30+ people:
- in any indoor space, allow between 1 watt and 15 watts per person, with more watts per person for fewer people, and less for more.
- in any outdoor space, allow 2 to 15 watts per person for a large sound system. The increase in the minimum wattage mainly affects large gatherings, and insures against the effects of wind and humidity on sound propagation.

When taking a domestic system outdoors, Martin Colloms suggests that with the aid of a substantial wall, four times the indoor power would be needed, and considerably more in a fully open space.

Naturally, the power decided upon has to be close to what's available from the power amplifier, into the nominal impedance of the speakers it is to be used with.

2.4.3 Loudspeaker vulnerabilities

It's a common experience that audio amplifiers can deliver signals that destroy loudspeakers. High performance domestic systems are the most vulnerable, as abuse ruggedness is not much called for. Regardless of whether the design is domestic or professional, the fact remains that nearly every protection system taxes sonic quality, if not directly, then by stealing from the finite design and component budgets.

Whether professional or domestic, there are two main causes of failure in drive-units:

1. **Thermal**. Alias 'burning out'. Applies solely to electro-dynamic drivers, including ribbon types. Caused by excess power (energy integrated over time). When cone and compression drive-units are somewhat over-driven, the high temperature-rated adhesive holding the voice-coil wire (or foil) together, melts and deforms. If this causes the coil to rub, there will be distortion, sometimes only at higher drive levels, effectively making the driver unusable. The wire may also fracture from chafing or impact, either at first, or eventually. Or turns can short, changing the driver's characteristic. These fates are common in bass-drivers.

   If the over-drive is harder, and particularly if its onset is abrupt enough, the glue is burnt to a crisp and the conductor, if copper, may be heated to incandescence, before snapping. Large, unwanted RF signals delivered from amplifiers often have this effect.

   Either level of thermal failure is most common in HF drive units.

2. **Mechanical**. Applies to most drive-unit types. This covers ripped cones, diaphragms and surrounds, snapped 'tinsel' leadout wires, and fractured voice coil wire or foil.

   A loudspeaker cone attempts to move further as it is driven harder. It also attempts to move further when it is resonating, and in most enclosures, increasingly further when driven by lower frequencies. Large excursions are a problem especially for compression and hf drivers (if driven down to their low end limit), and bass cone-drivers. If driven some way beyond the maximum *linear* excursion (X_{max}) rating (in mm or inches), damage will ultimately result to the drive-unit, later if not immediately. The excursion this occurs at may be specified as the X_{damage} rating (again in mm or inches). For a high power bass drive-unit, X_{damage} is typically 300% (3x) X_{max}.

   Mechanical failures can also result from the hugely high-g-forces that hf drivers are subjected to. G-forces in bass drivers can be far less yet they are commensurately stressful, in view of their higher moving mass.

   Mechanical damage is rare in midrange drivers.

2.4.4 High power, the professional rationale

It was the demands of rock'n'roll sound reinforcement rental companies and guitar instrument amplifier makers that revolutionised the maximum SPL and hence dynamic range of the electro-dynamic drive-unit family: cone, compression and dome. Between 1965 and 1995, the ability of such drive units to sustain power without frying, and to give useful output in near proportion, has increased over 30 times, from barely 30w, to as much as 1kW. The techniques needed to cope better with sustained high power and excursions were pioneered as much in the UK (notably by Fane, ATC and Celestion) as in the USA (by JBL, Gauss and ElectroVoice).

The development was co-spurred by power amplifier developments, as throughout Rock 'n' Roll's history, the ratings of loudspeaker voice coils have lagged behind the power available to cook them. In the decade 1984–94, professional power amplifier makers competed at supplying a given power for less bucks, while also shrinking size and/or weight per watt. The exploration (making cheaper, smaller, lighter) amplifiers proved worthwhile, but while the amplifier makers' backs were turned, some loudspeaker system developments crept up.

The majority of drive units built with modern materials can now handle music transients with instantaneous 'power' equivalent to at least 3 to 5 (and up to 10) times their AES (or EIA) power rating, depending on the music's *peak-to-mean ratio* (PMR). To experience the full dynamic capabilities of such driver without risk of damaging them through clipping, amplifiers capable of delivering 1 to 10kW are required. Even for quite compressed music program with a 10dB PMR (averaged over a period), the long term heating (comparable to the AES rating) will be about 1/10th of the transient maximum (Figure 2.17), so a 600 watt rated driver (say) meeting this signal would be in no danger when driven by an amplifier rated for 2kW – assuming it has no excursion problems and is never driven into clip.

Excursion capabilities have been extended in some high power drivers. The extra excursion may not be very linear but it is more linear than hitting the end stops or ripping the surround. In other designs, the pole pieces or other 'end stops' are made softer, so damage through over-excursion is lessened. Remaining excursion (X_{max}, $X_{destroy}$) limitations at the bottom-end of any driver's range may be handled by suitable dynamic EQ processing [9].

Why does anyone need all this power? In main studio monitoring, direct radiating and vented speakers with from 0.4% to at best 2.5% efficiency are the norm. This is better than some domestic designs, but no better than a quarter as efficient as the horn-loaded designs. Improvements in reproduction accuracy are usually at the expense of efficiency, so with continuing refinements, increasingly high power handling is necessary for the ideal single drive-unit per frequency band to continue to comfortably reproduce the highest SPLs in music, say $140dBc$-wtd_{SPL}.

In medium to large-scale PA systems, for a given maximum SPL, higher power rated drive-units and amplification spells fewer drivers and cabinets. In turn, fewer

Figure 2.17

The upper graph shows a speaker-level music signal momentarily peaking at over 110v, and regularly peaking at over 55v. A 1kW/8 ohm amplifier would be required to handle this signal without clipping, but with just 1.5dB of headroom. The middle plot shows instantaneous power, which momentarily exceeds 1600 watts. This bothers the active devices inside the amplifier, but not most drivers. The lower plot shows the average power, the rms value that mirrors voice coil heating. Here, after 1.5 seconds it has levelled out at barely above 160 watts, or 1/10th of the peak instantaneous power. With adequate amplifier headroom, a 200 watt rated driver could be safely used in this situation, and with the same music signal, a 600 watt rated driver would be safe with transient powers and voltages up to 3 times as large. However, an amplifier capable of about 220v rms (i.e. 3kW into 8 ohms) would be needed to realise this capability.

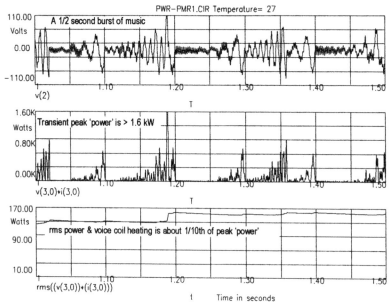

speaker boxes (subject to having enough for full coverage) generate a more coherent soundfield – as well as lowering operating costs. In the UK, Turbosound have made breathtaking reductions in array size with their *Flashlight* and *Floodlight* speaker systems. Amplifier makers may in turn be challenged to put more power than ever before in a medium sized box, rather than offer existing power ranges in ever smaller, deeper boxes.

Tannoy's System 15 DMT–II is a major refinement and ruggedization of their renowned 15" Dual Concentric monitoring speaker. Standing just over 3'/1m high, it's been used by the author to test amplifiers of all sizes. It has handled over 1300 watts of bass-heavy house music and reggae as well as Baroque and choral music, without distress, although nominally rated at about a quarter of this.

Increased power handling is equally about operating headroom and the effortless sound it permits, most appreciated by those who can afford it, and about avoiding the amplifier clipping that's so harmful to drive units and human hearing alike.

Reliance on limiters as the primary way to avoid output clipping creates a temptation to underspecify amplifier power, hence system headroom. This makes routine hard limiting more likely. The resulting compression may not be much kinder to ears and drivers than amplifier clipping. But *with adequate amplifier headroom*, limiters can be employed solely as safety nets, rarely to be heard.

2.4.5 Active systems, power delivery requirements

The considerations for amplifier power ratings are different for amplifiers that handle only portions of the audio band, because an active (or line-level passive) crossover is in use.

HF

High frequencies are where most of the peakiest transients are found relative to the average level. Amplifiers handling hf need (ideally) just as much voltage headroom (relative to the rms signal level) as mid and LF amplifiers. This is an important consideration when selecting amplifiers for multi-way active systems (Chapter 9). Dedicated HF amplifiers do not however need to contain the large and heavy parts that are needed for continuous high power delivery. It is worth remembering though, that some tweeters may be mainly capacitative (e.g. Piezos), and others may be driven at frequencies where capacitative loading occurs, which might cause unexpectedly substantial dissipation in a dedicated hf amplifier, with music having persistent high treble content, (e.g. metal, percussion) at the 'wrong kind of frequency'.

LF

This range, below 300Hz, ideally requires an amplifier having a higher, *sustained* power rating than the mid and hf. This is because the PMR of in some modern music forms, the PMR of subsonic bass signals in particular, can be less than 10dB, approaching that of a continuous sinewave.

2.5 Current delivery requirements

The currents that audio power amplifiers are required to supply cover a wide range. At the very worst, for almost opposite ends of the usable power capability scale, the difference in currents for a given impedance is just 100:1, as follows:

$$1w \Rightarrow 16\Omega = 0.25A \text{ rms x } (1.414 \text{ x } 6) \quad = \quad \textbf{2.1 Amperes}$$
$$10kW \Rightarrow 16\Omega = 25A \text{ rms x } (8.484) \quad = \quad \textbf{212 Amperes}$$

Note: These figures are based on Otala's caveat that a speaker's dynamic impedance *may be* as little as a sixth of its nominal rating, times 1.414 to show the peak (not rms) current value.

With live sound, and with the best analogue recordings, the smallest musically significant output currents will be around 120dB below this figure, or *one millionth* of these current levels, hence 2.1µA and 212µA respectively.

2.5.1 The low impedance route

Most professional drive units are nominally 8 ohms, and only a few professional power amplifier channels can yet 'swing a *continuous* k' ($1kW \Rightarrow 8\Omega = 89$ volts rms) or more into 8 ohms. After that, how do we get even higher power from one channel, hence without 'bridging'? Most modern, transistor amplifiers closely approximate a voltage source (see earlier), so increased output power is readily achieved by reducing the rated load impedance, down to 4, or even 2 ohms, not forgetting to ensure the output transistors, the power supply, the heat-sinking, and even the mains supply are rated to handle the extra current that will flow:

$$\text{0.125w} \Rightarrow 2\Omega = \text{0.25A rms x (8.484)} \quad = \quad \textbf{2.1 Amperes worst case}$$
$$\text{1.25kW} \Rightarrow 2\Omega = \text{25A rms x (8.484)} \quad = \quad \textbf{212 Amperes worst case}$$

Note: These figures are again based on Otala's caveat that a speaker's dynamic impedance *may be* as little as a sixth of its nominal rating, times 1.414 to show the peak (not rms) current value.

When power density is demand, ideally only to cover for breakdown(s), an amplifiers' 4 or 2 ohm rating is commonly utilized by connecting cabinets (or 2 or more drivers in one box) in parallel. While concentrating power delivery into fewer amplifier racks, this does nothing to concentrate power delivery into individual drive-units, so the system and number of sound sources shrinks less than it could. Also drive-units with higher power ratings are wasted, unless they have appropriately low impedances, which almost never happens.

Still, this way, it's long been relatively straightforward to attain notional kilowatts of power capability. It's how it was done in the first touring PA amplifiers of the early 70s, when there weren't any rugged BJTs that could reliably support a voltage swing much above 45v rms, as needed for $250w \Rightarrow 8\Omega$: Four cabinets were connected in parallel to a channel, so the amplifier was called upon to deliver a nominal $1kW \Rightarrow 2\Omega$. This is how it's still done in Brixton, London, world centre for alternative sound systems. Here, 'Charlie' produces monoblock MOS-FET amplifiers rated at 8kW into 1Ω, to drive twelve 15Ω bass-bins from one amplifier, for serious appreciation of the dub, with the minimum of equipment and expenditure.

Otherwise, a few PA amplifiers are still rated to drive 1 or 1.5Ω loads, a worthy reserve for handling impedance dips. But today it is recognised more widely that any low impedance drive capability in the amplifier should really be held in reserve *in case of* impedance dips when driving most cone speakers, particularly those reproducing bass. So a 1 ohm capable amplifier would have to be rated down to about 0.167Ω to meet high performance audio standards (again based on Otala's caveat that a speaker's dynamic impedance *may be* as little as a sixth of its nominal rating).

Loudspeaker makers *could* design drivers with a 2 or 1 ohm coil. With these, and enough parallel output transistors, adequate power handling could be had without using bridging or high voltage-rated transistors. But there would be tradeoffs, in magnetic field strengths (for EMC), and resistance losses including, skin and proximity

47

effect. Today it is also recognised that such a low impedance drive unit would be pro-rata more sensitive to cable resistance and inductance, a more purely reactive load, and not an efficient load for most amplifiers and their power supplies.

The exceptions to these high-current caveats includes ribbons, and ordinary moving-coil drive-units with conjugate passive crossovers, or copper caps, all of which present a nearly pure resistive load. Also, some of the high quality professional midrange and hf horns, and other drive units operated in 4-way or higher active systems, may be operated at those frequencies where they are mainly resistive. In these cases there are no reactive current peaks. *In all such cases, much or most of the usual low impedance driving reserve is not required and is available for parallel connection.* The most common example in touring PA is four 15Ω HF compression drivers driven from one 2Ω or 3Ω rated channel, which are loading it with a nominal 3.75Ω, that probably won't dip below 3.6Ω.

References

1   Duncan, Ben, *A versatile active crossover*, parts 1 to 4, HFN/RR, Feb to May 1981.
2   Otala, M, & Huttdnen, *Peak current requirement of commercial LS systems*, J.AES, June 1987.
3   Martikainen, Ilpo, Ari Varla, and Matti Otala, *Input current requirements of high quality loudspeaker systems*, preprint 1987-D7, 73rd AES, Mar 1983.
4   Norton, Thomas.J, *Questions of impedance interaction*, Stereophile, Jan 1994.
5   Duncan Ben, *Modelling cable*, EW+WW, Feb 1996.
6   Duncan Ben, *Black Box*, Hi-Fi News, Oct 1995; also, letter in 'Views', July 1994.
7   Hendricksen, Cliff, *Engineering justifications for selected portions of the AES recommended practice for specification of loudspeaker components*, JAES, Oct 1984.
8   *EIA/ANSI standard RS-426-A*, 1980.
9   Duncan, Ben, & Mark Burgin, *Dynamic loudness compensation*, RP-3, Proc.IOA, 1987.

Further reading

10   Colloms, Martin, *Basso Profundo - bass perception and low frequency reproduction*, Stereophile, Dec 1991.
11   Colloms, Martin, *Developments in loudspeaker system design, Acoustics Bulletin*, I.O.A, Nov & Dec 1995.
12   Duncan, Ben, *Monitoring, part 6 - drive unit power ratings*, H&SR, Apr 1984; and part 7 - *Drivers, safety zones and Xmax*, May 1984.
13   Duncan, Ben, *The spirit of bass*, EW+WW, Feb 1993.
14   King, Gordon, *Interface, 1, Amplifier to loudspeaker*, HFN/RR, Dec 1976.
15   Recklinghausen, Daniel.R.Von, *Mismatch between power amplifiers and loudspeaker loads*, JAES, Vol.6, 1958.
16   Woodgate, John.M, *Loudspeaker and amplifier specifications*, Maplin magazine, March 1988. (Newsletter of Maplin Electronic Supplies Ltd, UK).

The input port –

Interfacing and processing

3.1 The Input

For the user 'The Input' is often just a socket – often one groped for amidst a tangle of leads. This chapter untangles the details of the rarely recounted considerations that lie behind audio power amplifier input sockets, that enable the signal source to connect to the amplifier (and maybe to many amps) with the least loss of fidelity, and without introducing unwanted noise.

The amplifier is treated as a whole without considering the power capability or type of the output section.

3.1.1 Input sensitivity and gain requirements

Definition

Input sensitivity is the signal level at the input, that is needed to drive an amplifier up to its full capability, to just before clip, into a stated, nominal impedance, often 8 ohms. Clip may be defined as the onset of visible waveform flattening, or as a certain percentage THD+N distortion factor.

An older, less used definition (favoured in the 1978 IHF standard) is the signal level needed to deliver 1 watt into a given nominal load, say 8Ω. This is fine for comparing or normalising drive levels between amps having different power ratings, but as input sensitivity *per se* has no particular merit, the usefulness, for real amplifiers and speakers of widely varying power capabilities and sensitivities, ends there.

Description

Sensitivity is usually expressed as a voltage, either directly in V or mV (1/1000ths of a volt), or in dBu. Mostly, sensitivity figures are assumed to be rms values (cf. peak) and also specified with a steady sinewave, and for power amps in particular, with loading – all unless stated otherwise. If a peak (or any other non-rms) voltage value is cited, the maximum output to which it is referred must also be cited likewise – so like is being compared with like.

Variables

An amplifier's sensitivity depends (as defined above) on gain and swing. If an amp's output power rating, hence voltage swing capability into a given load impedance were to be increased, maintaining the sensitivity requires more gain from the amplifier. This is a consideration for the maker, and the installer who uses different sizes of a given design.

DIY gain resetting

For those uses with two or more different models and/or makes of amplifier, it is likely that sensitivities (however referred) will differ. Gain controls may not be present or it may be desired not to use them. If so, to align the system (ideally within a fraction of a dB), so for example all the amps enter clip at about the same drive level, the gain(s) of one type of amp will need changing. Usually, any gain controls are assumed to be at maximum. Then any 'accidental adjustments' can only cause reduced, not excess, gain.

In most well-designed, conventional high NFB power amps, gain may be easily changed up or down by changing one (global feedback) resistor per channel. The part being changed is usually in the output section. Changing gain by up to +10dB or down by as much as –6dB should have relatively little effect on sonic quality, assuming RF stability is not upset. But noise will be altered pro-rata.

In low- and zero-feedback designs, the availability of gain changing is far less, and effect on both measured and sonic performance of even a modest 10dB (x3) adjustment will be far more marked.

Gain restriction

In some power amp designs, gain changes may be unavailable – because they would upset RF stability, imperil a finally balanced gain/feedback structure; or violate some arbitrary %THD+N limit or other basic performance indication. Thus amplifiers from a product family spanning a range of output power ratings, may have either very similar gains (+ to –3dB); thus sensitivities (mV,V) almost commensurate with their ascending voltage swing. The upshot of this approach is (for example) a 2kW$\Rightarrow$8Ω amplifier which only provides 100w at normal drive levels (0dBu say). The +13dBu/3.5v rms input drive that is needed for full output makes it safer and more likely that the high swing will be kept in reserve as an inviolate headroom.

In other words, in *lieu* of increased gain when output swing is increased, such an amplifier will need to be driven harder. i.e. it is rated *less* sensitive. If the headroom achieved is ever used, then the higher input drive levels can cause increased distortion in the input stage. This effect will be most noted in esoteric amps with low feedback, but is still there in conventional high NFB amps.

Gain and fidelity

As noted, the positive side of having high swing amplifiers desensitised, by *not* increasing gain commensurate with the increased voltage swing, is that headroom occurs by default if the system's level/gain settings are not then altered. Reduced gain also reduces the risk of speaker damage by accidental loud blasts, dropped mics, styli, etc. Also, the audibility of the system's residual noise is lowered.

CM stress

In conventional power amplifiers with high NFB, 'common mode distortion', measurable as %THD+N [1], occurs due to *common-mode* voltage stress on the input stage, whether differential, or single-ended, the latter suffering CM stress if, as is common, it is non-inverting. The threshold voltage, 'V_{th}' – above which the input voltage to such an op-amp-type input becomes highly non-linear when open-loop may be sonically significant [2] [3]. These setbacks may not be revealed with conventional tests, notably %THD+N, which can contrarily show *lowered* distortion at high input drive test levels, because the noise (+N) may 'out-reduce' the rising common mode distortion [1].

Real figures

Every amplifier's sensitivity needs to match the *zero* (normal) *levels* of sources it is intended to be driven by. These vary. The upshot of all the factors is a spread of amplifier sensitivities that users know all too well.

Range of Input Sensitivities

Category:	in Volts:	in dBu:
Home Hi-Fi	30mV to 2v	−28 to + 8
Home Studios	100mV to 1v	−18 to + 2
Pro-Audio	775mV to 5v	0 to + 16

Ideally, there could be just one input sensitivity for all these uses. One that most could accept is the *de facto* professional standard of 0dBu alias 775 mV (millivolts). As a general rule, most lightweight domestic hi-fi and home studio equipment is likely to be more sensitive than 0dBu; and pro equipment likewise less sensitive.

However, as just discussed, a specific lower value, as low as 30mV, may be best (at least in high NFB circuits) from the viewpoint of circuit and device physics, for absolute best linearity [2]. However, the higher voltages that are mostly needed by de-sensitised high swing amplifiers (eg. driving 2v or +8dBu and above to clip) *confer the highest SNR,* hence dynamic range, *and also highest RF EMI and CMV immunity.* So the best of both these worlds appears not to be immediately reconcilable.

As most amplifiers are not pure voltage sources, when driven with continuous, high-level test signals into a real (or simulated) loudspeaker load (as opposed to an ideal, simple resistive load), the sensitivity (for a given clip level) can appear to increase at some frequencies, as the maximum output voltage with a conventional amplifier having an unregulated supply is reduced by typically by –0.5 to –2dB. The averaged shortfall is likely to be less with programme, at least at mid and high frequencies. *It follows that there is a complex frequency-conscious and dynamic peak-to-mean disparity in practical amplifiers' sensitivity ratings. The purer the voltage source, the less this can happen.*

Gain and swing

The following table shows the gain requirements both in dB, for some 'roundfigured' voltage swings, and how the nominal power then varies, into 4 and 8 ohms.

Table of Power Amp Gains

For 0dBu sensitivity @ clip   ⇒ means 'into'

Gain		rms Voltage swing	avg. Power ⇒ nom 8Ω	average Power ⇒ nom 4Ω
+24dB	= x16	12.5v	19w	38w
+30dB	= x32	25v	78w	156w
+33.5dB	= x48	37.5v	176w	352w
+36dB	= x65	50v	312w	624w
+40dB	= x97	75v	703w	1406w
+42dB	= x129	100v	1250w	2500w
+44dB	= x161	125v	1953w	3906w

For other sensitivities, gains are easily determined by appropriate subtraction or addition, e.g:

For +4dBu, *subtract* 4dB from the indicated gain(s).

For –10dBu, *add* 10dB to the indicated gain(s).

3.1.2 Input impedance (Z_{in})

Introduction

The amplifier's *input impedance* is the loading presented by the amplifier, to the signal source *driving* (or 'looking up' or 'into') it.

Impedances (often abbreviated 'z') are rated in ohms (Ω). As in this case the ohmic values are nearly always over one thousand, the counting is usually in thousands (k). 10k or 10kΩ ('10k ohm') are easier to say than 'ten thousand ohms'. When near a million or over, 'M' for 'Mega' is used, e.g. 1MΩ is 1000kΩ.

Common values

With ordinary, high NFB, power amplifiers, high input impedances (high Z_{in}, say above 10kΩ), to 1MΩ or more, are readily attained. For most sources, this is analogous to very light loading. But in most cases, power amp input impedances are commonly at the low end of this range, at between 10kΩ and 22kΩ. This restricts noise and buzzes when (particularly *un*balanced) inputs are left open, unused or floating, especially when cables are unplugged at the source end. This is less of a problem with short cables and in domestic environments.

The nominal values of amplifier input impedances vary widely. As a rule, professional equipment is the better defined:

Power Amplifier Input Impedances

Type of Power Amplifier	Z_{in} Range:
Domestic, sep. and integrated	10k-200kΩ
Hi-end domestic, esoteric	600-2MΩ
Professional	5k-20kΩ
Vintage Professional	600Ω

If balanced, Z_{in} is the differential mode Z.

An equipment's input impedance may be described as the source's *load* impedance. This is true enough at frequencies below 1kHz. But *load impedance* (since the signal source may be across a room, 100 yards down a hall, or even half way across a field) is the *totality* of loading – namely including all the cable capacitance – which takes effect increasingly above 3kHz.

Audio is not RF

Precise 'Impedance matching', where specific impedances (often 50 or 75 ohms) must be adhered to, is correct for radio frequencies, where cables above a metre or so act as a transmission line [4]. But at the highest audible frequencies (20kHz) even a 200m long input cable in a stadium PA system doesn't behave as transmission line.

Where the wavelength (the dual of frequency) is a fair fraction, say twenty or ten times greater than the cable, cables look mainly like the respective sums of their resistance, capacitance and inductance. As the ratio falls, the cable begins to behave increasingly like a transmission line.

Voltage matching

Since the widespread use of NFB (50 years ago), the majority of power amplifiers' inputs have been *voltage matched*. This means that the source impedance is low – much lower (at least ten times less) than the total destination, or load impedance [5,6]. The intention is to transfer the signal, which is encoded as a voltage 'wiggle', without significant loss of headroom, dynamic range or detailing.

The source's impedance – whatever's feeding the amplifier(s) – also has to be *low enough* and remain so at hf, to support a flat hf response into the capacitative loading of likely cable lengths. Voltage matching is defined by *de facto* industry practice, in the IEC.268 standard. Here, recommended *input* impedances are 10kΩ or over, and equipment *source* impedances 50Ω or less. This is easily memorised as:

Looking *Back* from amp: **Looking *Up* amp:**

$$\leq 50\ \Omega\ \Leftarrow \qquad\qquad \Rightarrow\ \geq 10k\Omega$$

With voltage matching there is no sharply defined 'right' impedance. Except that in high-CMR balanced systems and high resolution stereo systems alike, an amplifier's individual input impedances may be ultra-matched. Since with *voltage matched systems*, the wanted input signal is a voltage, the ideal, 'non invasive' amplifier input or load impedance would appear to be very high, say 1MΩ. Then only minuscule current would be taken from the source.

High impedances

Some high-end Hi-Fi makers have taken the high impedance route, claiming better sonics. This may be inseparable from the circuitry used to create the High-Z conditions, and not necessarily down to the High-Z conditions *per se*.

In power amps with low (or zero) feedback, and using BJTs in the input section, high input impedances (above 10kΩ) can be more difficult to implement consistently. On this basis, the early transistor amplifiers sometimes had their inputs rated in µA of input current drawn ! In contrast, there is usually no difficulty attaining impedances as high as 1MΩ or more, when the input stage parts are valves, JFET or MOSFET or any combination of these – whether loop or local feedback is zero, low or high.

When *un*terminated, such high impedance circuits are noisier (hissier) and far more liable to allow parts to be microphonic, than lower ('normal') impedance ones [7]. Demonstration is simple enough: try tapping the appropriate capacitors with an insulated tool, while listening with full-range speaker(s) connected. High impedance inputs can also be the cause of difficulties and compromises with direct coupling. Yet, unless the input is direct coupled, or is at least coupled via very large capacitors, LF and sub-sonic microphony and electrostatic noise pickup will not 'see' the lower source impedance, and will persist in accordance with the high impedance.

Low impedances

As input impedance is lowered, there is less microphony and electrostatic noise pickup when the amplifier inputs are disconnected - even with unshielded cabling. But loading is increased, as is ultimately the *susceptibility to magnetic field noise pickup - which is much, much harder to shield against.*

Loading

A single load of (say) 1kΩ may or may not compromise the source's performance. But two or a few of such loads almost certainly will, unless the source is rated appropriately (see below). Low impedance inputs are also the most easily damaged if one amp's output is accidentally connected to another's input. Added protection would add complexity, increase cost and likely degrade sonics.

In tandem

In professional (and even a few domestic) applications it is normal for each signal source to drive more than one amplifier input. The loading of amplifiers driven in tandem is cumulative: each added amplifier *reduces* the impedance (or *increases* the loading) *pro-rata,* in accordance with its impedance. Assuming conventional power amplifiers with 10kΩ input impedance, the reciprocal pattern is as follows:

No of Amps in tandem	Total Z_{in}
x 1	10 kΩ
x 2	5 kΩ
x 3	3.3 kΩ
x 4	2.5 kΩ
x 5	2 kΩ
x 6	1.7 kΩ
x 10	1 kΩ
x 15	666 Ω
x 20	500 Ω

Note that there are *very* few types and models of the likely sources (e.g. active crossovers, delay lines, preamps, etc) that are rated and able to drive impedances of below 600 ohms without degraded performance. Much pro-gear is rated and even spec'd for 600 ohms – but still gives its best measured and sonic performance into 2k or even higher.

For large tandem systems, existing equipment usually has to be retro-fitted with special *line-driver amplifiers*, or these are added as independent units, in line. Line-drivers used in live sound practice do not expand the allowable loading by much – usually down to 300 ohms, and possibly as low as 75 ohms. To be sure, only 50% of this rating would be used. The rest allows for tolerances, variables (see later), add-ons and the cable's capacitance loading at hf. In a major concert where 100 or more power amps have been required to handle just one frequency band alone [8], signal was split amongst up to ten line drivers, all *daisy-chained* off one line driver. This method is far preferable to having multiple crossovers, which might superficially simplify the signal path, but would also introduce near impossible set-up and band-matching demands.

Multi connection

When one signal has to feed many amplifiers, it is normal to connect the amplifiers by *daisy chaining*. To permit this, amplifiers made for professional use have both female (input) and also male (output) XLR (or other, gendered or ungendered in/out) connectors, linked together in parallel. '*Daisy chaining*' means physically, as the name suggests, that a short cable 'tail' carrying the input signal *loops* from one amplifier to the next in the rack or array. The signal being passed on is not really entering each amplifiers' input stage, but merely using the input sockets and case-work as a durable and shielded Y-splitting node. An alternative would be to make up a hydra-headed cable, ie. one splitting into *n* separate feeds. This would take up far more space and is far less flexible, but might prove the next best method if amplifiers without input 'link-out' sockets have to be used.

Ramifications

Professional power amplifiers, which are the sort most likely to have long cables connected to their inputs, and to reside in electrically noisy environments, mainly eschew impedances much above 10k. But if they're to be usable for live sound, their makers also can't welcome any much lower impedance, as this would further limit the number of channels that can be daisy-chained off a given line driver. In most multi-amp setups, the source that is being loaded is usually one of the band outputs of an active crossover, rated for 600 ohms with the NE5534 or 5532, 1977 IC technology that remains a *de facto* standard. In this common case, depending on the allowance for cable capacitance, between 10 and 15 amplifier channels (at most), should be driven.

Variables

As with other electronic equipment, input impedance is a function of electronic parts whose behaviour almost inevitably varies with frequency, and almost always depends on temperature. With unbalanced inputs, input impedance will also usually vary somewhat with the setting of the gain control (attenuator), if fitted.

Figure 3.1 shows how the input impedance of a typical, minimal power amplifier with an unbalanced input (Figure 3.2) varies across the frequency range. A 10kΩ gain control is assumed and is here backed off just 1dB. Note how the impedance in most of the audio band is almost constant at the scale used. Then notice how the impedance drops off (so the loading *in*creases) at high audio frequencies, and more so at higher, radio frequencies (Y). At low frequencies, if anything, the load imped-ance increases (X).

Figure 3.3 shows how the same input stage's impedance varies (without changing anything else) as temperature is changed from 15°c to 85°c. In other words, what can happen to the input impedance when an amplifier is 'cooked'. For the most part, impedance increases, which will do no harm. But in live work it *might* just alter a howlround threshold, as the higher load impedance allows the signal voltage to rise ever so slightly.

Figure 3.1

Input impedance (load) variation in an typical, simple unbalanced power amplifier input stage.

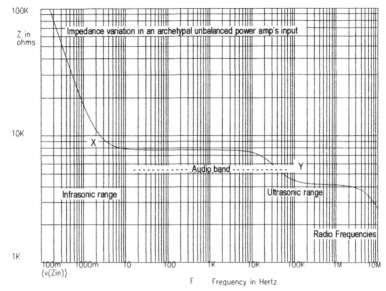

Figure 3.2

A typical unbalanced input stage.

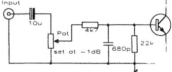

Figure 3.3

Impedance variation in a typical unbalanced power amplifier input stage as the amplifier warms up.

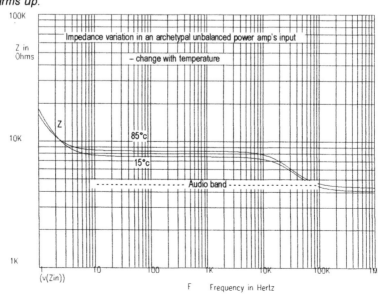

Figure 3.4

Impedance variation in a typical unbalanced power amplifier's input stage, as the gain control is swept.

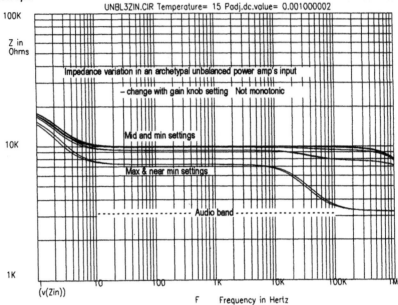

Figure 3.4 shows how the input impedance typically varies as the gain is adjusted. The change with each 30° rotation step is non-monotonic, so Z_{in} goes up then comes down, as you might expect. A 10kΩ log pot is assumed.

Ideally, an amp's input impedance would remain constant in spite of these changes. In unbalanced circuits, there is not much harm so long as any change in impedance is gradual and stays above certain limits, and anything which isn't like this happens well above (or even further below) the audio band. Staying constant is *far more* important in balanced circuits.

3.2 RF filtration

Introduction

Music starts out as air vibrations. These are not directly affected by EM (Electro-Magnetic) waves, *except while* they are passing through an audio system in the form of electronic signals. Planet Earth has long had natural EMI, in the form of various electric and magnetic storms; both those occurring in the atmosphere, and those occurring on the 'surface' of the Sun and Jupiter in particular. Since 1900, the planet has increasingly abounded in human-made EMI-babble, comprising electro-magnetic energy fields and waves, some continuous, some pulsed, others random. As stray signals nearly always have *nothing* to add to the music at hand, and most are profoundly un-musical, and as EMI permeates almost everywhere above ground unless guarded against, music signals require '*pro-active*' protection.

E-M waves used for radio broadcasting and communications mainly start in earnest at 150kHz (in the UK and continental Europe) and above, and continue to frequencies 10,000 times higher. However, special radio transmissions (for submerged submarines, national clocks and caving) may use frequencies below 100kHz, and even those below 20kHz.

Requirement

All active devices are potentially susceptible to EMI. BJTs, all kinds of FETs and also valves can all act as rectifiers at RF, demodulating radio transmissions. But this is very much more likely with BJTs, as the non-linearity of a BJT's forward biased base-emitter junction that gives rise to rectification, is triggered by considerably lower levels of RF voltage or field strength. All kinds of FETs and valves are relatively 'RF-proof' in comparison. *Caveat!* Oxidised copper, generally dirty contacts, crystalline soldered joints, or wrong metal-to-metal interfaces can all act as RF detectors as well, through rectification.

3.3 The balanced input

Balanced inputs – when used properly – can clean up hums, buzzes, RFI and general extraneous rubbish. When not used properly, the balanced-input's object may be partly defeated, but the connection will probably still improve the amplifier's and system's effective SNR.

Definition

To be truly balanced, a balanced input *and* the line coming in *and* the sending device must all have equal impedances to (signal) ground, to earth and to everywhere else. Also, the signal must be exactly opposite in polarity but equal in magnitude, on each conductor.

Real conditions

In practice, the signal is not of exactly opposite polarity. At high frequencies (and low frequencies in some poorly designed equipment) phase shifts add or subtract up to 90° or more, from the ideal 180° polarity difference. Otherwise the requirement for having a signal of opposite sign on each conductor is usually met. The exception is when one half of a ground-referred, balanced source has been shorted to ground. Not surprisingly, this degrades the benefits of balancing.

3.3.1 Balancing requirements

Input impedances

The norm in modern pro-audio equipment is 10kΩ across the line. This is commonly known as a 'bridging load'. It is also the *differential input impedance*.

The *common mode impedance*, what any unwanted, induced noise signals will see, is often (but not always) half of this, eg. 5kΩ in this case.

Considering the hum/RF noise rejection capability of an effective balanced input, input impedances much higher than 10kΩ, say 500kΩ, would seem feasible and useful in professional systems. However, if the input resistance is developed by the ubiquitous input bias path resistors connected from each input to the 0v rail, then there are limits to the usable resistance, before the input stage's output offset voltage becomes unacceptably high. Although low V_{oos} op-amps exist, a number of otherwise good ICs for audio have execrable DC characteristics, as IC designers do not appear to comprehend that good DC performance is a most helpful feature for high performance audio. In this case, input impedances above 15kΩ to 100kΩ are found to be impractical, depending on bias current.

A galvanically floating input, (i.e. the primary of a suitably wired transformer) has no connection to signal 0v (as it has no bias currents), so there can be a very high common-mode impedance, say 1M or more, up to modest RF. This aids rejection.

Conversely, differential impedances of less than 10k increase the influence of such random, external factors as mismatched cable core-to-shield capacitances.

3.3.2 Introducing Common Mode Rejection

Common Mode Rejection (CMR) is an equipment and system specification, that describes how well unwanted common-mode signals – mainly hum and RF interference – are counteracted, when using balanced connections.

Minimum requirements

At the very least, *all* the equipment in a system must have a balanced input (alias a 'differential receiver'). CMR can be improved and made more rugged when balanced inputs are used in conjunction with balanced outputs (alias 'differential transmitters'), but this is not essential.

What does CMR achieve?

CMR action prevents the egress and build-up of extraneous hum, buzzes and RFI when analogue signals are conveyed down cables, and between equipment powered from different locations – and all the more so in big or complex systems. CMR helps make shielding more effective by canceling the *attenuative residue*, the bit that any practical shield 'lets through'.

Sending the signal on a pair of twisted and parallel conductors ensures this latter residue and any other stray signals that are picked-up *en route* are literally coincident, and appear 'common mode', ie. equal to each other in size and polarity. A tight enough twist makes the conductors almost experience interfering fields as if they occupied the same space. This is true below high RF (200MHz, say), when averaged out over a cable's length.

By contrast, the wanted, applied signal from both balanced and unbalanced output sockets is distinguished while being no less equal in size by appearing *opposite* in polarity on each input 'leg', called *hot* and *cold*.

CMR also makes shielding more effective by freeing it from signal conveyance, enabling it to be connected at one end only, according solely to the dictates of optimum RF suppression and/or individual system practice. Breaking the shields' through connection also prevents (or at least lessens) the build-up up of the mesh of earth loops that causes most intractable hums and buzzes. CMR is also able to cancel differences between disparate, physically distant or electrically noisy ground points in a system.

Above 20kHz, even a modest CMR lessens the immediacy of the need for aggressive RF filtering. RF filtering can take place at higher frequencies and both the explicit and component-level effects on the audio may diminished accordingly.

Figure 3.4 shows the CMV that CMR helps the audio system ignore. Even when connection to mains safety earth is avoided by ground lifting (groundlift switch open), or by total isolation (switch open and groundlift R omitted), considerable capacitance frequently remains, through power transformers and wiring dress.

Overall, the rejection achieved (which is a ratio, not an absolute amount) is described in minus (–) dB. Often the minus is understood and omitted. In plain English, 'CMR = 40dB' means 'all extraneous garbage entering this box will be made 100 times smaller'.

Figure 3.4

Most of the Common Mode noise that CMR defends against is either RF and 50/60Hz fundamental intercepted in cabling (Vcm1), or 50/60Hz hum + harmonics caused by magnetic loop, eddy and leakage currents flowing in the safety ground wiring between any two equipment locations (Vcm2).

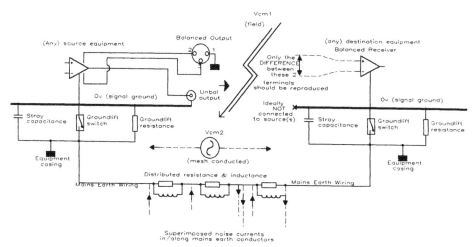

What CMR cannot do

Like the stable door, the one thing CMR can't do is remove unwanted noises that are already embedded, in with the music. It follows that just one piece of equipment with poor CMR and in the wrong place, can determine the hum and RFI level in a complex studio or PA path.

The ingress of Common Mode noise, called *Mode Conversion*, is cumulative, as each unit in the chain lets some of it leak through. So the CMR performance and/or interconnection standards of all the equipment in complex systems (e.g. multi-room studios and major live sets) must be doubly good.

The higher CMR of well-engineered equipment (80dB or more) provides a safety factor of 100 to over 1000 fold, over the minimum 40dB that's common in more 'cheerful' products. But the higher CMRs are more likely to vary with temperature and aging, as with all finely tuned artifacts.

Relativity rules

The size of common mode (noise) signals is not fixed nor even very predictable; they may range from microvolts to tens of volts. CMR is just a layer of protection. 40dB of protection is not much against 10 volts of CMR, but it is definitely enough for 1 microvolt.

Sonic effects of RF

RF interference is a Common Mode noise, and sources of RF go on increasing. In a competently wired system in premises away from radio transmitters and urban/industrial electrical hash, a modest rejection no better than 40dB has before seemed good enough to make inaudible induced 50/60Hz hum and harmonics, and the 'glazey' sound of RFI and RF intermodulation artifacts. Unfortunately, RFI artifacts aren't always blatant, and when any sound system is in use, they're the last thing that users are likely to be listening for the symptoms of. Yet even if there are no blatant noises, inadequate CMR can allow ambient electrical hash to cover-up ambient and reverberative detail.

System reality

The CMRs discussed are those cited for power amplifier input stages. The actual *system* CMR is inevitably cumulatively degraded by the cabling and the source CMRs. But it can be maintained by ensuring all three have individually high CMR *and* have highly balanced leg impedances. Lines driven from unbalanced sources give numerically inferior results, but often quite adequate ones (subject to appropriate grounding and cable connections) in low-EMI domestic hi-fi and studio conditions, where equipment connections are also compact; and even in outdoor PA systems, in open countryside.

Summary

Generally, 20dB is a low, poor CMR, 40 to 70dB average to good, and 80dB to 120dB or more, is very good and far harder to achieve in a real system. In a world where some audio measurements have had their credibility undermined, it's reassuring to know that with CMR, more dBs remains simply better.

3.4 Sub-sonic protection and high-pass filtering

Rationale

All loudspeakers have a low-end limit; their bass response does not go endlessly deeper.

Sub-sonic (infrasonic) information comprising both music content and ambient information, may occur below the high-pass 'turnover' frequency (or low-end rolloff) of the bass loudspeaker(s). It will not be efficiently reproduced.

Note: While potentially within humans' aural perceptive range, sub-sonic signals are 'below hearing' (strictly *infrasonic*) in the sense of being 'out-of-band' to, and only faintly or at least reducingly reproducible by, the sound system.

Loudspeakers vary in their ability to handle large sub-sonic signals. Small ones may or may not be heard but won't ever cause damage. Large sub-sonic signals are more risky with some kinds of loading. An approximate ranking of sub-sonic signal handling robustness is as follows. *Individual designs can vary widely* however.

Loudspeaker Subsonic Handling (Infrasonic Handling)

More Robust ⇑	Transmission lines.
	Differentially loaded cabs *
	Properly arrayed bass horns.
	Sealed boxes.
	Open-backed cabs.
	Large cone vented enclosures.
Less Robust ⇓	Small cone vented enclosures.

* alias bandpass or push-pull.

Sub-sonic stresses

Other than straining the speaker(s), if the amplitude of the subsonic (really infrasonic) signal(s) is large enough, then significant amplifier capability will be being wasted. At the very least, the unrealisable portion will cause unnecessary amplifier heating and electricity consumption.

If the amplifier is also being driven hard, the presence of a large sub-sonic signal will reduce the threshold for clipping and also thermal shutdown. The amplifier will behave as if rated at only a fraction of its actual power capability. There are broadly two approaches to the problem.

The pro approach

Sub-sonic filtering may be regarded as an essential part of editing and sweetening in recording. 'Sub-sonic' frequencies ('sub' here being rather loosely designated as any 'out of context/too-low bass information') are usually removed before amplification, by HP filters (HPF) with fixed, switchable or sweepable roll-off frequencies, usually available on each channel or group of a mixing console. Alternatively, HP filtering may even be available 'up-front' as a switch on some mics, or on portable, location tape machines.

Generally such filters are at least −12dB/octave, and more usefully the steeper −18dB/octave (Figure 3.6) or even −24dB/oct. They may be occasionally appended to professional power amplifiers, as well as to preceding active crossovers, on the basis of providing 'maximum' (read: brute force) protection at all costs. In this guise they are described as 'Sub-Sonic Protection' (SSP). Often this facility is superfluous and needlessly repeated, as the mixer and active crossover already do or can provide sub-sonic filtering.

Figure 3.6

Typical high-pass (sub-sonic protection) filter circuitry.

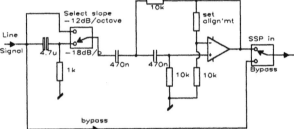

Logistics

The mixer can provide SSP most flexibly per channel, solely for those sources requiring filtration. The active crossover may provide overall back-up sub-sonic protection, in case a mic without HPF'ing on its channel, is dropped.

When sub-sonic protection is fitted to and relied upon in amplifiers alone, there will be an enforced and unnecessary repetition and diversification of resources, in any more than the simplest, 2 channel PA. If sub-sonic filter provision is made in an amplifier, it should be switchable (or programmable or otherwise controllable) so its action can be *positively removed* when not required.

Indication

A few power amplifiers have LEDs (often jointly error indicators) that indicate subsonic activity or protection shutdown arising from excess sub-sonic levels. This kind of protection is most common where the maker is also a speaker maker or where the amplifier is closely associated with a particular speaker as the protection's frequency-amplitude envelope that will allow the most low frequency action, is very specific to the cabinet and driver used.

Overall, in high performance professional power amplifier designs benefiting from modern knowledge, filtration and any HP filtering is avoided as far as possible or else minimised by adaptive circuitry [9].

Hi-end approach

In 'hi-end' hi-fi *and* professional power amplifiers, high-pass filtering is (or should be) depreciated or at least kept to the bare minimum, for two reasons:

First, because all practical HP filters progressively delay low frequencies relative to the rest of the music. Every added HP filter pole only adds to this 'signal smearing' [10]. Simulation in time and frequency domains shows this [11].

Second, because HP filters require the use of capacitors. Capacitors that are almost ideal for audio and not outrageously expensive and bulky, are limited in type and values. Capacitors which are *Faradically* large enough not to cause substantial 'signal smearing' are in practice medium type electrolytics, and not in practice nor in theory anywhere near so optimal for audio as other dielectric types.

For these reasons, even routine HP filtering (alias *DC blocking or AC coupling*) may be absent altogether. Figure 3.7 shows the points where HP filtering occurs in the majority of otherwise direct and near-direct-coupled power amplifiers. The output and power supply capacitors E and F are considered in chapters 4, 5 and 6.

Figure 3.7

High pass filter capacitor positions
The potential locations of DC blocking/HPF capacitors in the signal path of conventional transistor power amplifiers, assuming gain blocks (the triangles) are internally direct-coupled.

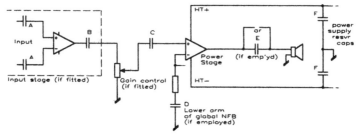

Low approach

In 'consumer grade' audio power amplifiers, HP capacitors are made as small as possible in value while maintaining what is judged by casual listening or first order theory to be an acceptable point for the bass response low cutoff frequency ($f3_L$). The result is considerable HP filtering, permanently engaged. Subsonic signals may then rarely pose a problem, but sonic quality may be degraded up into mid frequencies, while a great deal of the music's ambient cues are lost. See also 4.3.5.

3.4.1 Direct Coupling

When all HP filtering is removed, a power amplifier becomes direct- or 'DC' (*Direct Current*) -coupled. 'DCC' would have been better, but that now means something else.

Extending response to zero frequency, ie. 'down to DC', is readily achieved at the design stage with most transistor topologies. The advantages are sonic, and substantial, due to the excision of intrinsically imperfect parts *and* the removal of an intrinsically unnatural filtration, and the signal-delay and the possible charge accumulation on asymmetric music signals it brings. For this is the truth of all signal path HPF capacitors, both those in series, and in NFB arms. Whether DC coupling is safe or workable in a particular amplifier is a separate design question.

With conventional valve amp topologies, response to DC is not fully achievable, except in the few workable 'OTL' designs. But it is still possible to direct-couple the remainder of a valve amplifier, with global DC NFB taken before the transformer. In fact, the first precision DC amplifiers *were* valve op-amps.

DC management

With direct-coupled circuitry, unwanted DC 'offset voltages' will be amplified by the power amplifier's respective stage gains. Excess DC is of great concern and must be avoided. It can be:

(i) internally produced, by mismatches in resistor or semiconductor values; or by intrinsic topological asymmetry.

Or

(ii) externally introduced, from preceding DC coupled signal sources.

Internally produced DC offsets may be kept to safe levels by precision in design and component selection. This requires matching of two or three apposite parameters of the *differential pair* at the front-end of each stage, assuming some version of the conventional high NFB 'op-amp' type of architecture. The 'pair' might be BJTs, FETs or valves. And to ensure that the source resistances (at DC) seen by each input leg, are the same, or close, and not too high either, depending on bias current. If the resistor values then conflict with CMR, the latter should have priority, in view of EMC requirements, and the non-recoverability of CMR opportunity. DC balance may be restored by other means, e.g. current injection.

Externally applied DC, appearing on the inputs, due to essentially healthy but imperfect preceding equipment will usually be in the range of 0.1 to 100 millivolts. More than +/–100mV would suggest a DC fault in the preceding source equipment. Assuming a gain of 30x, this would result in 3v at the amplifier's output. Such a steady *offset* will eat up headroom on one half of the signal swing, so the clip level is lowered asymmetrically. A direct coupled power amplifier should not be harmed by this, and should also protect the speakers it is driving, but equally it is entitled to shut down to draw attention to such an unsatisfactory situation. In the most advanced designs of analogue path yet published [9], DC coupling is adaptive: if DC above a problem level persists at the input, DC blocking capacitors are automatically installed and the user is informed by LED.

Some low budget domestic power amplifiers have long offered part and manual direct coupling. The power stage may not be wholly direct coupled, but at least the DC blocking capacitor(s) at the input can be bypassed via a second 'direct' or 'laboratory' input. The user is expected to try this but revert to the ordinary AC coupled inputs if DC on the source signal is enough to cause zits and plops.

Autonulling

DC offset may be continually forced to near zero volts by a *servo*, which is another name for brute-force VLF and DC feedback, applied around an amplifier overall, or just the input or output stage. Servos have been *de rigueur* in US and US-influenced high-end domestic power amplifiers for some years. Alas, those who have designed them into high performance power amplifiers have clearly *not* thought through the consequences. Tellingly, servos are not usually nor likely to be found in amplifiers with truly accurate sounding bass.

The reasons are clear enough today: Servos cause the same or even wilder distortions in LF frequency and/or phase response, and/or signal delay vs. frequency (group delay). Figure 3.8 shows this.

Figure 3.8

DC servo-circuits cause at the very least the same phase and delay error as using a dc blocking capacitor conventionally. The upper graph shows the frequency response of a standard two pole Servo (2 x {1MΩ x 470nF}). The lower graph shows the phase shift, which is clearly non-linear below 85Hz - place a ruler against the line. The curvature indicates frequency-dependent signal delay, hence smearing (after Deane Jensen). An alternative, custom 3 pole compensating type (C3P) is plotted. This overcomes the smearing, as phase shift is much less than 0.1° above 5Hz, but the amplitude (upper) is now peaking below 1Hz.

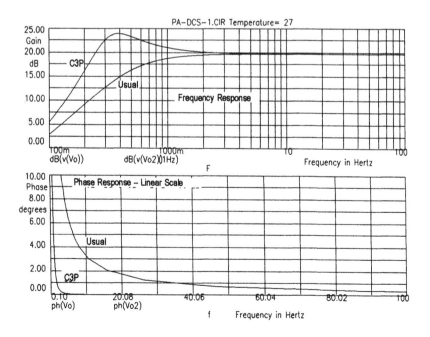

They also compromise the integrity of the circuitry they are wrapped around, by increasing noise susceptibility, while the capacitor imperfections that DC coupling is supposed to overcome, are reintroduced, since distortion-free DC servo action depends on an expensive, bulky, high performance capacitor for integration. In this way, the DC servo returns us to *before* square one, with the added cost and complexity. Worse, the original thinking behind servo'ing was to save money (!) on input transistor and part matching, as a servo will 'fix' any DC in its range, often up to +/–5v, including DC appearing on the equipment input. This is neat, but like so many 'smart' options, DC servo'ing is not *quite* suitable for audio.

3.5 Damage protection

The input stages of most audio equipment are unprotected. This approach appears to save on parts cost, complexity and sonic degradation, but in reality, it may indeed *cause* costs and degraded sonics. The inputs of power amplifiers are certainly amongst those most likely to sustain input voltages that may be damaging to the active parts inside.

Causes

Typical culprits include firstly, large signals from line level sources, and from amplifier outputs, experienced through accidental connections (see section 8.5.2). Here, excessive signal voltages that could be applied could range from a few volts, up to 230v rms, and from below 10Hz to above 30kHz.

Second, the outputs of crossovers or consoles, or misconnected amps, which are kaput and have DC faults, so the output voltage might range from +/–10v to up to +/–30v for line sources, and up to +/–160v DC for power amplifiers, but more typically +/–30 to +/–90v DC.

Scope

The parts most at risk from excess input voltages are the solid-state active devices, particularly discrete BJTs, and most monolithic IC op-amp input stages.

Valves are relatively immune to input voltage abuse. They are most likely to be harmed by gross overdrive conditions that bias the grid positive, so a damagingly high current flows.

J-FETs and MOSFETs are next most rugged. MOSFETs are most susceptible to gate-source overvoltage, but gate-source protection is straightforward and effective.

IC input stages are the most fragile. Due to IC structure, even FETs, when monolithic may have parasitic weak points. For long term reliability, currents flowing into or out of IC op-amp pins [12] must always be kept below 5mA.

Harmful conditions

There are *two* kinds of potentially damaging input voltages:

1) Common mode.

2) Differential mode.

Either may occur when a power amplifier is in:

i) the on-state, or

ii) the off-state

...giving four possibilities.

On-state risks

When an amplifier employing BJTs at the front of its input stage is on, powered up, and settled down, it can sustain relatively high differential (signal) voltages without damage. Generally, in high NFB op-amp and other dual-rail based designs, the max differential voltage is a volt below the supply rails, hence a maximum differential voltage ranges from +/−14v for input stages working from +/−15v supplies, up to +/−30v or even over +/−100v, where the input stage transistors operate from the same or else similarly high supplies, as the output stage.

Long before differential overload, the input stage will be driven strongly into clip. Provided the amplifier has clean recovery, an *overvoltaging* may pass un-noticed if the high differential voltage only lasts 1mS. Yet this is plenty long enough to damage a semiconductor junction. In BJTs, the most vulnerable junction is the Base-Emitter, when reverse biased.

Under the same powered-up conditions, common-mode voltages above +/−10v can damage unprotected BJT input stages. In large systems, the common-mode voltage can be this high, commonly comprising 50/60Hz AC and harmonics, and arising from differences in grounding or AC power potentials.

The input stage's supply rail voltage usually has a large bearing on the maximum safe DM and CM input voltages. Here, low supply voltages may do no favours.

Off-state vulnerability

When an amplifier using BJTs is switched off, both differential and common-mode voltages as low as +/−0.5v may be damaging. In chapter 9, users are advised to always power-up preceding equipment before the power amps. This is universal practice amongst informed users, both domestic and professional. But if the pre-powering of the source involves the passage of signals above 0.5v peak to amplifier inputs, then unless the transistors behind the sockets are protected before the amp is powered-up, they may well be damaged. This mode of subtle, progressive damage and sonic degradation to analogue electronics has yet to be widely recognised. It can be overcome without changing otherwise sensible practices, by suitably designed input protection.

Occurrence modes

Damage to input devices may be catastrophic, if the overvoltage causes high currents to flow. This is rare.

Otherwise, with BJT inputs, damage may be subtle. Transistor parameters are degraded but NFB action initially hides the worst. Telltale signs would be changed or reducing sonic quality, raised, increasing and/or intermittent noise, higher %THD, and possibly increased DC offset at the amp's output.

With ICs, damage may be cumulative, caused by peculiar metal migration effects occurring in ICs' microscopically thin conductors. This means an input stage can appear to handle abuse repeatedly – until eventually the catastrophic failure occurs when all the conductor has migrated away!

Protection circuitry

Power amps have been designed to survive likely levels of both CM and DM overvoltages by the use of some combination of the following:

1. Series input resistors, which may already be part of the input stage's RF filtering, will limit the current flowing into inputs. If the resistance between the input socket and the active device is 5kΩ, then above 25v DC or peak signal would be needed to get more than 5μA to flow.

2. Back-to-back zeners to 0v, working in concert with series current-limiting resistors (which may already be part of the input stage's RF filtering). Both CM and DM voltages can be clamped to any available zener voltage. Designers must allow for quite wide variations with tolerance and temperature, and possible sonic degradation. Programmable zeners may also be used; or zeners may be combined with BJTs.

3. Ordinary, fast diodes across the active differential inputs, in concert with series input resistors in both legs. Protects against DM overdrive only. Internal to some IC op-amps, eg. NE5534. External diodes with larger junctions may be used to enhance protection.

4. Clamping relays. Placed after the series input current limiting resistors, inputs are shorted to 0v until power is up on all rails. With suitably rapid action and power sensing, relays in this configuration can provide complete protection against both DM and CM input signals.

5. Birt [13] describes a method developed at the BBC, using VDRs, zeners and current sources, providing input protection to audio balanced line inputs (including power amps) up to 240v AC. Alas, sonic quality may be detracted from.

3.6 What are process functions?

When in use, an audio power amplifier is *always* but part of some greater system. In domestic audiophile and even recording studio systems, it is commonplace for power amplifiers to have no gain controls – and to be devoid of any processing functions.

But in professional music PA applications, by contrast, it is the exception to find power amplifiers *without* panel gain controls (really attenuators). This facility turns into a *system processing function* when the gain control element becomes remote controllable, most particularly when all the amplifiers in a system or grouping are so equipped and also when the rate of gain control change is fast enough for it to be used dynamically.

3.6.1 Common gain control (panel attenuator)

The most common, almost universal form of 'gain control' is passive attenuation, set usually via a panel knob, with a rotary *pot* or pot*entiometer.*

Characteristics

As 'voltage matching' is the norm for modern audio, pots are nearly always wired in the voltage divider mode, where the wiper is the output. At this point, the source impedance seen varies, up to a maximum of a quarter (25%) of the pot's rated value (i.e. the end-to-end resistance) at half setting. At the pot's maximum and minimum settings, the source impedance reaches a few ohms above zero, which is usually much less than the preceding signal source's impedance.

Common values

In audio power amplifiers, the pot's value is commonly 5 or 10kΩ in professional and audiophile grade equipment, and 20, 50 or 100kΩ or even higher in 'consumer' grade equipment. The lower pot values offer lower maximum impedances at half-setting, for example just 2500Ω (2.5kΩ) for a 10kΩ pot. This lessens the scope for noise pickup in the inevitably unbalanced and relatively sensitive part of the amplifier circuitry where the pot is placed.

Audio taper

These considerations are true for ordinary pots with an audio taper, i.e. those marked 'log' or 'B'. As shown wired in Figure 3.9(i), these normally sweep over the maximum possible range of level setting, from a purely nominal –∞ (hard CCW or 'shut off', really more like –60 to –70dB) up to 0dB (maximum level). The 'audio taper' alias logarithmic resistance change per ° rotation, makes the change in sound level reasonably constant with rotation. The full span and audio taper are relevant when a pot is needed to act sometimes as volume control, where output levels very much lower than the power amplifier's capability are useful. It's also relevant where a quick sweep to –∞ (infinite attenuation) may be needed as a mute – to turn off the signal in one speaker, say – without switching off or unplugging anything.

Figure 3.9

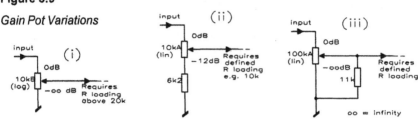

Gain Pot Variations

The right range

In many applications, the range offered by a raw pot is far too wide. In other industries employing pots, a vernier or a multiturn mechanism is added between the knob and shaft, to aid fine settings. But these are eschewed by modern professional audio operators, partly because of an ingrained fear of the loss of instant sweep control, and because of relatively high cost versus relative fragility. There is also the false sense of alignment is suggested by the verniers' 3 or 4 figure scale; scales on different amplifiers would be strictly incomparable, owing to most pots' poor tolerances, particularly good-sounding log pots. In the past twenty years, variations of 5% to 25%, (or 0.5dB to 3dB) have remained the *norm for the resistance mis-match* between different pots at the same mechanical setting.

Linear variants

Figure 3.9 (ii) shows how using a linear (A) pot and a fixed resistor, adjustment range is restricted to the 'top' 12dB, i.e., 0dB to –12dB. For system adjustment, this may be more usefully expressed as +/–6dB. This range of adjustment is preferable for active-crossover-based and arrayed systems, where the gain of individual amplifiers benefits from close adjustments, and only needs this limited range. In practice, switched (say) –20dB and –∞settings are then required. Note that the impedance vs. rotation relation is naturally slightly changed – the highest source impedance is here less at about 20% (rather than 25%) of the pot's value.

Returning to the full scale mode, a linear pot may alternatively be used (Figure 3.9 (iii)), with a fixed resistor used for 'law faking'. This converts the linear law to a log-like curve. If the pot and resistor values are kept within tight limits, this approach can give approximations of an audio taper that are at least more consistent than most log pots – which are made by butting *n* different-valued linear track segments together. Note that the pot's effective value is here a tenth of its rated value after the law faking resistor is included. So the pot shown in Figure 3.9 (iii) looks like a 10kΩ pot to the load. But the maximum source resistance is, as with the audio taper, at the 50% attenuation point, and is just about 10% from maximum.

Gain Control Identification

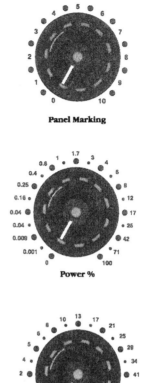

Panel Marking

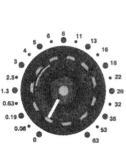

Attenuation (dB)

Power %

Voltage Gain (V Out/V In)

Voltage %

Voltage Gain dB

Input Clip Volts (PK)

Figure 3.10

Gain pot Settings

Six ways of looking at any power amplifier's gain control, in this instance the simplest and most familiar 'volume' control type. The final knob labelled 'input clip volts (pk)' scale is for peak levels, and is correct only for an amplifier that clips at 900mV rms. In reality, the point would depend on speaker loading, mains voltage, the programme, etc. The constant 9.6v peak reached at lower levels shows where the input stage clips, or else where zener-based input-protection clamping is operating. Courtesy Citronic Ltd.

73

Position

As in other analogue audio circuits, the placement of any gain control device requires careful considerations, in regard to considering trade-offs in headroom and SNR. But in power amplifiers having a minimum path, there is not much choice for location. They all end up after the input is unbalanced but before it is raised far.

Placement couldn't be contemplated after the point of signal passing to the input of the power stage, for example, as pots having film tracks (*cf.* wirewound) that are suitable for audio by virtue of low rotation noise are unsuited to high dissipation. In any event most power stage topologies don't have a place for inserting a single-ended, passive voltage divider, don't like having their gain widely changed, and are moreover wrapped around by NFB.

Adequate CMR (at the amplifier's input) demands good balancing and this in turn relies on resistance matching to better than *at least* 0.5% and since even makers of very expensive, high specification pots have problems maintaining matching between two or more sections to even 2%, over the entire travel, pots passing audio have to be placed *after* the input signal has been converted to single ended, i.e. after the debalancer (DTSEC). Virtually all power amplifier gain pots (or whatever other gain control devices) end-up thusly sandwiched. A few are used in active mode, where the pot is used in the NFB loop, of either an added line-level stage, or even a gain-change tolerant power stage. This seems smart but it has its own problems.

Fixed install

In amps principally intended for fixed installation, whether for a home cinema or public venues, and where power amplifier gain trims are needed or helpful for setting up, 'knobless' gain controls are welcomed. Here, shafts are normally recessed and can only be turned with a screwdriver. This avoids not just casual tampering, but knobs being moved (and settings lost) by accidental brushing, sweeping or knocking. A collet nut may be included. When tightened, the setting will then be immune to attack by screwdriver, as well as vibration creep. As a further discouragement to 'let's turn this up', such controls may be placed on the rear panel of the amp, or hidden behind cover plates.

3.6.2 Remotable gain controls (machine control)

Pots are mostly made to interface with human fingers, via knobs. When a sound system moves past the point where a single driver in each band can handle the power required, or where *Ambisonic* or other multi-channel sound is contemplated, remote control opens the door to 'intelligent' control of loudspeaker systems and clusters, including balancing and tweaking directivity, imaging, and focusing, by machines and via wires and radio links. The gain of an amp can be controlled by a variety of electronic means (Figure 3.11). The purely electronic means are fast enough to perform additional, true processing functions, e.g. limiting.

Figure 3.11

The family tree of electronically-controllable gain & attenuation devices.

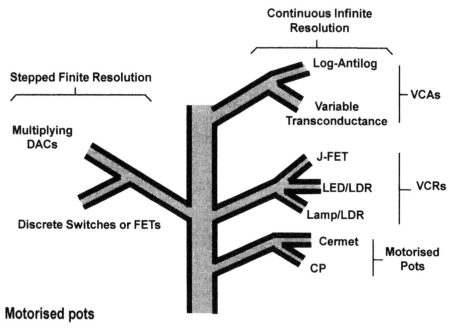

Motorised pots

Usually a motor connects to the same shaft as a knob, but the latter via slipping clutch. Either may override. This keeps the simplicity but shares the wide setting tolerance, sonic and some of the mechanical limitations, of ordinary pots, e.g. fragile shaft, relatively low setting speed. Which overrides the other depends on which way confidence most leans – toward human fingers, or computers! Control circuitry is needed to decode remote command signals which may be a variety of formats. Special driver ICs (e.g. BA series made by *Rohm* in Japan) make design and manufacture easy but might pose major replacement headaches to some owners in the future.

VCAs

Commonly called a Voltage Controlled Amplifier, but most are used as VC *Attenuators*. Usually a solid state and always an analogue circuit. Most are ICs based around one of a limited number of proprietary schemes, that are made (or licensed, e.g. *That Corp.* in US licenses, *National* in Japan) by one of three main patent holders, all in the USA [14]. Otherwise they6 are based on a discrete circuit; or on a consumer grade 'OTA' IC. Gain is accurately settable to within a fraction a dB, down to at least –70dB and even into positive gain with some parts. Gain is always defined by an analog control voltage (or current) that may be locally derived after decoding from a digital line or buss. Refined VCAs introduce considerable added circuitry into the signal path, which may defeat its own purpose. The simplest parts add two stages. They may boast low noise but it is at the expense of exposing the unnatural distortion patterns they create. The best performers add as many as five sequential stages, and more than five op-amps may be required. If part quality is not to be compromised, the added cost seems high. Operating speed with most types can be very high, under 1μS. In this way, VCAs *and all the following contrivances* are applicable to dynamic functions, up to the fastest meaningful audio *peak limiting*.

[LED+LDR]s

With this method, the control signal drives an LED (Light Emitting Diode), so that full brightness is defined as either maximum level or full attenuation. An adjacent LDR (Light Dependent Resistor) acts as the upper or lower arm of a passive attenuator. The intrinsic circuit isolation and physical separation that is possible makes LED/LDRs attractive in systems where isolation (of both grounds and common-mode voltages to 2.5kV or more) is important for safety or EMC. These parts provide remote control connections analogous to connecting digital feeds via opto-isolators.

Tolerance is an issue, and is dependent on the constituent parts, both semiconductors. The tolerance of both LDRs and LEDs is rather wide, and so manufactured combination devices are likewise broadly specified. The performance of both devices also varies widely with temperature. And in many circuits, there is no negative feedback loop to keep these variables within limits. Thus LDR/LED combinations are unsuited to system gain control due to inconsistencies of say +/–3dB. They *are* fast enough to be used as limiters for bass and even mid frequencies in active crossover systems, and sonic quality is regarded as amongst the best. However, the above gain variation (in a population) would translate as spectral imbalance making overdriven conditions in a large system unsafe and/or uncomfortable, as well as drawing attention to the limiter action.

An LDR may also be partnered with incandescent lamp. Even if small, the lamp is relatively slow to turn on and off, preventing its use for clean-cut dynamics processing, and lamp lifespan is more vibration sensitive, and so not as certain as solid state parts in road-going use.

J-FETs

J-FETs (Junction Field-Effect Transistors) are the lowest cost elements and can be made operative with little support circuitry. They are normally applied in the lower arm of an attenuator network. Without introducing complications of increased noise, noise pickup and other sonic degradation caused by introducing high ohmic value series resistors, attenuation is limited in range, and unless added circuitry can be justified, mild attenuation (around –6dB) produces high (1% to 10% but mainly benign, low order) harmonic distortion [15]. Low distortion control can be attained by placing the JFET in a control loop, comprising 2 or more op-amps and other active parts. However, as most JFETs' R_{on} is in the order of a few tens of ohms, attenuation is still typically limited to –20 to –30dB, enough for limiting, but not as a VCA gain and mute control.

M-DACs

Multiplying DACs (Digital-to-Analog Converters) involve a resistive ladder, usually binary, with semiconductor switches, usually small-signal MOS-FETs. They are the solid-state equivalent of a relay-controlled attenuator ladder (below). Types suitable for high performance audio must have dB steps –awkward in binary format – and special MOSFETs for low distortion and absence of 'zipper' noise. The latter undesired sonic effect occurs in low grade M-DACs; it is caused by step changes in DC levels or feedthrough from the digital control signal. Unlike the previous elements, an M-DAC has discrete resolution – just like a stepped ('detented') pot. At low attenuations, step size must be no more than 1dB for precise control; below –30dB, larger steps (2dB) are usually fine enough. To attenuate down to –70dB in 1dB steps 12 bit M-DAC is required .

R&R array

Comprising resistors and relays, this is the mechanical counterpart of the M-DAC, with relays opening and closing paths in a 'ladder' or other array of (usually) discrete attenuator resistors. Only high reliability, ATE-grade, sealed reed relays are suited for high performance audio, on grounds of both reliability and sonics. Such relays can act in under 1mS and have fast settling, but are still not really suited to dynamics processing ! Getting dB steps to act binarily with a resistor array takes some lateral thinking. Although the relays required are relatively expensive, by ingenious network adaptation to increment in binary dB, a mere seven can offer a 60dB range in 1dB steps. With suitably well-spec'd resistors, this type can offer the highest transparency of any gain control device.

Summary

Motorised Pots, Lamp+LDRs and Relay/Resistor arrays are good for remote- or machine-controlled gain trim and setting. The latter are the fastest and likely most reliable.

J-FETs and LED+LDRs are good for dynamics processing but attaining accurate, non-invasive performance takes from initial simplicity.

VCAs and M-DACs are the elements which can do both kinds of job well.

3.6.3 Remote control considerations

Computers regularly feign precision that is only virtual. Until gain control elements become self-checking, self-calibrating and self-aligning, they require careful specification.

Temperature

Pots (particularly conductive plastic), JFET, LDR and particularly VCA elements are quite temperature sensitive. Unless designed with very low *tempco*, then when used in 2 or more channel amplifiers, they must be placed iso-thermally, i.e. co-sited to be independent of all the major temperature gradients, dependent on drive patterns, siting and even amplifier and rack orientation – since a hot gas usually rises upwards relative to the earth's surface. This is true even with amplifiers employing forced venting, when small signal parts are not in an airpath, and are left to cool by microconvection, conduction and re-radiation.

Without such precautions, differences in channel gains of 2dB have been observed in an amplifier employing VCA-controlled gain, when driven up to working temperatures. This is enough to cause howl round or upset spectral balance.

Repeatability

Remote gain settings must not drift or have repeatability errors which can accumulate to cause more than (say) +/–0.15dB total error. This may seem stringent, yet on top of an initial tolerance of another +/–0.15dB, it allows a worst case total difference between speakers of 0.6dB. Other errors (cable losses, driver mismatches) are of similar order and add to the differencing toll, so there is no room for complacency. Least is best.

Conclusion

M-DACs and relay-resistor-array attenuators have the highest stability against temperature and time. Other types may prove acceptable with ameliorative engineering. Setting precision should not be taken for granted.

3.6.4 Compression and limiting

Compression and limiting (*comp-lim*) are gain reduction, alias dynamics processing techniques, that are employed (amongst other things) to protect speakers, ears and amplifiers from excess, distorted signal levels. In professional, active-crossover-based systems, they are usually embodied within the active crossover. This is the best position for logistics in traditional large systems, with only one comp-lim per band to worry about. Positioned within the filter chain can also be the best location for sonics.

Where power amplifiers are driven full-range, or where active crossover filters sections are integral to the power stage, compression and limiting functions may take place within individual power amplifiers.

Compression must be used sparingly, otherwise average power dissipation in the drivers will be increased, potentially part-defeating the object, as speakers may then suffer burn-out. Paradoxically, the compression threshold (at least for bass frequencies) should be *increased* if the gain reduction exceeds about 6dB. Also, attack and release times require careful setting to avoid pumping on strong low bass.

Limiting is a higher ratio, more brute force (many dB-to-1) gain reduction. Its *raison d'etre* is to catch fast peaks, hence '*peak limiting*'. Attack times that are useful for protecting most loudspeaker drivers are in the order of 10µS. Faster rising peaks that 'get through' rarely cause damage to hardware, but may be efficiently reproduced by metal-diaphragmed drive units (*cf.* paper cones), and perceived and found highly unpleasant by the ear. Hence faster-acting peak limiters may enhance sound quality under many real conditions of 'operator abuse'.

3.6.5 Clipping (overload) considerations

Driving any power amplifier with excessive input results in clipping because the output's excursion is finite. Amplifiers offering higher power into a given load impedance provide a higher voltage swing into that impedance, so clipping for a given sound pressure level is less likely to arise. But linear increases in power give only under-proportionate, logarithmic increases in headroom (in dB), and cost linearly ascending amounts of money. At some point, whatever more swing could be afforded would make no difference, and a limit is set. Exceeding this is clipping. For short periods it can be benign but else it is unpleasant and potentially damaging to hearing, and positively damaging to hf and bass drive-units in particular. Moreover, considerable overdriving, into *hard* clip, as can happen at any time by accident, even with domestic systems, can heavily saturate and thus vaporise, the BJT output stages of inadequately designed power amplifiers.

3.6.6 Clip prevention

Destructive and anti-social clipping may be prevented with comparatively simple circuits performing like a dedicated, fast limiter. There are as many names as there are makers. Some example are:

ARX Systems	*Anticlip.*
Carver	Clipping eliminator.
Crest Audio	*IGM* (Instantaneous Gain Modulation).
Crown (Amcron)	AGC in PSA2 (Automatic Gain Control).
Malcolm Hill	*Headlok.*

In these and related schemes, clip prevention does not occur until a dB or so of clip. Using the 100w analogy, the usual low %THD does not rise until the signal passes above about 50 to 70 watts. If headroom is adequate, this point should hardly ever be reached with the majority of recorded sound. With live sound, it may be reached quite often, but the fact that the deeply unpleasant point only 1dB higher is *not* crashed through, is of *far* more importance.

3.6.7 Soft-Clip

'Soft-Clip' is a feature that aims to defeat the suddenness of the onset of hard distortion above the *clip level* in conventional, high NFB power amplifiers. It may be provided as a fixed or switchable option. Unlike compression and limiting, there are no time constants, no settings, and no attempt to avert serious distortion of a sinewave. But the clipped waveform does not readily square off, and retains some curvature (dV/dt) even with heavy overdrive (e.g. at +10dBvr). This greatly reduces the massed production of unpleasant, high harmonics and intermodulation products of hard clipping. One apparent (but not necessarily actual) snag is that because hard clipping is a real limit, soft clipping has to begin to occur up to –10dB below full output (–10dBvr). This is tantamount to saying that distortion (%THD say) with a 100 watt amplifier begins rising from above about 10 watts, as opposed to rising very abruptly above exactly 100 watts, while remaining extremely low *up to* this point. Here is one difference between low and high *global* feedback amplifier behaviour.

Soft clipping restores the more forgiving behaviour of low-feedback to a high NFB amplifier. The extent to which it undoes all the high feedback's other benefits is unqualified. At least the high NFB is in operation for most of the time, for with proper headroom allowance, most of the musical content should lie below the –10dB threshold or so, whence the soft clip is inactive. Usually soft-clipping is arranged to be symmetrical. This may not create the most consonant harmonic structure – see chapter 7. Figure 3.12 shows a classic circuit.

Figure 3.12

A typical soft clip circuit - as used in the Otis Power Station amplifier.
Copyright Mead & Co 1988.

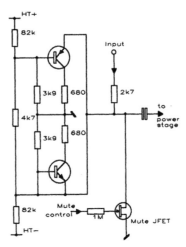

3.7 Computer control

Computer control of audio power amplifiers has been slow to develop. This is because amplifiers have not been a useful place, in most instances, for physical control surfaces. With a virtual control surface, the traditional limitation vanishes. In turn, installation setup, constant awareness of status, and troubleshooting of amplifiers in medium to large installations are all enhanced. One person can 'be in six places at once'.

The Dutch PA system manufacturer Stage Accompany was a pioneer of the computer-controlled and monitored PA system, in the mid-80s. But the first widespread commercial system that wasn't a dedicated, integrated type, was Crown's *IQ*, running on Apple Macintosh (1986). The second was Crest Audio's aptly named *Nexsys*, running on PC. Most subsequent systems have been IBM-PC compatible types, running under Microsoft's *Windows*. Every system is different yet offers similar, fairly predictable features; there is no clear-cut choice. At the time of writing (1996), some 'future proofed' universal, non-partisan, networkable system contenders that seem most likely to become industry standards appear to have priced themselves out of consideration. Instead, makers continue rolling their own. Recent examples include the IA (Intelligent Amplifier) system by *C-Audio*; the MIDI-based interface used by MC² (UK); and QSC's *Dataport* system.

Today's computer-control systems theoretically offer:

(i) the remote control of many of most of the facilities and controls considered here and in other chapters.

(ii) the flexible ganging, nesting and prioritization of these controls.

(iii) the transmission of realtime signal, thermal, rail voltage, or PSU energy storage data, monitoring, logging and alarming. May even include a measure of utilisation. For example, if a particular amplifier's swing is largely unused due to an overspecification.

(iv) the remote, even automatic, testing of amplifiers, speaker loads, and their connections.

Thus far, most computer controlled power amplifiers require an interfacing card to be plugged-in. Some types have integral microprocessors.

A well designed computer control interface must not affect the analogue systems grounding, nor compromise mains safety. These requirements are met by the fibre-optic, opto- or transformer-coupled interfacing, familiar enough in digital audio. Such systems must also not only meet EMC requirements, but also, in real world conditions, not radiate, nor introduce EMI to the power amplifiers. The system must also be able to recognise faults in its own connectivity to power amplifiers.

References

1   Ball, Greg.M, *Overlook THD at your peril*, letters, EW+WW, Aug 1993.
2   Cherry, Prof. Edward, *Ironing out distortion*, EW+WW, Jan 1995.
3   Jung, Walt, *Audio applications*, Section 8 of System Applications Guide, Analog Devices, 1993.
4   Penrose, H.E and R.H.S. Boulding, *Principles and Practice of Radar*, Newnes, 4th ed., 1953.
5   Duncan, Ben, *Black Box,* HFN/RR, Oct 1994.
6   Bohm, Dennis, *Practical line driving current requirements*, Sound and Video Contractor, Sept 1991.
7   Duncan, Ben, *AMP-01 parts 3 and 4*, HFN/RR, July and Aug 1984.
8   Duncan, Ben, *Building the world's biggest PA*, Lighting and Sound International, Oct 1988.
9   Duncan, Ben, *A state of the art preamplifier: AMP-02*, Hi-Fi News, March 1990.
10 Duncan, Ben, *Delayed audio signals*, EW+WW, May 1995.

11 Duncan, Ben, *Signal chain*, Studio Sound, June 1991.

12 Buxton, Joe, *Input overvoltage protection*, pp 1.56 to 1.73 of section 1, of System Applications Guide, Analog Devices, 1993.

13 Birt, David, *Electronically balanced analogue line interfaces*, Proc.IOA, Vol.12, part 8, 1990.

14 Duncan, Ben, *VCAs investigated*, parts 1-4, Studio Sound, June to Sept 1989.

15 (Nameless), *FETs as voltage controlled resistors*, FET data book, Siliconix, 1986.

Further reading

16 Augustadt H.W., and W.F.Kannenberg, *Longitudinal noise in audio circuits*, Audio Engineering, 1950, reprinted J.AES, July 1968.

17 Fletcher, Ted, *Balanced or unbalanced?*, Studio Sound, Nov 1980.

18 Fletcher, Ted, *Balanced or balanced?*, Studio Sound, Dec 1981.

19 Huber, Manual, *Conceptual errors in microphone preamplification*, Studio Sound, Apr 1993.

20 Ott, Henry,W, *Noise Reduction Techniques in Electronic Systems*, Ch.4, John Wiley, 1976.

21 Perkins, Cal, *Measurement techniques for debugging systems and their interconnection*, 11th AES conference, Oregon, May 1992.

4

Topologies, classes and modes

4.1 Introduction

Topologies

There are seemingly infinite ways to connect parts to form a high performance audio power amplifier's *output* and *drive* stages. In the broadest sense, these 'ways' divide into different *topologies* (unique electrical connectivities or *identities* in schematic form). As in any other branch of engineering, the development of new topologies has been driven by the need or desire to:

(i)   achieve certain performance ratings or goals, e.g. X watts or Y% efficiency; or symmetrical limits, ability to drive a given speaker; nil manufacturing adjustments, and/or many other goals, often simultaneously.

(ii) achieve certain performance ratings or goals, with given available parts; or to overcome intrinsic or present day limitations in components and other materials, particularly with early, developing types of a given component.

(iii) make more profit by making (i) or (ii) simpler to achieve.

(iv) get around the ownership by competitors of patents on topologies which do achieve the above (i,ii,iii).

(v) simply exercise unstoppable human ingenuity and inquisitiveness.

Uranus, the Greek god of the electronic arts, is too scattered and busy with new projects in the expanding cosmos, to hand out perfect topologies for mundane human needs. Further, there are none so simple as 'good' or 'bad' topologies, although some are more elegant or convoluted than others. All topologies are like people: complex mixtures of largely undocumented chemistry you partly do and partly do not want. In turn, different designers and engineers find they have an affinity with particular topologies. An individual designer's focus and determination can potentially make any topology shine.

Classes

Classes 'A' and 'B' are well known to anyone who reads Hi-Fi magazines, but these and the other classes have been deeply and widely misunderstood – even by major makers who should know better. *In this book, class is taken to be a second tier of identification, since most amplifier topologies and sub-topologies are employable for most of the classes.* Of course, a few circuit topologies *are* specific to the amplifier's *class*; the class precedes the topology in such cases.

There are at present just five distinct, recognised *classes* of audio power amplification. With the approximate first date of commercial application in brackets, they are called:

(1) Class A (1917) and variants (>1960), *Sliding, Super,* and *Sustained Plateau*, etc.
(2) Class B, A-B and variants, ($\cong$1945) eg. *QPP*.
(-) Class C is used for RF. It is *in*applicable to audio.
(3) Class D – 'digital' or *PWM* ($\cong$1963).
(-) Class E is un-specified for audio.
(4) Class G (1977) and,
(5) Class H ($\cong$1983) are both adding onto classes A, B or their variants.
(-) Class S (*sic*) has been used to classify a kind of valve biasing unconnected with modern audio power; and later, unrelatedly, for a control method, which is erroneously titled. See section 4.10.2.

At heart, all these five class designations describe is fundamentally different kinds of output stage current behaviour in the controlling devices, labelled very nearly in the order they were first put to work. Beyond class A (which was not named as such until there was a class B) the principal *raison d'etre* of the other classes is an improvement in power conversion efficiency. These will be examined later in this chapter, beginning at 4.6. For now, it is sufficient to be aware of the conventional classes A and B.

Modes

This is a tertiary (3rd) layer of meaningful difference to with which to categorise different kinds of high performance audio power amplifiers. It embraces methods of control or error-correction, the best known being (Negative) FeedBack (NFB). There are many variations of NFB alone, many of them inadequately explored, and beyond, several more ways (such as feedforward) to help an amplifier control itself *and* the speakers' moving parts.

4.1.1 About topologies

In electronics, *topologies*, giving identity, allow hundreds of seemingly complex and disparate circuits to be classified into family groups, and the higher relationships grasped better as a result. It is an *ad-hoc* task, for while botanists are still cataloguing new plants (and there haven't been many new models for in the past 10,000 years), electronics designers are busy developing more circuits than there will ever be plants, without a thought as to classification.

In all analogue audio power amplifiers, *output stage* topologies are *at heart* about the way that speakers are connected to *power sources*, in series with *active devices* that modulate the power delivery as an *analogue* of the music's pressure waves (Figure 4.1). There is of course more to the story: the active devices have to be stably biased to be reliably active; and they have to be fed with a signal; and the operating conditions must be within ratings of existing parts. It's also important that the controlling active devices must give *gain*, i.e. must be controllable by fewer amps and volts than they are themselves controlling. And much, much more.

Figure 4.1

Mother of all Power Amplifiers.

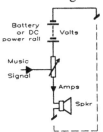

Before beginning, a recap of some basic details and also conventions:

o) The words '*transistor*' and '*device*' or '*power device*' are used interchangeably and as synonyms, where the part being referred to is always some kind of transistor, that may be BJT or MOSFET, or even IGBT, as appropriate.

i) There is only one polarity or gender for valves; but for transistors, whether BJTs or MOSFETs, there are two. This hugely increases the possible permutations. For simplicity, all the initial schematics in this section will show BJTs. Historically, most did use BJTs, but all *could* use MOSFETs. Figure 4.2 is a reminder of how the two BJT genders, *npn* and *pnp*, are shown and distinguished. Note with pnp the arrow points inwards – 'introverted' is a possible mnemonic.

Then in the topological drawings that follow:

ii) All circuitry is simplified to help counter '*not being able to see the wood for the trees*'. Makers and designers are naturally tempted into hiding or gilding those commonplace building blocks they may have used. Their schematics can sometimes be hard or impossible for any outsider, even a master topologist, to readily comprehend as a result. Here, nearly every drawing is uniformly presented. 'North' (top') is always more positive where feasible. Coupling transformers and capacitors are shown where they are essential to the discussion. Biasing and local feedback resistors (R_e where shown) are largely omitted in early drawings, but are shown with increasing frequency later as the eye accommodates the other details.

In practice, while not shown, negative feedback is almost always in use, either just locally, in the manner of the just discussed as *degeneration* resistors; and in many cases also connected globally, in a loop. The potential connections of this latter form are arrowed 'NFB'. The final section beginning 4.10 deals with modes of control including feedback.

Figure 4.2

OPS Topologies, Symbols Used.

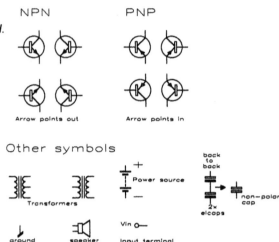

iii) Power sources are shown for simplicity as batteries. Figure 4.2 shows the symbol. Where the batteries are not connected to the same node as the ground, a floating supply is implied. The batteries may be taken also to represent the supply reservoir capacitors. The series impedance ('through') the battery (or capacitor) is very low at audio frequencies. *In other words, for music signals, the power supply source is transparent, a short circuit.* The indicated voltage range is a constant reminder of the range of real needs – with the notionally maximum 300v single supply offering just above 100v rms or about 700 watts into 15 ohms. Where two ('split') rails are shown, the total voltage required does not increase; it is simply a translation, and the voltages are halved accordingly as a reminder of this.

iv) *Ideally*, but not always, the grounded point will be the common point between (a) the input, (b) the output in most cases; and (c) not necessarily, between output and one or more power supplies. It is useful for makers, practical users and repairers, to have the ground node the same as one or other of the output transistors' main terminals, depending which terminal (Collector or Emitter; or Drain or Source) is connected to the case. This point is never mentioned in textbooks.

v) The input terminal is called 'Vin' (Input,Voltage). Ideally it is referred to 0v (ground) and if so, 0v appears as the open-ended line under Vin. Where an input connection must be floating, two pins are shown with 'Vin' between them. In a few topologies, two floating inputs are required, hence there are four input pins.

vi) All topologies are shown 'positive side up' (+ve towards the top of the diagram) so as not to faze the reader. This is at least consistent, as it follows on from valves. This is not the case in many 60's circuits, including the handful abstracted later in this chapter for their contemporary flavour. Do not be alarmed if these seem to be inside out or upside down. In the days when all transistors were pnp, it was evidently more helpful for readers to see emitter legs placed on the 'low side', than have positive 'above' – a sort of topological typesetting convention.

Subsidiary to the topologies, there are many so called 'building blocks' in electronics. A number of these are central to the development or workable existence, of the more sophisticated topologies. In the coming sections, they are described as *sub-topologies*, and are introduced in their approximate historic order of appearance in audio power amplifiers. There is even a place later for a higher-level *super-topology*.

There are several reasons for logging topologies. First, it has not been done before, and in the vacuum, there has been a great deal of haphazard topological information, much of it incorrect. There even appear to be some patents granted in the USA (where else ?!) on topologies that are not new at all, but mere re-arrangements with a new, often fancy name. Second, by discussing topologies in a light and readable way, amplifier users who are *not* circuit designers or *topologists* will be able to recognise what kinds of circuits they are really listening too (by comparing the maker's schematics with the simplified schematics in this book), and communicate with others informedly. Third, a solid, coherent comprehension of the significance of topological features, and why high performance power amplifiers are like they are, would likely prove impossible without imposing some structure on the vast diversity. Without it, there could only be an *ad-hoc* ramble.

4.2 Germanium and early junctions

The first transistors were point-contact. They were not for audio, or much else, but they did establish the potential [1]. The first junction transistors were npn, but later, most were pnp. So the tables were reversed (from what is familiar today), with *npn* (or *n-channel*) being the gender limited in availability, expensive, and so preferably avoided in design.

All early transistors principally employed germanium. They had high temperature sensitivity, readily developing potentially fatal leakage currents (I_{ceo}). Their absolute maximum *junction temperature* was 100°c – which is far less ample than it sounds. Even if the signal did not heat up the junction enough to cause failure, *thermal runaway* could. Runaway could be triggered by marginally poor ventilation, or a bias voltage that did not reduce with increasing temperature to compensate for slightly hard-driven transistor junctions.

The early germanium junction power transistors also had very low f_Ts (transition frequencies), so low they were well within the audio band, e.g. 7kHz. They should never have been used for audio, at least not with global feedback, nor for full-range use.

4.2.1 Out of the vacuum-state

The first transistor audio power stages followed existing valve topologies. Yet these topologies were developed *because of* the limitations of valves, namely (i) no opposite polarity *complement* devices, and (ii), limited heater-to-cathode voltage differences, necessitating many floating heater supplies if the topology was at all adventurous.

Figure 4.3

Topology 1: Single-ended, Transformer coupled.

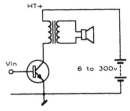

The simplest transistor power amplifier output stage is single ended (Figure 4.3). Here, a single transistor is driven by the incoming signal, and its collector is loaded by the speaker, reflected back through the transformer winding. The impedance 'step-down' arrangement lowers the voltage swing across the speaker but did enable a relatively low current-rated transistor to drive the speaker more 'stiffly' more than it otherwise could. This stage can operate in class A only (see section 4.6.1) and is not very efficient – except by the standards of class A amplifiers. The transformer output coupling is needed; otherwise the speaker would have to pass the bias current (DC), which in class A, is large. But the transformer's primary still has to pass a DC bias current. Though smaller than it would otherwise be, this saturates the core, degrading the transformers' performance. We are left with an amplifier requiring an expensive, weighty, oversized core for any chance of fidelity, and which is practically suited only for low powers, below a few watts. Worse, this transformer alone severely limits the amount of overall ('global') NFB that can be applied, without ringing or oscillation at both high *and low* frequencies.

Biasing for topologies like the one pictured in Figure 4.3 was critical when there were only germanium transistors. It relied on an *ntc thermistor*, ideally clamped to the power transistor, or its heatsink. The resistance of the thermistor would progressively reduce with increasing temperature, so reducing the bias voltage or current, or even the gain.

4.2.2 Push-pull, Transformer-coupled

With valves, power stage designers had long ago overcome the inefficiency of single-ended class A, going as far as class B, A-B and related schemes (see 4.6 *et seq.*). Valves were arranged in push-pull, i.e. they were driven and delivering differentially, or in 'phase opposition'. This ideally removes any DC biasing of the transformer core. Push-pull operation, essential for class B, could be used for, and benefit, class A operation. In all these schemes, transformers were not just used for impedance matching, *but also to produce and combine currents and/or voltages in phase opposition*. Without them, symmetrical push-pull would have been problematic, without complement devices. Push-pull action also ideally cancels those even harmonics that are self-generated. So it reduces distortion – but *doubly* selectively, i.e. (i) only if even harmonic and (ii) only if self-generated.

Figure 4.4 shows a double transformer coupled scheme, a direct translation or descendent of classic valve topology. The input transformer is ultimately no good for accurate audio, but it simplified the circuitry in low-performance amplifiers, and *to*

Figure 4.4

Topology 2: Push-Pull Transformer-coupled, CE type.

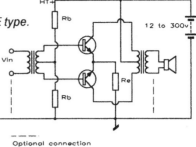

Optional connection

those simple-minded enough to believe that the map is the territory. In reality, making a good input (*interstage*) coupling transformer is no easy task. Operation could be class A or B. But in class B, the output transformer gave problems, as the stored magnetic energy was left to ring each time one side cut-off. The ringing caused voltage spikes that could cause the transistors to experience C-E voltages that were much more than the rail voltage, and their own ratings. Death was instant and there were no warnings. An unexpected 'free lunch', due to the maths of dissipation, or really, the high efficiency of a push-pull transistor output-stage operated in class B, is that transistors rated at 10 watts each, could deliver 49 watts into a (perfectly resistive) speaker before exceeding their ratings! This was later exploited in Sinclair's infamous *Z30* DIY hi-hi module and relatives (1969), which pushed cheap, low power driver transistors to their limit.

Figure 4.5

Topology 3: Push Pull TX Coupled, CC type.

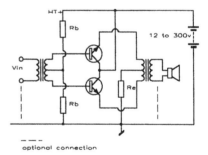

optional connection

Figure 4.4 employs the output BJTs in the *common emitter* (CE) configuration. This offers voltage gain, but lower current gain. The transformer's step-down ratio would have compensated for this, offering one excuse for employing a transformer. Figure 4.5 shows the alternative *emitter follower* (alias common collector) configuration. This configuration is possible with valves, but wasn't uppermost, as it stirs up trouble with heater-to-cathode voltage limitations, or else demands the expense of multiple floating heater supplies. In this configuration, voltage gain is just below unity (x0.95 typically), but current gain equals beta. Alas, beta is very variable with individual BJTs and with junction temperature.

4.2.3 Sub-topology: the Darlington

Early on, the cascade connection of two BJTs to achieve very high current gain, was discovered and patented at Bell Labs, birthplace of the BJT (see Appendix 1).

Figure 4.6

Sub-topology 1: The Darlington

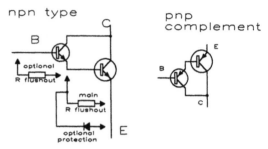

The particular connection, shown in Figure 4.6, has long been known as the *Darlington*, after its inventor. The two transistors may be identical, but for most utility, the left hand (input) device may (and often need only) be a 'smaller signal' part with a lower current rating (but no less high a voltage rating), while the second or 'final stage' device would be rated commensurate with the maximum current. To suggest the likely difference in maximum current and power handling capacity, the transistor symbols are sized accordingly on this occasion.

In early transistor amplifiers, where inadequate current gain (*beta*) was quite a problem, the Darlington reduced distortion, and also production variability, caused by the naturally wide spread of beta in early and any unselected BJTs. In modern high performance power amplifiers, it is still used and is often still made (like the original) from two discrete transistors, despite the existence of many *Darlington transistors*, which are monolithic ICs of a power and driver transistor, in the Darlington connection. These types are not much favoured, as one or other of the constituent BJTs may be sub-optimal for the designer's requirements; may be unsuited to audio, inadequately rated, bad sounding, or they may have the wrong value and quality of *flushout* resistor connected internally. The flushout resistor is essential to ensure the second, larger BJT turns off. Other resistors and diodes may be required to ensure clean, benign operation under all conditions.

There are other forms of two-transistor marriage able to create a high gain supertransistor – all loosely termed 'Darlington'. One, often called 'the compound pair' but far better named after its Americanized-Japanese inventor, Sziklai [2], is particularly effective [3,4] (Figure 4.7) as it has self-feedback.

Figure 4.7

The Sziklai (Compound Pair).

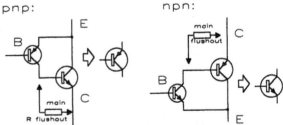

4.2.4 Transformerless push-pull (transistor OTL)

Elimination of the output transformer was a major step forward in high performance audio. This had long been recognised, since the 1930s, when designers first struggled to make *output-transformerless* (OTL) valve amplifiers [5]. Valves have the difficulty that they are intrinsically high voltage, low current devices. The amount of current available is out-of-step with the requirements of ordinary electrodynamic speakers of nominally 16Ω and below. The output transformer was a logical and theoretically elegant way to compensate for this. The reality is very different though. Even today, with computer design, good power stage output transformers that do not wreck measured and/or sonic performance are complex and expensive, both to design and manufacture. They can account for as much of an amplifier's weight and size as the usual 50/60Hz power supply transformer.

Figure 4.8 shows prototypal OTL transistor topologies. The small differences in where the ground is placed, and thus how the two, floating push-pull (differential) input signals are connected, has a large effect on the behaviour – with or without global NFB. The first configuration (a) is half-common emitter, which has moderate voltage and current gain, and also a moderate output impedance; while the other (b) is half-emitter follower (common collector), with unity voltage gain, but potentially high current gain, and a low output impedance. In both cases, the transistors might be Darlingtons. On the other side of the coin Figure 4.8 (a) might require a substantial current from the preceding *driver stage*, while (b) will in all cases require an input signal voltage slightly larger than the anticipated output swing.

Figure 4.8

Topology 4: OTL Types.

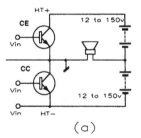 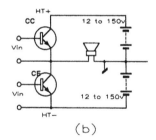

(a)   (b)

4.2.5 Sub-topology: diode biasing

Biasing is essential, but omitted in the surrounding topological discussions for simplicity. A snag in the early, elementary class B (and A-B) transistor power amplifier topologies where capacitor input coupling replaced the interstage input coupling transformer, was that music's asymmetry caused a DC voltage to build up, as the output or drive transistors' junctions rectified the signal [99]. Extra parts, typically a BE reverse-connected, parallel diode with a series resistor, were needed to prevent this. Long after coupling capacitors were abandoned in output drive stages, the same diode connection has since reappeared in cautious designs to protect BJTs; once again to do with charge control under overload.

Jones and Hilburn demonstrated the uses of diodes as biasing elements in the late 50s. They used a silicon diode followed by a voltage divider, to attain the optimum bias voltage[99]. Later on, series strings of two or more diodes were used, sometimes with ad-hoc shunt resistors. The *in*ability of a small, glass-encapsulated junction to follow the junction temperatures of the power devices, alias 'thermal lag', was recognised in 1957, with a paper that described a diode bonded to the base of a power device [6]. Forty years on, there still aren't more than a few very specialised power MOSFETs, and no BJTs, with integrated junction temperature monitoring. The world has more or less given up waiting for semiconductor makers to launch a range of power devices incorporating this most elementary engineering improvement.

4.2.6 Complementary push-pull OTL

Beginning in the early-to-mid 60s, the arrival of complementary (matching *npn* and *pnp* sets of) power transistors overcame the need for floating push-pull (differential) drive. Even if two drive signals were still required, at least they could now be referred to a common point, eliminating the argument for an interstage transformer. Figure 4.9 shows the all-emitter follower (alias Common Collector, CC) connection with single (a) and dual (b) supplies. Note how symmetry is lost in (a). Around 1962, while the most practical OTL topologies were still being worked out, there was even a proposal to retain symmetry and still have only one power source, by using a special, 3-wire speaker with a centre-tapped voice coil, but this didn't help speaker efficiency, nor did it catch on [7].

Figure 4.9

*Topology 5: Complementary
OTL, all-CC Type.*

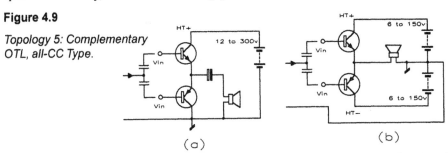

(a)   (b)

Figure 4.10 shows the all-common-emitter (CE) type, with a single supply (a) and with dual supplies (b, c) with different grounding arrangements. With the superficial similarities, confusion is easy. To distinguish between CE and CC (follower) stages, look to see if the speaker is (or would be) *connected between the emitters of the output transistors and the input signal reference* (ground). If so, the stage is a follower (CC); if not, it is CE.

Figure 4.11 is an example of how the basic follower-connected BJTs could be replaced by compound devices, either Sziklais (a) [8] or Darlingtons (b).

Figure 4.12 shows a complementary pair of Sziklai connected in CE format, the natural development of Figure 4.10 (c). With the Sziklai, don't forget that the effective emitter terminal is the actual collector terminal of the final BJT. In this circuit, some of the real world elaborations have crept back in: R1a,b are flushouts. Re1, Re2 provide local feedback, ideally for thermal stability. Other descriptions are possible: Cherry calls the configuration '*Push-Pull folded totem pole*', while Roddam describes it as comprising '*Cascaded complementary pairs*'.

Figure 4.10

Topology 6: Complementary OTL, CE Type.

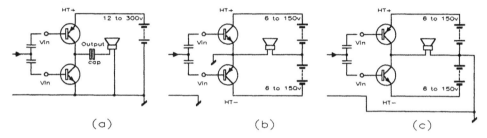

(a)　　　　　　　　　(b)　　　　　　　　　(c)

Figure 4.11

Topology 7: Complementary OTL, Compound OPS variants, CC type.

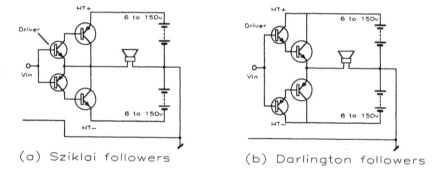

(a) Sziklai followers　　　　(b) Darlington followers

Figure 4.12

Topology 8: Complementary OTL, Compound CE type.
E = effective emitter!

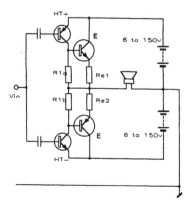

4.1.8 Quasi complementarity: the faked match

In the event, many makers eschewed these perfectly symmetrical schemes, preferring for example Figure 4.13, a possible and once popular arrangement of the infamous *quasi*-complementary scheme. The latter was developed as a smart solution to the lack of good (or *any*) pnp complements, in the early days of silicon transistors. Those pnp opposites that existed usually had a lower 'speed', ie. transition frequency (f_T), and they were not so rugged as their npn sisters. The outcome was (i) asymmetric (*dis*-complementary) performance at high frequencies, and respectively, (ii) sudden silence – except for a loud buzzing.

Figure 4.13

*Topology 9: Quasi Complementary
Compound, CC type.*

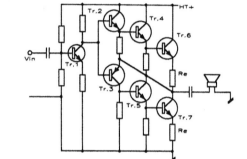

This time, the preceding, CE *driver stage* (TR.1) is shown. TR.2,4,6 act straightforwardly as a giant Darlington, or *triple* (section 4.3.4). For the lower half however, TR.3, the only pnp transistor in the circuit, acts as a so-called 'complementary phase splitter'. Really, it is more a level shifter, and certainly the first part of some kind of three-stage Sziklai (TR3,5,7), which may be seen also as a conventional Sziklai (TR3,5) combining with a final follower (TR.7). Keynote reasons for using circuits such as this one (*Ca.* 1960 – 65) were fourfold, three of them interwoven. First, *all power* transistors were at the time high cost, metal-canned items – there was no plastic packaging. Second, the saving in not having to buy the pnp type, and third, instead being able to buy *twice* as many of the cheaper npn type, added up to a tempting increase in profitability. Also, far more so than today (1996), silicon power transistors of any kind, let alone pnp complements, were only available in a limited range of voltages and currents, particularly for the output stage.

4.1.9 Sub-topology: paralleling

Output stage transistors devices have long been paralleled to increase current handling. As brute high-current handling does not square altogether with speed or high f_T, parallel connection is essential beyond a point. Without it, there would still be no high performance, rugged audio power amplifiers with output power capabilities much above 100 watts. Paralleling may even be beneficial when not essential, since '*n*' smaller 'lighter' parts will likely have a faster response time, and wider bandwidth. The outcome can be a species of supertransistor with a heightened *bandwidth* x *current-handling* product. Figure 4.14(c) shows the symbol invented by the author to denote one such, the 'Triplington'.

Figure 4.14

Paralleling BJTs.

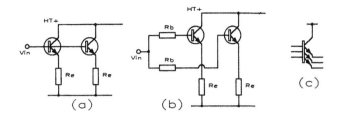

Paralleling is not problematic with valves, just expensive if it's not needed. But it was messy with the first output stage transistors, which had to be germanium types. These had *bad* problems with leakage current at working junction temperatures. Paralleling was at the same time often much needed, as current capability was most lacking and also easier than voltage rating to boost-up, see 2.5.1. Figure 4.14 (a) & (b) shows how even silicon paralleled BJTs must be *ballasted* with small degenerative (local feedback) resistors R_e, and base stoppers, R_b, both to aid current sharing and the latter to prevent possibly destructive, parasitic RF oscillation.

4.3 Silicon transistors (Si)

Germanium transistors were not only putative, but also expensive, as they had to be made as individuals. In 1959, Fairchild Instrument in the USA pioneered a method of mass-producing transistors, using lithography, and silicon as the substrate. This was called the *planar* process. In 1962, it was augmented with the *epi-taxial* process, where 'epitaxy' is the successive deposition of crystal layers. The resulting *Silicon planar epitaxial bipolar junction transistor* alias *silicon BJT*, is what we take for granted, today.

Silicon transistors were the opposite of germanium, in that it was easiest to make good *npn* ones; *pnp* types were intrinsically slower, less highly voltage rated and tended to be more expensive. And they differed in other ways. The base-emitter voltage needed to put them into *forward active mode* alias 'turn them on', was about double that of germanium (Figure 4.15) at between 500mV and 700mV (0.5 to 0.7v). This meant they could not be directly 'dropped into' existing circuits designed for germanium devices; every circuit required substantial redesign or at least rebalancing to get the right bias conditions. This probably gave designers an impetus to throw away the circuits they had been using, and begin a fresh with new ones, better adapted to silicon. The higher V_{be} (active threshold) also brought bad crossover distortion effects to light when the Lin circuit (see next section) was first employed *en masse*. It has been noted that crossover distortion had likely not been so very apparent when the Lin circuit was initially made with germanium transistors, owing to their gentler turn-on characteristics.

Figure 4.15

This room-temperature comparison of the 'turn on' voltages of germanium and silicon junctions shows not only how silicon required a higher voltage under all conditions, but how it turned on more abruptly. This made it better at generating crossover distortion. But silicon epitaxial parts were also cheaper to make, faster and more reliable when run hot.
Courtesy John Linsley-Hood.

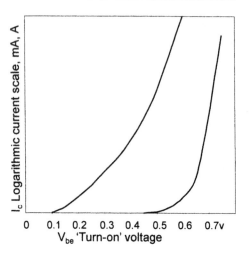

4.3.1 The Lin topology

Mr.Lin at RCA (see Appendix 1) proposed a transformerless, non-push-pull output stage topology for BJTs in 1956. It was elegant and simple, but it took around about five years, until the early 60s, and complementary, mainly silicon transistors, and the aid of a coalescence of sub-topologies covered in the subsequent sections, before it was put to major use. Then within five years, this most ubiquitous transistor circuit (Figure 4.16) had made most earlier forms obsolete. It was first popularised in the UK in a DIY Hi-Fi design by Tobey and Dinsdale [9].

Figure 4.16

An early but classic realisation of the Lin topology, with quasi-complementary OPS, diode-resistor biasing, and nearly all pnp, germanium transistors. C6 is the bootstrap capacitor.
Drawing from Transistor amplifiers for audio frequencies, P. Tharma, Iliffe, 1971.

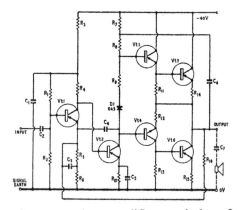

Today, the *Lin topology* is still the basis of many modern amplifiers, and also of classic, 3 stage, 'voltage-feedback' IC op-amps. For cheap, direct-coupled audio power, with high NFB, it seems elegant. Alas for quality audio, it is deeply flawed by asymmetries [10].

The early Lin circuits followed the scheme in Figure 4.17 (a). Just as in Figure 4.13 the upper output BJT is a *Darlington* while the lower is like a *Sziklai*, giving the same quasi-complementary kind of output. But now the input is a single node, referred here to ground, and *without* any coupling capacitors. This is one important feature; with it, *overall direct (DC) coupling of the signal path becomes possible –*

Figure 4.17

Topology 10: The Lin, Quasi-Complementary, OTL.

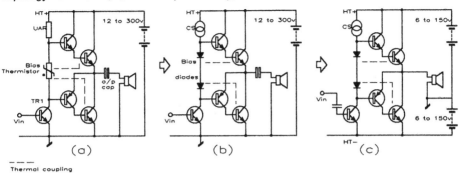

– – –
Thermal coupling

even if early versions retained the output capacitor. Another important feature is that a low, direct-coupled impedance, free from capacitor charge, can exist between the base and emitter of both the upper and lower driver devices, and here, just the upper output device, without loading the signal source. From the viewpoint of BJT voltage ratings, this condition, once called 'V_{ces} mode' by RCA, helped flushout excess drive charge from the BJT bases, and by doing so, increased the output BJTs' safe voltage to the V_{cbo} rating, commonly at least 30% higher than the V_{ceo} rating. Even better, the Lin topology enabled the devices operating in class B (or like) that were not conducting during the majority of the opposite cycle, to receive a reverse bias or 'back bias', to turn them hard off. None of these features is unique to the Lin topology – but it did offer these characteristics along with apparent simplicity, and was in the right place at the right time.

Crossover distortion (discussed later, in 4.6.7) was worsened by the use of two different kinds of transistor sub-topology (Figure 4.18). The symmetry was 'corrected' (certainly improved) by adding a diode [11,12,13] and passive components in line with the Sziklai's *effective* emitter leg. Later, with the arrival of 'matched' (read: 'mostly similar') npn and pnp transistors, it was possible to draw a perfectly

Figure 4.18

The asymmetry of a Sziklai mated with a plain Darlington, as in the basic quasi-complementary pair, shown enlarged. Shaw and Baxandall (both UK) later introduced simple modifications that greatly improved the symmetry.

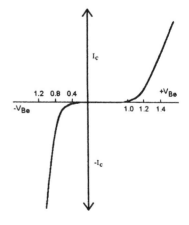

symmetrical circuit of a '*pure complementary*' stage, with *either* Sziklai *or* Darlingtons in both positions, although this is past what Lin envisaged! This type was popularised in a DIY design by Bailey [14].

Returning to Figure 4.17 (b), the upper resistor originally needed to be made small to get an 'equal swing'. This sets a high current in the chain through UAR, the bias resistor and TR.1, making the circuit unworkably inefficient. The circuit eventually developed so the upper resistor (UAR) was replaced by a current source (CS). This was one way to overcome the asymmetrical clipping that otherwise prevailed, as the upper Darlington is limply driven. Before the current source had been invented and had also 'rippled-through' from academic journals and its first use in monolithic ICs, the standard way was to a add *bootstrap capacitor* (C6 in Figure 4.16), always a large valued electrolytic, to the midpoint of the upper arm resistor (R7,8), and connecting the other end to the output. The capacitor voltage was then able to 'pull itself up by its own bootstraps', to be above (more +ve than) the upper arm transistor, so that it could be turned hard-on. This technique was widely used in the past. But like all tricks employing large capacitors in audio signal paths, it is unsuited to high accuracy.

First, program asymmetry will cause some charge accumulation, which is liable to upset any DC coupled circuits' optimum operation – quite apart from the transmission errors arising from the imperfections of large electrolytic capacitors. The benefit of the current source (which comprises a discrete circuit or IC comprising semiconductors and resistors only) is that it achieves what the bootstrapping-capacitor did, but more overtly and with greater transparency. By the late 1960s, silicon diodes were cheap compared to ntc thermistors, thus the bias compensating thermistor vanished in favour of sensing diodes, shown thermally coupled to the final stage pair.

Then in Figure 4.17(c), the supply is split so the output capacitor can be omitted, in order to realise the direct coupling potential of the Lin circuit. A small snag is that the ground to which the input was referred, has now become a noisy, unregulated HT supply [15]. The necessary ground-referred input was initially achieved with one level shifting transistor. Not long afterwards, it was replaced by the classic *VAS* and *LTP*.

4.3.2 Sub-topology: the long-tailed pair (LTP)

The familiar 'differential pair' or long tailed pair (LTP) circuit (Figure 4.19) is not a power stage topology, but is the most important element used to control power stages, wherever loop or global feedback is used. It was invented in England by Alan Blumlein in 1936 (see Appendix 1). This circuit played a key part in the world's first public TV service, and then in RADAR, and in the related electronic developments of World War II and through the Cold War, as the heart of valve and later transistor *op-amps* that formed the analogue computers [16], that enabled weapons to be designed and aimed, in the days when digital computers were still in their

Figure 4.19

Sub-Topology 4: The Long-Tailed Pair (LTP).

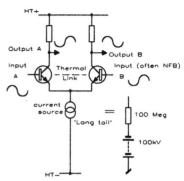

infancy. But for some reason, the LTP did not see much use in audio power stages of any kind until the early '60s. Then it was used as much as a phase splitter, which is a key sub-topology for valves, but not with transistors, once you have *npn* and *pnp* types. The LTP began to see extensive use as the ideal power stage input, when RCA popularised it in their application notes for the new silicon power transistors of the mid 60's. It has never looked back, and nearly every high performance transistor audio power amplifier has used the LTP since, as the input stage. It has even (re)appeared in valve amplifier designs [17].

The LTP is widely used because it is simple, yet free from most of the vices that afflict simple transistor circuits. For example, the input needs no messy bias connections to the supply rails. Instead it is biased to 0v (or the supply midpoint, see 4.3.5) and resides close to 0v DC which allows it to be direct-coupled. Supply rail rejection can be extremely high, and the two transistors can balance out each other's drifts.

4.3.3 Sub-topology: the V_{be} multiplier ($V_{be}X$)

Bipolar transistor biasing is naturally temperature sensitive. Silicon transistors only reduced the slope of the effect, and expanded the maximum junction temperature. The V_{be} (Vee'bee'ee) multiplier, alias '*the amplified diode*' and comprising one BJT and as few as two resistors (Figure 4.20 (a)), has appeared in power amplifiers with BJT output stages, since the late 60s [18]. It is a simple, elegant, temperature-dependent, 2 terminal floating voltage regulator.

The keynote is that the temperature dependency of the voltage is set by the transistor material (Silicon or Germanium). It is closely akin to that of other transistors of the same material, in this case –2.2mV/°C, with silicon. Compared with the diode

Figure 4.20

Sub-Topology 5: The V_{be} Multiplier.

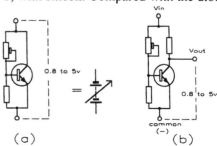

strings that had been used before, it was no more expensive to build, and offered fine adjustment of the bias voltage without altering the slope – as happens when diode strings are partnered with variable resistors to set precise current levels. It was also easier to force a plastic transistor (or bolt it) into close contact with a heatsink, than a glass-bodied diode. Thermal tracking was later improved with a third resistor [19], and a subtle topological change, increasing the pin count to three (Figure 4.20(b)).

4.3.4 Sub-topology: the triple (compound BJT)

Long before their use in audio output stages handling tens or hundreds of watts, triples are documented in the late '50s, as designers first explored the direct coupling of transistors. Figure 4.21 (a) shows an early, elegant scheme for interconnecting two gain stages, melding voltage gain with a low output impedance. Connection was soon extended to three stages (Figure 4.21 (b)). This raw triple combination had latch-up problems if overdriven, but the quest had begun.

Figure 4.21

Sub-Topology 6, Triple precursors.

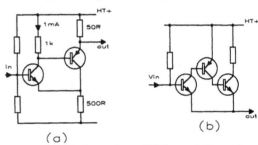

The *triple* (tripp'ull') is a mega-transistor built from three BJTs, and the natural progression of the Darlington. Figure 4.22 (a) shows the most obvious format – as still used by Crown. The sizes of the transistor symbols are a symbolic reminder that even more than with the Darlington, the transistors in a triple are cascaded in their current handling and capability, i.e. small, medium, then large.

But the triple was first brought to world attention in the more sophisticated form used in QUAD's 303 amplifier [20]. Figure 4.23 (a) shows the QUAD triple of 1966.

Figure 4.22

Sub-Topology 7: Triples, Pt.A. Upper half of OPS only.

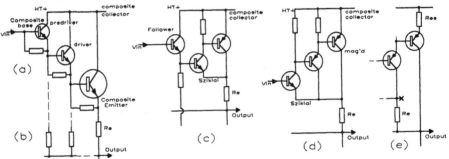

Figure 4.23

Sub-Topology 7: Triples, Pt.B

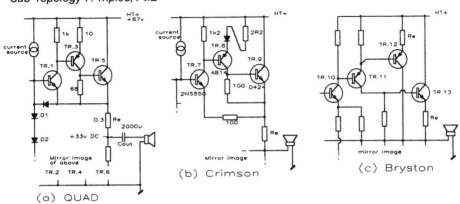

(a) QUAD

(b) Crimson

(c) Bryston

This particular topology has been instrumental ever since in BJT output stages of high sonic repute. One more of already several possible permutations (Figure 4.22 (c),(d),(e)), it is almost the same as Figure 4.21 (b) but with some major improvements, including being enclosed in a negative feedback loop, which not only linearises it, but prevents the latch up or 'sticking' with overdrive. Such 'dwell points' are quickly fatal to output devices, and are not good for sound either.

The QUAD triple also has increased local feedback (note the added resistors) giving better DC and thermal stability – both gained in conjunction with the use of silicon transistors. The arrangement of the triples used in the QUAD 303 circuit also allowed the small, cool-running pre-driver transistors (TR1,2) to set the bias, in comparison to the voltage across (hence current in) the ballast resistors, R_e, with the normal bias diodes (D1,2) giving a reference voltage that ideally tracks the same ambient temperature that TR1,2 are experiencing. This was an early attempt to overcome transient crossover distortion problems, covered subsequently in section 4.6.7 and 4.6.8.

Returning to Figures 4.22 a,b,c and d, the behaviour of triples has been documented by Bongiorno [21]. The raw type (a) has a notorious problem with the overcoming of RF instability and possibly with crossover distortion – as there are three junctions of different sizes for the signal to 'pass through'. Crossover distortion in class A-B (or B) may be ameliorated by the Crown scheme, or similar, where small signal currents near zero volts, up to a few tens or hundreds of milli-amps, are passed directly to the speaker by the predriver and driver, via resistors (Figure 4.22(b)).

The type in (c), called 'Darlington Reverse Amplifier Emitter Follower' (Darlington RAEF) by Bongiorno, may be seen as a follower, driving a Sziklai pair. This type is also tricky to make stable. In both cases, the need for unconditional RF stability demands measures that will compromise amplifier speed and increase hf distortion. The next type, (d) is like a Sziklai with the follower part magnified, Darlington-like. RF stability is again a problem, but with only one B-E junction, class A-B (or B) biasing is as straightforward as any. Figure 4.22(d) shows a variation where as in Figure 4.22 (b), the penultimate device is able to pass current directly to the output. Note also how the local feedback resistor, R_e, has moved. Consideration of local feedback in triples greatly increases the possible permutations.

Continuing with Figure 4.23 (a), the 4th type of triple (called *RAEF Darlington* by Bongiorno) and that chosen by QUAD in 1966, is relatively easy to make RF stable. Intrinsic distortion is relatively low, and biasing is stable. It was subsequently developed by Crimson Elektrik, a British domestic maker and DIY module supplier, in 1976-81. By then, the single rail and output capacitor had been done away with, and Figure 4.23 (b) shows again direct, 'progressive drive transfer' connections from driver and pre-driver, to the output; and the insertion of the Shaw-Baxandall diode above Tr.8, to enhance complementarity. Last, Figure 4.23 (c) shows an elegant development by Bryston in Canada. Having already employed the QUAD scheme, they improved complementarity by adding Tr.12, which is in parallel with the usual Tr.13. The output stage thus comprises paralleled *npn* + *pnp* devices, and vice-versa in the lower half complement. Aside from canceling non-complementarity (as both *pnp* and *npn* transistors are conducting at all times) an unexpected benefit is that the driver (penultimate) transistors work less hard; the current demanded of them is more spread out. Overall, and when used in class A-B (or B), distortion – the high harmonics associated with crossover-distortion in particular – are reduced, all the more so when there is hard drive into low impedances. *By repute, this topology, more than other BJT output stages, allows class A-B amplifiers to challenge the sonic behaviour of class-A*

4.3.5 Sub topology: Dual supplies (+/–Vs)

In some of the topologies seen so far (in Figure 4.8 through 4.12), two batteries, alias dual or 'split' (+/–) supplies, are already shown. Early makers of anything other than battery powered amplifiers fought shy of this. First, before the widespread use of the LTP as the input stage, dual (+/–) rails for the output stage would not have suited any directly-coupled drive stages.

Second, makers must have been wary of a transistor failure causing DC (the full rails) to appear across the output, then burning out the speaker. Rather than design DC protection, nearly all elected to use the variants where only a single supply is needed (Figures 4.9 (a), 4.10 (a) and 4.13), but where *what would have been* the negative supply rail's reservoir capacitor, has been moved round to be in line with the speaker as an 'output DC blocking capacitor'. This, charging via the speaker at turn on, was the cause of the universal, substantial *Pluop!*, a characteristic of countless rather low-performance amplifiers created with this blatant topological asymmetry. The output capacitor was least welcome in those days, when large valued electrolytics gave poor performance and were still rare, bulky and expensive. As a result, it was minimally sized, often just 2000µF for a nominal –3dB at 20Hz when driving a nominally 4 speaker, when in fact 100,000µF could and should be specified today, if such a capacitor is to be used at all. *The tight value most unfortunately tested the weak drive stages of the time, at the time when it was least welcome, by causing substantial phase shift on low bass and subsonic signals.* This may account for the obsession of some amplifier designers over the years, for throwing away all trace of subsonic frequencies !

Dual supplies, alias 'split' rail supplies, are based on practice developed since the 1940s for op-amps and analogue computing [16]. Ground (0v) is centered between approximately equal and certainly opposite voltage rails, alias symmetrical supplies. These rails are in a sense balanced. When resting, the output (e.g. at the junction of the R_e's in Figure 4.12; or the hot side of the speaker in Figure 4.17(c)) is kept at or very close to 0v by NFB, and DC blocking just isn't required *until* a transistor blows and shorts. A major advantage of split supplies working in combination with the LTP is that the ground can be almost solely a reference point that can be kept free from contamination by large-signal currents. See also chapter 6.

Having made their first appearance in battery powered transistor amplifiers in the late '50s, dual supplies re-appeared in high performance amplifiers in the early '70s, after it was noticed how much the absence of series capacitors (and transformers) improved sonics [22].

Figure 4.24

Topology 11: Intermediate Lin.

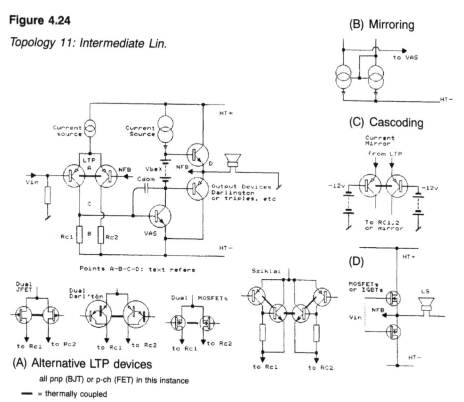

103

4.3.6 Sons of Lin

The Lin topology absorbed the sub-topologies just discussed, so that by 1972, the standard, intermediate format had emerged, shown simplified in Figure 4.24. Over the years, further topological improvements came flooding in, and were bolted-onto the Lin topology. These include:

** *LTP altered* to employ various Darlington-connected BJTs; or J-FET or MOSFET input matched dual pairs. While slightly increasing noise (hiss), these could respectively enable slew limit to be increased, and reduce DC offset voltages, both initial, and with heating up – amongst other benefits (Figure 4.24 (a)).

** *Improved constant current sources*, and variants (Figure 4.25).

** *Current mirror* loading, one of the varieties in Figure 4.26, replaces Rc1,2, doubles A_{vol} and lowers distortion (Figure 4.24 (b)).

** *Cascode connected transistors* stand off HT from LTP's collectors, to make high-quality but low voltage-rated BJT or FET input pairs usable with high voltage supplies.(Figure 4.24 (c)).

Figure 4.25

Sub-Topology 8: Constant Current Sources

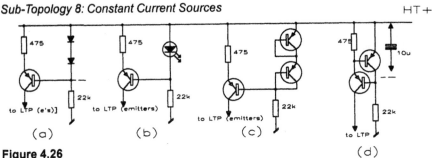

(a) (b) (c) (d)

Figure 4.26

Sub-Topology 8: The current mirror.

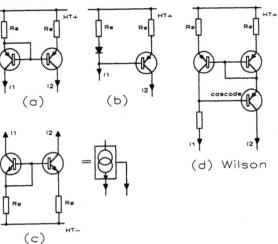

(a) (b)

(c)

(d) Wilson

** *LTP altered* with cross-connections to further increase gain, and cancel common-mode currents or related error signals. Outside of textbooks and IC equivalent circuits, BGW have used such schemes [23].

* *VAS altered* to a Darlington, Sziklai, or beyond.

** Output stage alterations. Other than the permutations shown previously (viz. replacing the original quasi-complementary stage with an honest complementary stage made from two straight Darlingtons, two Sziklais, two triples, or two of something else), audio power MOSFETs, which first appeared around 1976-8, may be used instead with only minor changes (Figure 4.24 (d)) [24]. The benefits of MOSFETs are covered in chapter 5.

The double asterisk ** is a stark reminder that all but one of these changes were no less applicable to many other, subsequently developed, and completely different topologies. And that more than any other part, what distinguishes the Lin from other topologies, is the VAS, with its ill-defined referral, and attachment to a noisy and high voltage negative rail.

The individual and combined role of each of these 'bolt-ons' in reducing steady-state %THD to a few parts-per-million, is discussed in a series devoted to 'Gilding the Lin' [3]. Experienced listeners are not so impressed, as they recognise the onset of monolithic IC disease, where in the words of Nelson Pass *"...the complexity of the circuit is used to create the gain necessary for the feedback loop to correct for the greater distortion of the more complex circuit"* [25].

4.4 True symmetry: the sequel

It may seem ironic that complementary BJTs were one of the developments that enabled push-pull driven output stages to be abandoned. And all the more so at a time when %THD and %IMD (distortion) figures were believed to have a strong correspondence with sonic quality, since push-pull stages do cancel some of their own distortion products, principally the more benign, even-order harmonic ones. It would also, by implication, cancel far less benign intermodulation products caused by those same even-order harmonics.

So while the Lin circuit was followed by some designers and makers, others worked on developing the visually elegant symmetry of the topologies in Figure 4.9 (b), 4.10 (b and c), and 4.11 (a and b) [26, 27]. Since the mid 60s, there has been an increasingly wider choice of better matched, complementary BJTs, in an increasing range of voltages and currents (see Appendix 3 for some cumulative examples), and pnp parts are today relatively less costly. The *next* challenge was to develop a symmetrical driver stage, while still employing the LTP. The best known topology, part invented by Borbely, part by Lender, employs 'back-to-back' LTPs and is completely symmetrical (Figure 4.27 (a)), at least on paper. As an example of bolt-ons, the cascode section is not essential but reduces distortion. Note that the first two stages are working in push-pull. Later, connections between the first and second

Figure 4.27

Topology 12: Borbely-Lender, all-symmetrical types.

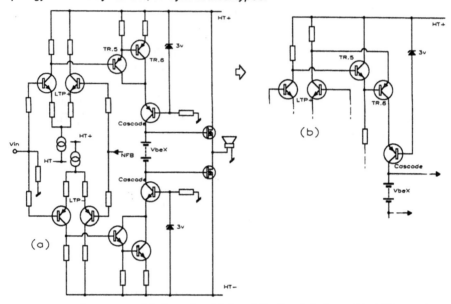

stage were made more sophisticated, with local feedback being added (Figure 4.27 (b), to enhance linearity.

4.4.1 Later topologies

As early as 1973, Otala [28] was presenting a relatively complex, all direct-coupled, and partly symmetricised driver topology that employed three cascaded LTPs (Figure 4.28). Although this kind of circuit could easily exhibit intractable RF instability when built without full knowledge of acceptable transistor substitutes, when it did work, the distortion, bandwidth and speed specifications were of a whole new order. The 'part PP' label is a reminder that here too, the first two stages are working in push-pull. About a decade later, this kind of 'more and more gain is better' topology had filtered through to a DIY design (by Tillbrook, in *Electronics Today International*) where the output devices were by then lateral MOSFETs.

Jumping back to 1977, someone working for Hitachi in the US 'downsized' to the simpler, *double LTP* (alias two differential stage) drive topology, that with the simple addition of Hitachi's new lateral MOSFETs, gave fair measured results all-round, as good as and in many ways far better than any all-BJT-based circuit at that time, and with considerably fewer parts (Figure 4.29 (a)). This topology, in its earliest known form, had appeared in Bob Carver's Phase Linear 700 amplifiers in 1972. Now with MOSFETs' huge current gain, it was possible to have only three stages causing phase shift in the onward signal path. So RF stability was easily attained and kept, and less compensation was required.

Figure 4.28

Topology 13: Cascaded LTP Driver, part PP,
generic OPS.

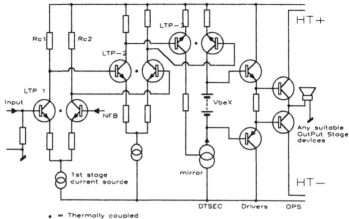

Figure 4.29

Topology 14; Dual LTP Driver-Stage, part PP.

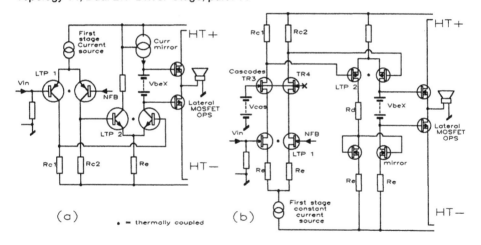

This same topology could also 'mate' with BJT output devices, either Darlington or preferably triples – but then the path brevity and the ease of hf compensation would have been lost. Hitachi subsequently published a variant employing all FETs, but this was not at all widely taken up. The circuit (Figure 4.29 (b)) is an upside down type (compare to Figure 4.29 (a)), and had MOSFETs for the second LTP (LTP 2) and current-mirror below, but employed J-FETs in the first LTP and for its cascode (TR.3,4), which is compulsory in view of JFETs' limited voltage rating. This small added complexity (it might alternatively have been used with a BJT input pair – refer back to section 4.3.6) may have been offputting, along with many maker's dislike of parts which cannot be multi-sourced, which may well have been the case with the specified J-FETs and MOS-FETs at the time.

4.4.2 IC power

Note: ICs are not limited to particular kinds of topologies; this section is a necessary connective interlude.

By the early 70s, monolithic IC makers began to have success with making audio
power amplifier ICs with capabilities up to a few watts. The first monolithic and
expressly audio power IC was a British development, made by Plessey in 1968. It
offered about 1w. It took Clive Sinclair to make a success of it second time around,
in 1971, as the 'Sinclair IC12', with an alleged 6 watts of output [29]. But Sinclair
then turned his efforts into pioneering pocket calculators. The main makers from
then on were National Semiconductor, SGS-ATES (today SGS-Thomson) and Philips
(as Mullard), later joined by Japanese makers, e.g. Hitachi, Rohm and Toshiba. For
the most part, the many audio power stage ICs that have been made by these companies are highly successful for undiscerning use, but well beneath the scope of this
book. The Japanese types have often not even been readily or at all reliably available in many Western countries, to quality makers who do not order in millions, nor
either for public testing.

Faced with the typical twenty to thirty transistor circuit that's arisen after many
bolt-on improvements to the classic Lin and BJT-based power stage, inquisitive
makers and users are bound to look in askance at analog ICs, black boxes that
contain about as many transistors and similar circuitry, at less than it costs to just
test or fit twenty discrete transistors. This is the superficial attraction of ICs. Yet
after nearly thirty years and $1000,000,000s of monolithic power stage 'develop-
ment', all that has been demonstrated is that monolithic IC audio output stages with
few exceptions, have intrinsic limitations, and can only be worth employing if all-
out miniturisation, the pursuit of profit, and/or manufacturing simplicity outweigh
or precede audio quality.

One keynote engineering problem is that a large area of silicon is required for high
power and ruggedness, and that past a point, the cost of silicon 'real estate' increases extremely disproportionately. If the chip designer gets at all generous, the
IC won't be able to drop far in price, so manufacture with discrete parts will 'cost
out' cheaper, and then there'll be insufficient numbers of the IC being ordered for it
to be viable. The upshot is that the most is forced out of the tiniest area of silicon.
The other keynote difficulty follows on from this: that of thermal interactions. The
power stage transistor junctions are close enough for their pulsating heatedness to
modulate and upset the conditions of the surrounding, cooler, lower power drive
circuitry [30]. In the absence of an elegant way of preventing this, it remains true
that if using an IC is a must, then far better performance can be had with a 'two
black box' solution, where the IC requires only discrete power devices to be added,
which may also be integrated amongst themselves.

The *dis*incentives of IC power amplifiers to makers and users of high performance
audio systems, are first, that like all ICs, if they misbehave or sub-perform in any

way, likely not discovered until 1000s have been bought, and sent out in amplifiers, there is usually nothing the maker or user or anyone can do about it. All-IC power stages are largely untweakable, and the ability of most IC makers, being large companies, to ignore a user's feedback (where they are a 'small' customer' taking under 1000,000 pieces per year) particularly their discovery of a fatal flaw after 'mask revision', has been well documented. Second, measured audio performance has not improved; the 'learning curve' is haphazard. Recently launched (1994) ICs from one maker, for example, develop a large-signal, asymmetric RF oscillation when mildly overdriven. Another, by the same US maker, picks up hum on its output, if speaker connection wires above a few inches are used [31], while other parts with the theoretically ideal, 'spiKe' (sic) protection system suffer from this intruding on peaks, and producting asymmetric RF oscillation if the output is clipped. What these ICs otherwise achieve is staggering for almost no money, but at what cost ?

Most audio power ICs have only primitive protection against shorting and other adverse loads and most sporadically omit the half-dozen other most basic housekeeping requirements (see chapters 3 and 5), all of which are manifest in long proven circuits which require little silicon, and should have been long integrated by now. Worse, the ability to plug-in a new IC has been thrown away in the cost squeeze. *If audio power ICs were pluggable like some power transistors, it wouldn't matter so much if they had no protection.* What *has* happened is that power has crept up, and price down. An IC capable of delivering 40w now costs less than a restaurant tip, at around £0.07 per watt, whereas the original Sinclair/Plessey IC12 cost (in 1972 money) about 65p per watt, or in real terms about fifty times as much. Way above these prices, some hybrid ICs made today in the US by Apex, have met with some approval by a number of quality audio makers and users. Here, the power stage is mounted inside a pluggable metal can, and comprises discrete but tiny MOSFET chips, which are micro-wired to a surface mount IC. Thermal interaction is greatly reduced. The other obvious remedy just hinted at, is to be content with integrating just the majority of small parts, namely a driver stage IC. There have been surprisingly few of these – the occasional monolithic drive-stage IC has been for European and Japanese 'consumer audio' and none are considered high performance. Again, Apex (who are not an audio maker) have pioneered PA42, the first high-voltage IC with good enough specifications to be considered for high performance audio.

Figure 4.30a

Apex's hybrid ICs combine discrete, hardwired MOSFET chips with a monolithic driver chip. Although more costly than an ordinary monolithic IC, a major hurdle to sonic accuracy is overcome. Courtesy APEX.

Figure 4.30b

Integrated High Voltage Audio Drive Stage

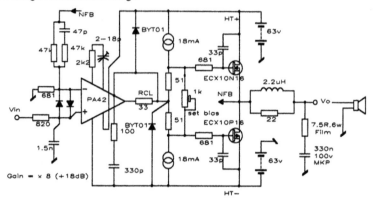

Figure 4.31 shows a test circuit used to demonstrate the PA42's application as an audio power-stage driver [31]. Clearly, the integration could be taken further. At the time of writing, the protection of audio power amplifier ICs against adverse loading still remains an issue, as *comprehensive protection is never included*, if not on-board, then in some separate protection IC.

Figure 4.30c

Apex's hybrid true high voltage IC op-amp achieves what ICs are uncompromised at - simplifying and shrinking small signal parts. When coupled to conventional power devices, a true high performance, yet highly compact amplifier can be realised.

4.4.3 The Op-Amp topologies

While integration of the complete input to power stage has not really made the grade, the less greedy approach of employing an IC op-amp as a partial manufacturing simplifier has had a long history, with continuous performance improvements. Losmandy was one of the first to publish a circuit showing an IC op-amp driving a power stage [32]. Topology No.15 (Figure 4.31 (a)) had all the classic problems of the day, e.g. asymmetrical clipping, inadequate slew rate, and bad crossover distortion. In time these have been whittled away by vastly improved IC op-amps and BJTs. One not intrinsic restriction is that the output voltage swing is limited to the op-amp's supply rails. The absolute maximum supply limits of most IC op-amps have remained in the +/−18 to +/−22v range since the late 60's due to process limitations, let alone limited dissipation (mainly due to class A-B bias current) in what is a *small-signal* part, usually in a plastic or ceramic package.

To increase voltage swing, hence power delivery into a given speaker impedance, topology 16 (Figure 4.31 (b)) shows how the op-amp is 'wobbled' (UK) or 'bootstrapped' (US) allowing it to swing up to a maximum of twice its supply rails, without popping. The zener diodes may be added to clamp the op-amp supplies at

Figure 4.31

IC Op-Amp Drive Stages.
Topologies 15, 16, 17 and 18.

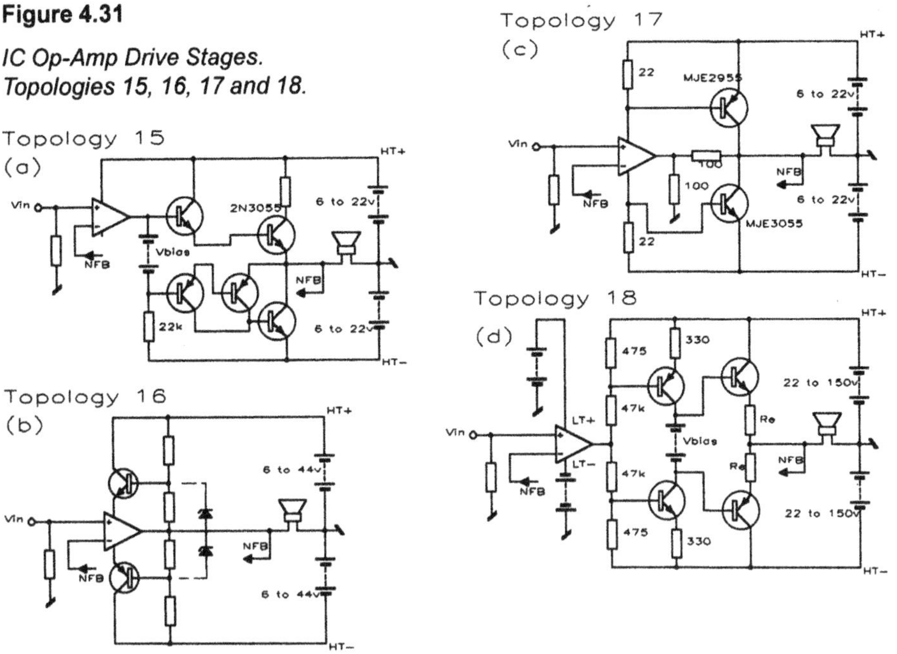

safe limits if the supplies are able to approach or exceed double the op-amps' rated voltage. In this drawing, the op-amp may be *any* power stage with split rails and a differential input. If only a low power op-amp, a current gain (output) stage would be interposed. If a power stage, we have a *supertopology*.

In topology 17 (Figure 4.31 (c)), the op-amp unusually drives current into the ground, while output devices are driven 'high sided' by the voltage-drop across the current sensing resistors in the supply leads. Although BJTs are shown, MOSFETs may also be used [33]. Swing is not increased, but it is one way to get a push-pull output from an op-amp. Note that the op-amp can also drive low current directly into the output, as first employed in the triples (Figure 4.22 (a) and 4.22 (a,b,c)).

Next, in topology 18 (Figure 4.31 (d)), the path is amended to handle high voltage unlimited by the op-amp's own ratings, by re-deriving signal from the op-amp output and interposing level-shifting resistors. As ever, the added followers boost current-handling capability [34] and for high performance, the individual BJTs would be uprated with Darlingtons, Sziklais, triples or MOSFETs.

Moving on to Figure 4.32, here the four-supply LT/HT topology of topology 18 has since developed, by moving the ground to change the power stage sub-topology to CE. Variants of this topology with both BJT and MOSFET output transistors have been employed by several professional power amplifier makers. One version, with an extra inner feedback loop, called the 'Super-Nova', was patented by Jim Strickland (see Appendix 1). Another variant with inner feedback was originated in the UK by Tim Isaacs. In the generic 'grounded source' stages illustrated, the lateral MOSFETs

111

Figure 4.32

*Topology 19:
Op-Amp-driven,
Common Source types.*

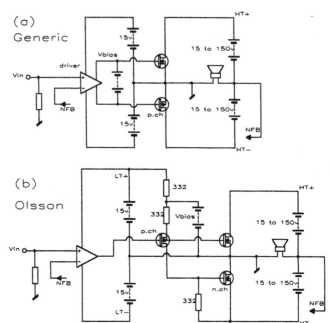

are working again in 'CE' or strictly, common-source (CS) mode. Compare with 4.10 (b,c). In the CS mode, and with global feedback, the MOSFETs are forced to provide all the voltage gain that is required, up to clip, as well as high current gain. The drive stage may be simplified to an IC op-amp and with appropriate design finesse, a fast op-amp, and MOSFET output devices, *slew rates commensurate with the highest voltage swings are attainable*, e.g. if swinging 150v rms, *however transiently*, then 300v/µS is just comfortably above the minimum safe value at 20kHz. Lastly, in Figure 4.32 (b), after Olsson [35], the grounded-source topology has been modified to eschew complementarity, and return to an all *n-channel* (or even *p-ch* or *npn*) output stage. This same scheme can also be applied to more conventional topologies. Other than banishing all aspects of crossover distortion caused by mismatch, the Olsson scheme allows (for example) 'V' or 'D type' switching MOSFETs to be used, whenever p-channel 'complements' (or *vice-versa*) are scarce, too expensive, or dubious. Note how the drive stage employs a small (but HT voltage-rated) p-channel 'phase splitter' MOSFET that is essentially perfect and without problems compared to its BJT counterpart. Notice also the asymmetric biasing.

4.4.4 Power cascades and cascodes

When transistor main terminal ratings (e.g. V_{ceo} or V_{DS}) are inadequate, transistors may be connected or 'stacked' in series to achieve the desired swing. The two methods are known as cascoding or cascading. Either may be employed in lieu of bridging to achieve high voltage swings without even-harmonic cancellation, for which reason they are to be sonically preferred [36]. In Figure 4.33 the simplified circuits show the cascading of BJTs and MOSFETs. The older *'totem pole'* arrangement is similar. The compensation capacitor 'C' may be required to adjust HF response, particularly with BJTs. Suitable matching is also required. If one device switches out of sync or exhibits higher leakage, the approximately 50/50% voltage division on which both devices rely, will go astray with potentially explosive results.

Figure 4.33

Sub-Topology 10: Cascades & Cascodes.

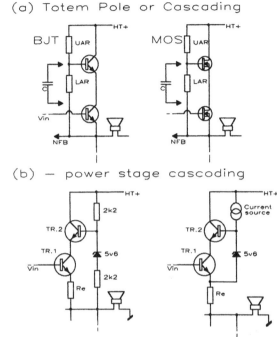

(a) Totem Pole or Cascoding

(b) — power stage cascoding

Figure 4.33 (b) on the other hand, shows *cascoding*, already discussed in the context of the LTP in Figure 4.24. Here, the upper-tier power device sustains the majority of the voltage, and all the voltage swing. Pass [25] refers to it as 'shielding' the lower, lower-voltage-rated, more linear device. One possible benefit of this arrangement is that the lower device may be a very fast switching part, easily obtained in low voltage ratings. In an ideal situation, the lower device is rated to (just) sustain the full voltage. In this case, it can be a faster switch than the upper part, sustaining the full voltage only for nano- or micro-seconds, while otherwise sustaining a steady 5v (say).

4.5 Introducing bridging

Bridging, as nearly every professional power amplifier user or owner knows, is a free 'tweak' switch. Move it across, and a conventional stereo amplifier (or pair of channels) condenses down to mono, but in return, swings more volts. If the output is stiff and able into the load, voltage swing and slew limit capability both double (+6dBr), relative to the voltage rating of the individual transistors and the power rails. If so, power delivery capability potentially quadruples – always assuming the speaker's voice coil resistance doesn't rise, and that the power supply and transistors can handle the doubled current and the quadrupled dissipation that also results.

In practice, if we are bridging because we are dealing with increasing already high swing capability, then *even if* the amplifier were fully capable, thermal compression knocks back the anticipated increase in SPL, from +6dB to more like +4 to +3. In practice, with most audio power amplifiers, continuous average ('rms') power delivery is doubled. Conversely, for a given power output and safety margin into a given impedance, bridging halves the transistors' voltage rating requirement.

What is less well known is that some amplifiers are already bridged per channel; and much else besides. *Bridging* is really a higher level of topology or 'supertopology', as it is *a different way of connecting two output stages that you already have*. It dates back to the second half of the 50s, at least, and has seen service in valve audio amplifiers, in various forms, most notably as the Wiggins/EV *Circlotron* amplifier (the topology was even drawn as a circle).

Terminology is a good place to begin. In most of the topological diagrams presented so far, two opposing output transistors (it does not matter what they are, BJT or MOSFET, and whether push-pull or complementary) are arranged to control the passage of bi-polarity (or alternating) current through the speaker. This general arrangement is called a *half-bridge* in electronics (Figure 4.34 (a)), but not often in audio. On the right (aa), it is drawn in the most familiar form. It is sometimes called a *'single-ended'* stage, but this is then misleading, except while used in the immediate context of 'non-bridged'. What is certain is that audiospeak's *'Bridge mode'* is called the *'full bridge'* by electronics designers. Figure 4.34 (b) shows this viewpoint (analysed in 1959 [37]). Again, Figure 4.34 (c) translates it back into the more usual perspective of audio users bridging two separate amplifier channels. Another perspective is that bridging could be called the seriesing of two output stages. These

Figure 4.34

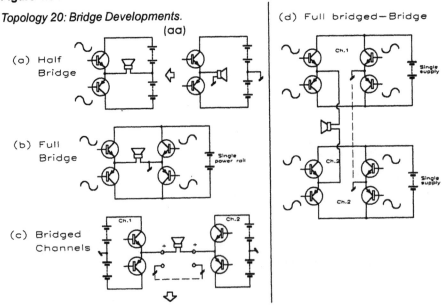

Topology 20: Bridge Developments.

(aa)

(a) Half Bridge

(b) Full Bridge

(c) Bridged Channels

(d) Full bridged—Bridge

are also driven differentially, as indicated by the sinewave symbols – else there would be no effect. So a third perspective is that we have a balanced (or at least, differential) output also. One feature that is subtly different about Figure 4.34 (b) compared to (c) is that it has just one, floating, power supply. In other words, the bridging of a stereo amplifier with the usual independent, dual supplies is not necessarily the most elegant example of *bridgology*.

Bridging was in use in transistor power amplifiers by at least 1961. It was one of the tricks that engineers had up their sleeve, for achieving what were then regarded as PA level power deliveries (like 30w to 100w $\Rightarrow$8Ω) which were way beyond what the voltage ratings of output transistors would then allow. Voltage rating could be doubled by using the *beanstalk* sub-topology [38], alias cascoding, Figure 4.35. But bridging was also in use (Figure 4.36). In either case, any added current handling required could be met by straightforward paralleling of transistors.

Figure 4.35

The cascode or 'Beanstalk' configuration was used pre-1967, to get higher swings before remotely high voltage output stage devices were available. It meant stacking up pairs of output BJTs, e.g. V4 and V6. V4 is arranged to take a fraction of the voltage, normally 50% at all times. Note the biasing or voltage setting chains, and the interstage transformer.
(From Transistor amplifiers for audio frequencies, Thomas Roddam, Iliffe, 1964.)

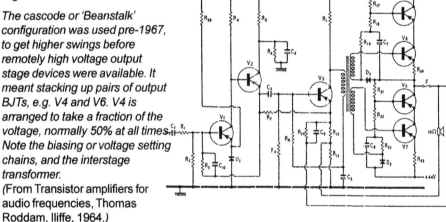

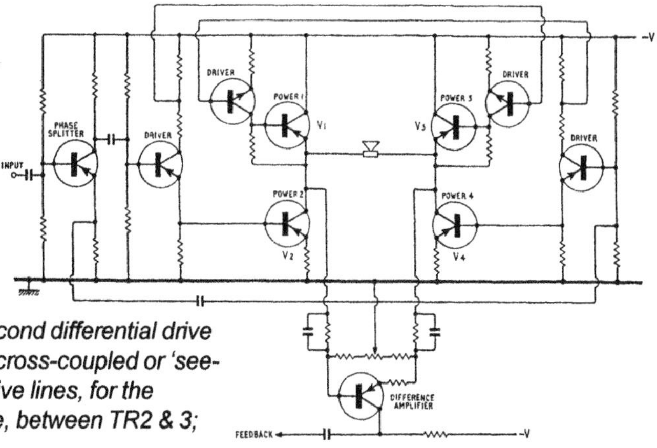

Figure 4.36

Bridge in 1963. This early '60s circuit shows a single BJT used instead of an LTP, as a difference amplifier, for NFB control (lower centre) - a long lost sub-topology. Note the 'phase splitter' is also the input stage, and unbuffered, and that a second differential drive is then developed by the cross-coupled or 'seesaw' connection of the drive lines, for the necessary push-pull drive, between TR2 & 3; and between TR.1 & 4.

One reason to choose the bridge was the recognition that as with a push-pull stage (in the normal half-bridge), bridging cancels asymmetric differences – which create *even-order* harmonic distortion products. The foremost cause of such differences, at the time, was that between nominal *pnp* and *npn* 'complements'. Bridging excised these bugging little asymmetries. It seemed elegant at the time.

Amplifiers capable of developing over about 50v rms i.e. 300w into 8 ohms (and pro-rata, i.e. >1.2kW⇒2Ω, etc) were increasingly sought from the late '60s onwards for concert sound and even studio monitoring, let alone for industrial uses. By doubling to quadrupling power in one step, bridging continued to appeal to many pro-amplifier designers with a 'linear power mentality'. It was also seen as sensible considering the ruggedness of low voltage output devices versus the ease with which non-bridged high swing amplifiers used to blow-up; to those who continued to need more power at all cost; and to others out to make a loud bang for a small price. The fact that bridge outputs are floating above ground is not normally a problem, as nearly all speakers are electrically floating also, in their wooden boxes.

Returning to conventional switched bridging between two channels, Figure 4.37 shows how drive schemes have developed. In Figure 4.37 (a), from 1965, the drive circuit is a push-pull type just retired from driving the pre-Lin topologies. Now each output 'half' drives a separate non-push-pull type of output stage. In Figure 4.37 (b), dating from the early 70s, channel 1's output is returned via an invertor (after due attenuation) and fed into channel two. This technique ensures relative balance (if Ch.1 is loaded 'down', Ch.2 will follow) but channel two output is also delayed (differentially), passes through twice as much signal path, and accumulates twice the distortion. In amplifiers with grounded emitter or related topologies, high

Figure 4.37

Sub-Topology 11:
Bridge Drive Schemes

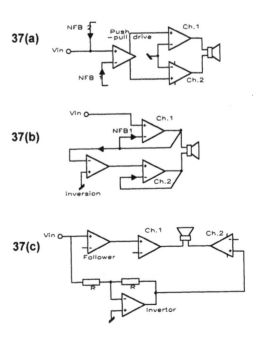

currents will pass through the ground system, and distortion may worsen and will certainly change, as a result. Figure 4.37 (c) shows the more refined method, where the push-pull or differential signals for both sides are derived in parallel, and at line signal levels, where contamination is less likely.

Driving a speaker differentially from a bridged amplifier has an important characteristic in professional use, particularly when the sound source involved must be particularly fail-safe. That's because if the amplifier on either side (not both alas) has to be shut down (for maintenance) or has shut down because of a problem, or dies, the music will continue, albeit at a 6dB or so lower SPL. We have in other words, *redundancy*. This assumes the off or dead side has a low impedance path through to ground. This might not be the case if the amplifier is very ill or dead. In Crest's amplifiers, the bull has been grabbed by the horns by arranging the channel output muting relay to connect the speaker end to ground, when muted.

4.5.1 Bridging the bridge

In 1985, Crown, a maker of industrial amplifiers, introduced a development of their grounded-output topology, at least to the audio world. It had actually first seen use in their model M600, in 1974. *Each* channel in *Crown*'s Macrotech 1000 (and many subsequent models) is bridged yet it is conventionally ground-referred. Crown overcame the *floatingness* of traditional bridged amplifiers' outputs by floating each channel's power supply, then grounding one side of each bridged output (Figure 4.34(d)), two amplifier channels, driven differentially, can be safely bridged to double the swing again. The result is called not unexpectedly the 'Bridged-Bridge'. It was first implemented for audio by Bongiorno and colleagues in the 1970s, who formed GAS (Great American Sound). Figure 4.38 shows Bongiorno's own, complete and highly symmetrical topology [21]. A passing resemblance to the Borbelly-Lender topology (Figure 4.27(a)) is to be expected, since Bongiorno was a collaborator.

Figure 4.38

Topology 21: All Balanced Bridge.

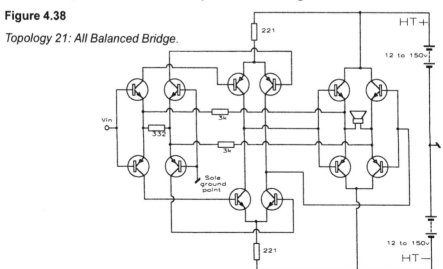

A key benefit is a given, high voltage swing using four times as many low voltage transistors, which are ideally disproportionately cheaper. For example, for +/− 80v rails, a quite scarce, complementary set of transistors rated at (at least) 180v are needed. The bridged-bridge allows 50v transistors to do the same job. Or no less common 75v transistors could be used, giving a far larger than normal safety margin at little added cost. Another benefit, cited by Bongiorno is that each side of the speaker has 100% NFB control. But does this really amount to double damping?

As with all 'smart' techniques, the bridged-bridge has limitations. One is that there are now four output-stage devices with complex metal junctions in series. Another is that even if 50 volt transistors are cheap, they still need mounting, and properly and this can be costly in skilled time. A third negative consideration is a reduction in reliability, arising if one part is not fastened properly, as failure of one device in 8 (rather than in 2 or 4) will cause total breakdown.

Other kinds of bridge have been proposed [39], but none have seen so much use as the Bongiorno/Crown circuit.

Bridging is further discussed in Chapters 6 and 9.

4.6 Class-*ification*

As mentioned at the beginning of this chapter, the class designations to be considered in the following section, describe fundamentally different kinds of output stage current behaviour *during each wave cycle*.

After *class A* (the original), the principal *raison d'etre* of all of them is an *improvement in power conversion efficiency in the output stage. Higher efficiency* can translate into better use of resources or a cost saving across the board. In the past fifteen years, this has been most important to large-scale touring PA, where even a little weight and space saving can have some disproportionate knock-on benefits. In the future, if not already, efficiency should be of increasing concern to those manufacturing studio monitoring and domestic systems with high swing (power) capability, to ease EMC requirements and reduce AC power consumption.

4.6.1 Class A

Perspective

The first power amplifiers operated in class A. The subsequent classes (B, D, etc) are not inferior, 'lower' classes as might be implied, but were invented later and named (almost) in order. While not often referred to as such, class A operation is the norm for most small signal stages at line levels and below, in preamplifiers and processors.

Basis

By itself, class 'A' indicates a continuous conduction cycle, i.e. 360°, in which none of the output devices are cut-off. For symmetrical clip, the output stage is arranged to 'tick over' at 50% of its intended maximum output current. This condition is ideal for linearity in analogue electronics.

Class A formulae

Peak current (before exiting class A) = $2I_q$

Max power in load = $[(I_q{}^2 \times RL)/2]$

Maximum signal dissipation - as above.

Max DC input power = $I_q{}^2 \times RL$

No signal dissipation = as above.

Where I_q = quiescent current

RL = nominal LS ('Load') resistance or impedance (e.g. 8Ω)

Costs

Many loudspeakers require voltage swings equivalent to hundreds of watts (see chapter 2) to cleanly handle occasional or potential peaks. With class A, *twice* the rated power is necessarily dissipated *whether there is any signal requiring it or not*, with the environmental costs of many kW-h of wasted electricity. This must be discharged in the form of heat. For noise free cooling, weighty and bulky convection heat exchangers are required.

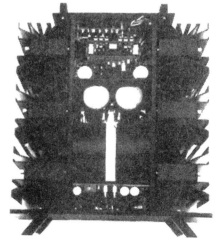

Figure 4.39

Behind the facade of this classic high-end domestic class A amplifier produced by Mark Levinson, is a large area of heatsinking (sides), and a huge power supply (centre) with two transformers. Signal handling circuitry with elderly metal-can transistors, is visible at the rear. © Hi-Fi News

Efficiency

Class A efficiency is highest at an ideal 50% when the output stage is push-pull. This could be one of several of the topologies shown earlier in Figures 4.9, 4.27 and 4.28 for example. It is second best at 25% if the output is single ended with an active collector load. Figure 4.40 shows both BJT (a) and MOSFET versions (b).

Figure 4.40

Class A, single-ended type, with active load.

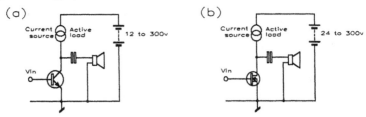

Without the active load (as in simple followers used at line level), efficiency falls to 12.5%. Also, efficiency only approaches these figures under the ideal circumstance, when the output is (i) a continuous sine wave (ii) delivering full power (iii) into a particular optimum speaker impedance. *With music's varying PMR and signal asymmetry, efficiency in real use is nearly always much less than these theoretical figures. typically no more than quarter to a fifth as much,* e.g. 10% for push pull; and 2.5% for a raw single ended output.

Excepting transformer coupled outputs where there are matching options, the speaker 'ohmage' at which class A efficiency peaks is a specific impedance (usually 8 or possibly 4 ohms) that the manufacturer has decided upon. Above or below the optimum load impedance, efficiency again falls off.

Power range

Class A amplifiers with conventional NFB behave as near perfect voltages sources – as much as any other amplifier. However, as the load impedance is reduced below the rated figure, and power output increases, the output stage current will pass into either a non-linear range (if single ended) or if the topology is push-pull, into 'overbiased' class A-B, wherein the current in one half is temporarily cut off, i.e. common-mode conduction becomes less than 360°.

Application

Conventional class A amplifiers above a watt or so are increasingly impractical for battery powered audio. They are no less impractical for large-scale sound systems. In a mid-sized 50kW system there would be an average 95kW of static heat generation, not falling below 50kW on signal peaks. Extra, factory-sized air conditioning plant would have to be budgeted for, and in a fixed installation, the cost of the electricity over several years would soon overtake the original cost of the entire sound system. Finally, with the weight and bulk per unit output capability being at least double that of other amplifier classes, class A would also add impossibly to already large show running costs.

Solely non-adaptive class A amplification of any power rating is mainly restricted to 'high-end' Hi-Fi and the occasional 'no-compromise' recording studio. Honest power ratings above 100 watts into 8 ohms (which must imply a transient capability down to 400w into 2 ohms) are rare. With these sort of ratings, the dissipated energy should be recycled by ducting the hot air to supplement room heating, pre-heat cold water, or dry clothes!

Sonic benefits

Class A is widely advertised as the sole keeper of audio quality. *In reality, it positively overcomes just a couple of the many distortion mechanisms that audio power amplifiers suffer from.* The following achievements of refined class A operation are not trivial but it is important to keep them in mind, in the context of a much longer list.

1. No seams

Within the rated load impedance, there are no 'seams'. Potentially highly toxic crossover and related slope distortions (the effects of which are audible in *parts-per-million* by some listeners) are absent by definition, rather than just made very small or fleeting. This relieves those individuals who are most sensitive to crossover distortion, and removes one of many layers that mask or destroy subtle musical detail.

Without crossover distortion to battle against, the signal path circuitry may be simpler. Also VAS loading (in a Lin-based circuit) can be linear and those distortions caused by misplaced feedback connections and beta (β) mismatches in the output BJTs are absent [3].

2. Constant draw

Within the range of rated load impedance, class A power supply current draw is constant. With the crude, unregulated supply rails that have long been the norm for audio power amplifiers, the constant current drain prevents the creation of signal-related but grossly distorted half-wave signals, on the output stage supply rails and power supply wiring. The possibility of injection of such garbage into low level stages or sensitive nodes is thus averted. When a power supply is shared between two channels, the constant current draw of class A also prevents interchannel crosstalk, at least through this route. It is worth noting that where wideband, tightly regulated power-stage supplies are employed, the other classes are ideally no worse, i.e. this advantage of class A is quite circumstantial [40]. Turning to power supply ripple noise, class A does not prevent this *per se*, but rejection of it in push-pull and in those single-ended configurations employing current sources, is relatively good.

3. Simplicity

The simplest conceivable audio power amplifier (Figure 4.40 or the earlier Figure 4.1) has to operate in class A. *Simpler signal paths assist focusing on component and material quality.*

Bias benefit

The optimum biasing of output stages operating solely in class A differs from class B and relatives. Stability of biasing is somewhat simplified after initial equilibrium is gained, as the heatsink temperature is likely well above ambient and remains constant with signal. Figure 4.40 shows a 'self-biasing' or 'active bias regulation' scheme. Without this or similar feedback biasing control, the device junction temperature takes time to reach equilibrium. Here is one reason why class A amplifiers are commonly switched on well in advance of use. Active bias regulation for push-pull class A requires thoughtful design. In this design by Allison, published in Wireless World in 1972, gross distortion that would be caused by simplistic sensing of the (signal+bias) current – with any practical degree of music-removal filter – is avoided. Here, the bias current alone is sensed by the series sensing connection across $R_{e1} + R_{e2}$. If the bias drifts too high, it is restrained as regulator transistors Tr.1,2 turn-on equally, so the drive current is diverted.

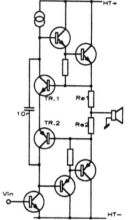

Figure 4.41

Class A bias regulation. Bias in this class A push-pull stage is automatically regulated without detriment to sonics by TR.1,2, which sense only the bias current, not the signals' apportionment of it. This scheme was published by Allison in Wireless World in 1972.

This same scheme also offers protection against short circuits, required with this push-pull output stage topology. Note that no such protection is required for the single-ended schemes that are biased by a constant current source – just so long as the current source controls itself against temperature.

Social benefit

There is a belief amongst technically uneducated purchasers that the 'class A' label means it is '1st class', as in travelling. The sheer size, weight and cost of honestly specified and adequately-rated class A amplifiers reinforces the belief.

'A' overview

Overall, noxious crossover and power supply distortion artifacts may be made very small, %THD can be the lowest of (or as low as) any class, and the harmonic structure can be the most benign. As a result of high intrinsic linearity, global feedback may be reduced or excised altogether, yielding amplification with low or zero-global feedback, while THD still remains below 1%. The reduced feedback can simplify the signal path, reduce the translation of low order harmonics to higher orders, and improve hf/rf stability margins when driving into practical cable-lengths, while the absence of global feedback may enhance sonics, considering the back-EMFs of dynamic speakers and some passive crossovers.

Reliability

While active devices operating in class A dissipate at consistently high temperatures compared to the other, more efficient classes, the MTBF of the output transistors may be less compromised than expected, if the equipment is run continuously, *as output devices are freed from the peak and cyclic stresses that might otherwise be the foremost factor in their demise.*

Summary

Class A is a benchmark, against which the sonic quality, reliability and ecology of all the more advanced power amplifier topologies may be contrast. The fact also remains that some listeners are highly-sensitive to crossover distortion and related, highly dissonant or masking perturbations that occurs in some form (however slight) in the other classes. As little as 5ppm (parts per million) may be audible. In practice this is an rms measurement, and as crossover distortion is highly spikey, and the real, peak level could be 10 or 30 times higher, so its minutest audibility may be slightly less remarkable.

4.6.2 Class A alternatives

To engineers, class A operation is a frustrating mixture of idealism, simplicity and brute waste. In turn, a number of schemes that 'bounce off it', and attempt to square the circle, have been devised.

4.6.3 Class A sliding bias and 'Π-mode'

This is the earliest [41] [42] development of class A aiming to redeem efficiency enough to make it more usable for high power transistor amplification.

The basis

The instantaneous signal voltage is arranged to modulate the quiescent current *above a level*. Small signals experience class A conditions. When a large signal is present, the bias (distinguished, at least in a push-pull output stage, as a common mode current that flows through the power devices but *not* through the load) increases to keep just ahead of the instantaneous signal current.

Efficiency

When signal drive is absent or just small, the bias current can be left to idle at moderate levels, dissipating a fraction of pure class A. With a continuous test signal, sliding the bias *halves* normal class A dissipation for a given power rating. With music, the reduction is still greater. Efficiency can thus approach that of class B and relatives.

Basic issues

Derivation of the correct bias voltage is dependent on suitable sensing and rectification of the output current. Establishing 'instantaneous' bias with a music signal takes time, particularly if the rectified signal is filtered, and unless very fast rectifiers are used. In early designs, the output stage was in this way under-biased on transients. The outcome is a transient increase in a steady state distortion. Fortunately, human ears can be surprisingly tolerant of momentarily increased harmonic and intermodulation distortion product. Moreover, the lag can be far, far less than those occurring in the thermo-mechanically coupled biasing systems used for class B and A-B biasing. On the other hand, with percussive and 'choppy' music, the transient distortions might be repetitive enough to be noticed. A prolonged release time counters this; or the bias release time may be made auto-adaptive.

Historic practice

In pro-audio at least, the use of sliding bias to make class A qualities realisable, is the attribute of a historic rock'n'roll amplifier lineage. Originally designed by Malcolm Hill in 1970, it employs 'sliding bias' under the alias of 'π-mode operation'. After intensive use in the company's PA rental system, it appeared on the world market in *Hill Audio*'s DX series in 1978. Looking at the simplified circuit in Figure 4.42, the input signal is fed directly into a low power amplifier (a 10 watt IC power amplifier, TDA2030 was actually used), which provides some voltage gain. It also has to have sufficient output current capability to drive a low impedance. This is necessary because its output is fed into a 1:1:1 transformer, which provides isolated bi-phase drive, in series with each output transistor's bias chain. Considering the topological journey we have traveled so far, the transformer coupled drive seems idiosyncratic. It is there to provide the two individually floating drive signals that are needed, with the least complexity. It also prevents the chain destruction that

Figure 4.42

Class A, π-mode 'Sliding Bias', after Hill.

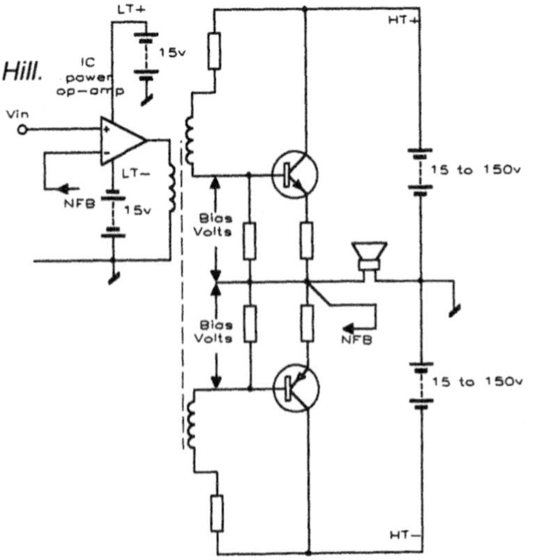

had plagued most direct-coupled amplifiers in professional use in the early 70's, when the circuit was conceived. The transformer also keeps any DC offset voltages (from the driver stage) out of the output stage, and thence from affecting the biasing. The alternative would be four blocking capacitors.

Real behaviour

In common with other diagrams, output transistors are shown singly for simplicity; in practice several are nearly always in tandem (paralleled). With nil signal, they are biased here (as is the case with most class A-B amplifiers) to around 25mA. For inputs above –40dBvr, the bias current then increases in proportion to the output signal voltage – up to a maximum of 1 Ampere. This is a practical limit, above which the signal current oversteps the bias, to allow high drive levels into low impedances. Nonetheless, the output is still operating in class A (by one definition at least), since both 'halves' of the output are *substantially* conducting at all times. Equally, 1 Ampere is about $^1/_8$th of 500 watts (63v rms) into 8Ω. Assuming system headroom is adequate at this level, the majority content of an *uncompressed* (or not unduly compressed) audio program is therefore experiencing class A. It follows that the 1 Ampere limit is a surprisingly *un*important compromise in amplifying some kinds of live rock'n'roll, big-band and other percussive music (even military brass, say), where PMR is at its highest, at 18 to 30dB; or any recordings managing to capture similarly wide dynamics.

In the Hill Audio designs, efficiency is at best a respectable 64%, when driven with audio. Counter-intuitively to the average professional user, but like the class A amplifiers that many domestic users have experience of, Hill's design runs hot when ticking over and cool*est* when driven hard. Unlike class A, the efficiency isn't 'tuned' to a specific load impedance, because of the bias current's dependency on the output signal's voltage, rather than current.

History presents a broad measure of the essential similarities in efficiency, and hence power density, between everyday class A-B amplification and Sliding-Bias class A: When Hill's DX 700 was introduced in 1978, it produced a maximum combined output of 1500 watts into 2 ohms from an enclosure just 2U high and barely 12" deep.

Figure 4.43

In 1978, Hill offered unrivalled power density to the world in a 2u case. Unlike its rival, the S-500D, it didn't blow up and even had substantially class A, 'π-mode' operation. Those using them in rackloads nick-named it the "Hill heat generator". © Hill Audio

In fact, for some years, there wasn't a class A-B amplifier which worked reliably and approached this power density. In 1990, Malcolm Hill (later operating as Malcolm Hill Associates, and latterly, as Malcolm Hill Audio) repeated this by launching *The Chameleon.* Despite operating on more conventional class A-B principles, its power density too has challenged even the highest-tech competitors with up to 1500 watts/ch, in a 1u ($1\,^3/_4$" high) enclosure.

4.6.4 'Super Class A'

In 1978-9, 'Super-Class A' and similar schemes was sold to the public by oriental Hi-Fi manufacturers as a snake oil which would cure the ills of conventional class A-B amplifiers without the costs of pure class A. The Technics SE-A1 for example, offered 350 watts in class 'A+B'. It comprised a low voltage class A amplifier, dissipating a few watts, connected in series with a floating high voltage supply, with the supply polarity and level arranged to keep up with signal, this being controlled by a class B amplifier, also in series, and rated at high voltage but having low dissipation (Figure 4.44). Some aspects of class 'A' purity are in part lost. On the other hand, maximum dissipation is lowered, so that heatsinking and power supply ratings may be a fraction of those required for real class A [43].

Figure 4.44

Super Class A.

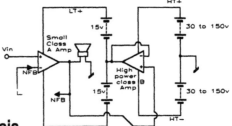

4.6.5 Dynamic biasing and Stasis

Nelson Pass patented 'Dynamic Biasing' for class A in 1976. With it, the Threshold 800A could deliver 200w/ch but when idling it dissipated only 50w/ch. This is a natural developments of the sliding-bias ideal. Subsequent to this patent, Pass conceived a dual to cascoding. He called it *Stasis,* or 'current bootstrap'. Here, a high current-handling device 'shields' a low current, highly linear transistor, from current variations, while leaving it to control the quality of the voltage swing. This is an elegant and no less fundamental attack on one of the root causes of non-linearity which is changing semiconductor gain over the small-to-large signal span of voltage and current. With it, Pass achieved *0.03% distortion without any global NFB,* which is 10 to 300 times less than the open-loop distortion of most circuits in the audio midband.

4.6.6 Sustained plateau biasing

Sustained plateau biasing was patented in the early 90s by Krell (US). It appears to be a further development of Dynamic Biasing, and something like it is prior art in dozens of amplifier designers' notebooks. When an amplifier using this technique is standing by, the bias level is low, but still enough for signals at 'average' listening levels not to cause transistors on either side to turn off. A signal transient that is about to demand a current higher than the quiescent bias level, turns up the bias, via very fast acting circuitry. The bias then stays up at the high level (there may be several gradations), for a period dependent on program history. If subsequent signals do not demand high bias, the bias levels reverts to the quiescent level. This is a sensible eco-measure, as it can save a lot of heat and electricity, but only circumstantially; it does *not* allow a purist class A amplifier to be made cheaper with much smaller heatsinking and power supply, as the loud playing of highly percussive music will put the biasing into the high mode and keep it there!

4.6.7 Class B and A-B

Introduction

The *class B* category of operation is the starting point for the vast majority of power amplifiers. Class B was originally used to save battery power in early portable audio amplifiers both valve and transistor, in the 1950s. In days of 'consumer-grade' valve amplifiers, special valves were even developed to operate in a form of class B called QPP.

Definition

Pure class B operation is defined as conduction of an active device (in practice, an output transistor) over 180°, ie. for only half a sine-wave cycle. For audio, class B necessarily implies a minimum of two opposing devices operating on *alternate* half cycles. Compared to class A push-pull (the two may use very similar circuitry), the devices on each side are cut-off about half the time, i.e. there is little or no common-mode conduction.

'B' efficiency

Pure class B efficiency can theoretically approach 78½%, a 157% improvement on class A's best efforts. In everyday figures, every 500 watts of class B power produces at least 137 watts of waste heat. Figure 4.45 shows clearly how dissipation is highest when the signal voltage (or current) is −3.9dB below clip, or 64% of full output, where efficiency has fallen back to 50%. At this point, *power* output is $(0.64^2) = 40.7\%$, or just above a third of maximum. It is important to note that efficiency is a potentially counter-intuitive notion when taken out of context. For while class B efficiency reduces towards 0% with lower level drive, so does the commensurate power delivery. *The efficiency that matters most is that in the top 6dBvr of the output span*, i.e. the 25 to 100 watt part of a nominally 100 watt amplifier.

Bias need

Pure class B operation is an anathema to pure sound as discontinuities in the signal waveform around zero volts arise because the output transistors in each half (whether

Figure 4.45

Dissipation and efficiency in Class B (and A-B, etc) amplifiers.

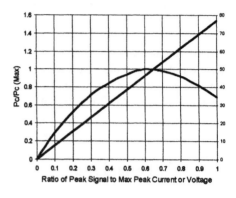

On the left, $P_c / P_c(max)$ is a dissipation scale, normalised to be 100% at the point where dissipation in the output stage devices (which may be BJTs, MOSFETs or Valves) is greatest.

Along the abscissa (below), the ratio of peak to max peak voltage (or current) may be translated into –dBvr, where 1.0 = 100% = 0dB = clip; 0.64 = –3.9dBvr; and where 0.1 on the left would be –20dB(vr) below full output, etc. The rounded curve 'Pc/Pc(max)' then plots the dissipation peak. On the right, % efficiency is scaled, and its straight-line slanting 'curve' is seen to rise in linear proportion to linear drive level, hence logarithmically in terms of dBs and approximate human perception of loudness.

After graph in Transistor audio amplifiers, Tharma, Iliffe, 1971.

they're BJTs or MOSFETs) need to see an input voltage above a certain threshold (namely V_{BE} or V_{GS}) to operate in their active region. The application of global NFB causes impossible feats to be attempted (comparable to HT electricity jumping a spark gap), and results in a predominance of high, odd-order harmonics, which are the constituents of sharp crossover spikes, seen in the distortion residue. That global NFB helps so little should be unsurprising, since linear circuit analysis starts with the presumption that all transistors in a circuit's forward path are biased sufficiently into their active region to exhibit a useful degree of transconductance. The unpleasant sonic effect of the discontinuity is notorious enough as *crossover distortion*. It's universally overcome by biasing the output devices into their active region.

Significance X

Tiny deficiencies within the zero-crossing point of audio waveforms can seem insignificant, particularly when a signal of any size is viewed (as on a scope) with a linear scale. When music is plotted in time with a dB (log) scale, (easy in simulation but rarely available on instruments) the zero 'point' becomes a seemingly infinite chasm. Then the amount of sonic detail and 100dB or so of dynamics residing in what seemed an insignificant place, can be glimpsed, and the substantial damage done by even very small errors in the crossover region understood. Added to this is the 'unreal' sonic quality of crossover distortion. With it, the overall system distortion behaves *non-monotonically* – meaning to say that distortion not only rises at high levels (for all the usual reasons, especially in all *reproducing* (cf.recording) sound systems – i.e. those with speakers), but with crossover imperfections (*nothing* to do with speakers' active or passive crossover *circuits* incidentally!), distortion *also* rises in the opposite direction, as levels diminish. Even if the distortion were benign, it is to be expected that the ear would still find the effect unnatural.

A-B variety

With BJTs in the output stage, the quiescent current needed to overcome crossover distortion is small, at between 10 and 40mA, even for a high power amplifier. With any kind of power MOS-FETs, the bias current needed (at between 20 to 100mA *per pair* of devices) is commonly more to much higher. The total current is high enough at between 100mA and up to 800mA per amplifier, to create class A conditions for small to medium output signals, i.e. up to a few watts or even tens of watts. Some BJT output stages have long managed similarly high bias currents.

Definition

Class 'A-B' is formally defined as any output stage having a conduction angle which is beyond 180°, but less than 270° (where the conduction angle for class A operation is 360°). A-B$_1$ and A-B$_2$ are sub-variants with different bias levels, rarely mentioned today. Class A-B$_1$ was defined in the 1950's, when the only audio output stages where class mattered employed valves. *It is worth noting that these suffer no less from crossover 'notch' and 'switching' distortion, which are not purely 'transistor diseases'*! And that in valve amplifiers, class A-B$_1$ meant that onset of (positive) grid current set the upper limit to output (anode or plate) current, rather than anode (plate) dissipation. Transistors don't share this limitation.

Overbias ability

Topologies differ in the way they respond to 'overbiasing' in A-B, this being past the point where the power stage halves' forward transconductances meld with the smallest or at least the most graceful, least kinked, deviation [44]. The CE mode is one optimal sub-topology. For topologies (mainly specific BJT types) that show a distinct optimum, *excess bias current may produce just the same sort of highly objectionable harmonics as under-biasing.*

Bias states

Figure 4.46 (overleaf) shows over-, under- and optimally-biased states, for BJTs. In BJT-based output stages, biasing may also interact with bandwidth and speed. This behaviour applies also to amplifiers employing MOSFETs, where there may even be two or more optimum bias levels. Fortunately the point of optimum bias is far less critical.

Bias vs. °C

For BJT-based output stages, the desired bias voltage is commonly developed either with diodes, or more universally with a small signal transistor configured as a V_{BE} multiplier (Figure 4.20). A few makers (e.g. QSC) have success with the older and slightly simpler alternative of a thermistor and preset resistor. The trouble common to all these schemes is tracking the output device's −2.2mV/°C change in V_{BE}, as junction temperature fluctuates with fast attack and slow release. It is being presumed that instantaneous changes in output devices' junction temperature are directly experienced by the biasing components. At best, they are loosely coupled to

Figure 4.46
The crossover point magnified. Dashed lines show the tailing-off of the currents in individual 'conjugate' halves, transistors TR.1,2. Characteristics in a, b & c show respectively, under-biasing, nominally correct and over-biasing current settings, for a particular BJT-based output stage. Class A-B & related biasing behaviour, still hotly disputed by engineers after nearly thirty years, is inter-dependent on largely unpublished topological subtleties as much as on active device type.
Based on drawing in Transistor audio amplifiers, Tharma, Iliffe, 1971.

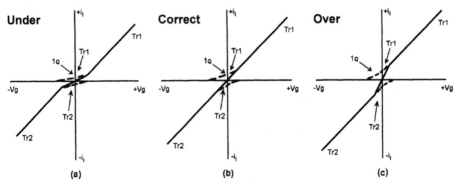

the heatsink, so there's a delay together with a shortfall in the peak temperature sensed. *This leaves the output stage over, then under-biased, during and after loud passages.* The overall scene is fairly complex, since the preceding tiers of driver transistors are also subject to thermal delay, and with smaller junctions, their transient thermal response is different again. The fact is that circuit developments aimed at overcoming the devious facets of crossover distortion are still being postulated, more than a quarter of a century after the problem was first recognised.

HF distortion

Secondary crossover distortion is rarely mentioned. It appears in BJT-based output stages, again as spikes and notches about the zero crossing. It is caused by the delay in sluggard BJTs turning on, then flushing out, current. With MOSFETs in output stages, and also with faster output-stage BJTs (f_T>10MHz), this kind of distortion need no longer be significant, at least below low ultrasonic frequencies. If it occurs, it involves uncontrolled common-mode conduction, which might develop catastrophically, i.e. leading to output transistor destruction.

Practical efficiency

Figure 4.47 charts typical real world class B waste heat (as a % of output power) and efficiency statistics against percentage output power, for sine waves and for varying kinds of music drive of the same average ('rms') power. Music drive appears more efficient, because above about 8dBvr, alias 20% of max output power, the peaks are being clipped, which lessens heat dissipation. As well as demonstrating variations with PMR, hence program density, the 'envelope' curve also encompasses small differences in efficiency arising from different output stage topologies.

Figure 4.47
Class B & A-B Efficiency.
The top curve (or edge)
refers to sine waves with
their low 3dB PMR, with
PMR increasing below, to
that of moderately
compressed live program,
portrayed along the bottom
of the envelope.
Copyright QSC, reproduced
with permission.

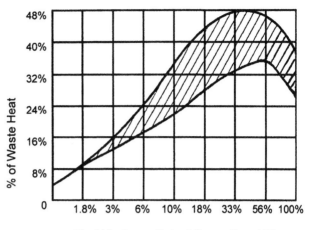

% of Maximum Output Power Capability

When driven at their rated maximum power, real class B amplifiers are plausibly efficient at 74%, compared to 78.5% predicted by first-order theory, which is ignorant (for example) of bias current. The graph also clearly shows how continuous sine-wave testing at about one third of full power delivery is the acutest bench test of a class A-B amplifier's heat dissipation capability.

Out in the real world, the amount of waste heat varies as widely. It approaches 50% when a monitoring or PA amplifier is driven with heavily compressed or hard limited programme averaging –3 or –4dBWr (i.e. about $1/_3$rd full power). Without compression, normal programme's average (so called 'rms') power is nearer 10% to 20% of the instantaneous output. Then when a class B (or A-B) amp is driven just short of clip by an uncompressed recording (live, or in a studio), waste heat falls off to between 20% and 35% of rated output. Heated amplifiers are one penalty for squeezing maximum 'volume' into a recording or performance.

4.6.8 Class A-B, developments and ameliorations

Resumé

For the first two decades (until 1976), virtually *all* quality transistor class A-B amplifiers were made with BJTs. Due to a positive temperature coefficient, leading to *positive thermal feedback*, BJTs are liable to act suicidal if the bias ever gets too high. As described earlier, they are fussy about biasing for optimum sonics, while the right bias hinges on their junction temperature, which in turn depends on drive and load conditions. Optimum class B or A-B biasing thus requires iteration and is kindly described as 'fiddly'. The time and preparation needed to get biasing right defied time-efficient production. Thus some designers sought to make optimum biasing automatic, or at least more robust.

Blomley

An early variant on conventional class A-B biasing was designed in 1969 by Peter Blomley, at Plessey (UK). The patented circuit was published in *Wireless World* in 1971 [45]. The scheme was not taken up commercially and was largely forgotten except by a few designers. It has been mentioned more often of late [27,46]. Looking at the simplified circuit (Figure 4.48), the output stage transistors are optimally biased at all times. When driven, the current through them can only increase, as they are arranged to respond to 'common mode' ('same') signals, and to ignore differential (opposing) ones.

Figure 4.48

Simplified Blomley Topology.

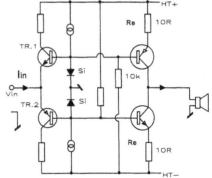

The purpose of this arrangement is to overcome switching problems with the larger output stage BJTs at high frequencies. Dividing the signal into clean halves, alias 'Conjugate Splitting', is achieved by the current driven pair, TR1,2. As small signal devices, their switching speed and *complementarity* can be exemplary. In turn, errors at crossover are greatly reduced. This ironically allows feedback within a circuit to work more effectively – further improving performance. In turn, the optimum bias may be lower than in ordinary class A-B amps. Even better, Blomley's scheme can be free from setting up of bias. In practice, to be satisfactory, this requires some sacrifice of power in the high ohmic value ballast resistors (Re). The critical splitter transistors might also need selecting. Blomley's technique was the first of a series, that would later be known as *non-switching class A-B*.

Current-dumping

The next well known development in a series of proprietary techniques which *did* change the status quo appeared in 1975. When QUAD's 405 arrived to replace their model 303, it had something special, called *Current-dumping*. Looking inside (Figure 4.49), it comprises a low power amplifier (A1) that can run in class A without significant energy waste or heating. In practice this is an IC op-amp. At low levels, and low frequencies, it initially drives the output directly. If and as the output current approaches the limits of the small signal driver, the high-current-handling 'dumper' transistors switch in, to take charge of higher output levels. These work in pure class B, and are relatively non-critical. Conventional, global negative feedback is replaced by 'feedforward error-cancellation' (see section 4.10.2).

Figure 4.49

Current Dumping.

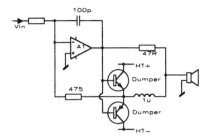

This is an over simplistic explanation. Noting the capacitor across A1, making it into an integrator, and the choke between the dumpers and final output, the circuit may be redrawn slightly to show it as the 'frequency conscious bridge' it really is, to aid comprehension. And yet while current dumping is unusual for having quite lucid descriptions of the circuit action by both of its designers, the late Peter Baxandall and QUAD's founder, Peter Walker [47], these descriptions are highly arguable, e.g. [48,49].

Current dumping is said to be more effective than ordinary global negative feedback at deleting crossover distortion and associated thermal modulation in BJTs. Alas, in being patented, it has not seen innovative commercial use. And while the same current-dumping circuitry has been employed in all of QUAD's models subsequent to 405, up until 1995, the subsequent models have reverted to conventional circuitry.

By bypassing the 'deadband', current dumping, as with Blomley, permits the optimum bias current to be smaller than it would otherwise be. Its undeniable benefit was in production, where the output stage bias required no setting for a decently low distortion residual, say <0.05%, and remained less fussy about drift with aging, something which happens inevitably with plain class B and A-B biasing. And by reducing bias current, QUAD's technique saves a few watts of static dissipation.

Non-switching A-B

Distortion caused by over-biasing was mentioned earlier. In otherwise apparently refined BJT amplifiers (like the QUAD 303), and particularly in some class A-B bias conditions, a large signal driving high current into a speaker could turn off *all* the bias current in the side of the OPS that isn't passing signal – due to the voltage drop across the emitter feedback and thermal stabilising resistors. This results in reverse-biasing, and hence hard switching of large, rather sluggish BJTs, and a step voltage shift that the feedback has to accommodate [50]. A few years later, Japanese designers almost overcame the problem by adding a secondary V_{be} multiplier ($V_{be}X$) type biasing circuit, which increased the bias on the 'non-conducting' side of the output stage, as the conducting half passed a large current. This involved a kind of positive feedback, and carried the risk of blowing-up the output stage ! It also relied on the switching of small diodes, which ameliorated and displaced, rather than overcoming the switching problem. The first schemes also involved as many as 11, admittedly small signal, extra transistors, and the effects on sonic quality are not documented. By 1980, Tanaka [51] at Sansui had developed a simpler bias control system which used both positive and negative feedback, and only two transistors (Figure 4.50). It is worth comparing to the class A bias regulator (Figure 4.41). The benefits of Tanaka's method are evident as a large reduction in %THD particularly at high frequencies (>10kHz), with an increasing distortion reduction with diminishing levels, i.e. a reversion to monotonicity.

Figure 4.50

Class B, A-B Bias Regulator.

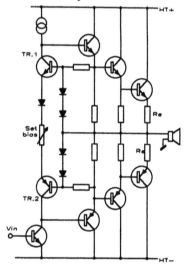

Subsequent work by Margan [52] in the former Yugoslavia, has documented what is going on. Margan's conclusions, based on real-time analogue measurements, using subtractive techniques, and applying to amplifiers with global NFB, are important enough to be individually tabulated.

Margan's conclusions

1. An ordinary, optimally biased class 'B' output stage generates crossover distortion until the output signal current falls below a level X Amperes. This level is set by the ratio of the speaker's impedance to the main emitter degeneration resistors (R_e), and also by the quiescent current setting.

2. When crossover spikes occur, a phase error is also generated. The amount of error is inversely proportional to the open-loop bandwidth of the voltage gain stage. It is also dependent on the ratio of amplifier output impedance, to the speaker's impedance. Output impedance is significant because we are dealing with a condition during which NFB is inactive; so the raw (or 'open-loop') impedances of the OPS components applies.

   With a resistive (dummy) load, and an output stage having wide gain-bandwidth, the phase error may be as small as 0.05° at midband frequencies, rising to 5° to 10° at 20kHz, or so. Thus with music, the phase-frequency relationships become non-linear. In turn, parts of the music signal that give spacial information and definition are lost, masked or blurred.

3. The envelope of the phase error signal stays in fixed proportion to the output signal envelope until a threshold is passed. It then disappears abruptly. This effect may be called *switching phase modulation.*

The conclusion is that non-switching class B (NSB) can overcome these effects and give results as good as class A without the dissipation – though only in terms of overcoming crossover-distortion.

New loop

The latest work on circumventing class B biasing problems in BJT power stages comes from Holland [53]. Here, a non-linear common-mode feedback loop (separate from the signal feedback) sets and regulates the quiescent current, preventing thermal runaway or drift, and crossover distortion after loud transients [54].

MOSFETs

The increasing use of MOSFETs in output stages, since 1976, has sidestepped the principal class B and A-B biasing problems, which are down to limitations in large ('power') BJTs.

Both lateral and vertical (large, power) MOSFETs switch much faster and much more cleanly. Thus reverse biasing and consequent switching is a relative non-problem. Complementary matching is better in most ways, than the best BJTs. Thermal drift isn't a problem either, at least with lateral MOSFETs. In theory, these exhibit an unstable positive temperature coefficient (ptc) when biased below about 80mA each. In practice, and with good thermal coupling, they can be stably biased in some topologies, down to as little as 20mA. The *ptc* region for vertical MOSFETs reaches up higher; they are potentially thermally unstable up to an Ampere or so. Again, bias regulation or 'servo' schemes can come to the rescue [35,55]. In both cases, a *negative temperature coefficient* (ntc) region follows. So the resistance of MOSFETs ultimately increases to set limits on their power dissipation.

Additionally, biasing of MOSFETs, at least lateral types in certain topologies, is less critical (as regards crossover distortion) than the biasing of BJTs in most topologies (that do not have B or A-B bias regulation). It follows that any thermally-derived compensation for bias (e.g. a V_{be} multiplier) may be quite loosely coupled, as there will be no explosions or sonic nastiness if the bias lags behind the output MOSFET junction temperature(s) for a few seconds.

In 1996, Linear Technology (US) were the first semiconductor maker to produce an IC expressly for general purpose class A-B bias regulation in MOSFET output stages. Their LT1166 contains independent voltage and current control loops. The current loop senses the output stage's bias current independent of the signal, and gate-to-gate bias voltage is continuously adjusted to maintain a appropriate voltage across the source degeneration resistors, R_e. The voltage loop keeps the levels 'centered'. In those D-MOS circuits where it is applicable, the IC claims to automatically take care of temperature tracking and part matching, making both trimpots and fiddly trimming unnecessary – without going down the topological and evolutionary cul-de-sac of 'current-dumping'.

The relatively high bias currents needed per individual lateral MOS-FET (compared to BJTs) can lead to significant heat dissipation on standby, particularly in amplifiers which aim to deliver high power into 8 ohms and therefore support high rail voltages. For example, with a nominally 550w into 8 ohm amplifier, with +/–100v rails and 8 MOSFETs (4 per 'half') biased at 50mA, static dissipation totals 40 watts. In this way, MOS-FET amplifiers may technically exhibit poor efficiency, but only under small and no signal conditions – where it least matters.

Unlike BJTs in all except *non-switching* and related class topologies, MOSFET output stages can cope with any amount of overbiasing, without any special considerations. other than adjusting hf compensation. With four MOS-FETs biased altogether at 550mA (for example), signals do not experience anything other than class A operation up to 2½ watts (assuming a nominal 8 ohms). With efficient, horn-loaded speakers, this may be enough for adequately high SPLs, possibly 120dB (peak, C-wtd) or more – all you need, in other words! Under the same conditions, but with a typical 40mA quiescent current, an equivalently power-capable BJT output stage ceases to operate in class A above some 25 milliwatts, a hundredth of the power, or –20dB lower.

Summary of B and A-B

Pure class B is not used for audio. *Where mentioned, class B is always presumed to be optimally biased for lowest crossover distortion, or an attempt in this direction.*

In amplifiers with BJT output stages, bias current is commonly 5 to 40mA, depending on the output topology and component values. If the bias is more than (typically) 50mA, it may also be described as operating in class A-B. But A-B 'overbiasing' is restricted to enabling class A operation up to a few tens of milliwatts, unless suitable topologies, extra heatsinking and bias regulation is used.

With MOSFET output stages, class A-B bias levels commonly sustain class A operation up to hundreds of milliwatts, and may even range (heatsinks and power supply permitting) up to a substantial fraction of the full power output capability.

4.7 Introducing higher classes

In 1972, Robert Carver, then president of Phase Linear Corporation, published a paper [56] outlining Phase Linear's research into sonic differences between power amplifiers. This was by way of explaining why his company had launched a model producing what seemed an *outrageous* 350 watts/ch. Back then, it was described as a '700 watt' amplifier by adding-together the power capability both channels, or else by citing 'peak' power.

Re-appraising power

In his 1972 paper, Carver cites a series of measurements which drive home a fact that was then less well known than it should be: that the 'loudness' capability of a power amplifier can be substantially affected by the supply rails' regulation and voltage. This is at least one way in which, for a given loudspeaker, the highest undistorted sound level (whether experienced or measured) doesn't necessarily follow the amplifier's continuous rated power.

PSU dynamics

Carver noted that this was particularly true if the amplifier's power supply was fairly 'soft', i.e. one exhibiting significant voltage sag on continuous drive, thanks to an undersized transformer, or reservoir capacitors. For example, an amplifier rated for a continuous 100 watts (into 8 ohms) with a badly soggy supply dropping from +65v to +45v on load, could sound louder on music passages, than a 200 watt amplifier (again into 8 ohms), with tightly regulated rails of + 57v. Although Bob Carver wasn't the first person to stumble across the effect, judging by what he implemented, he clearly glimpsed consequences that others didn't. Subsequently, the effect was recognised as 'dynamic headroom'.

Dynamic headroom

For domestic Hi-Fi in the US, the IHF (Inst. of Hi-Fidelity) defined dynamic headroom nearly two decades ago, in their IHF 202 specification. It's the maximum average (or short term 'rms') power available for just 20mS every 500mS, beyond the rated continuous power. Sometimes it's called 'burst power'. Since 1987, there's been documentary evidence [57] of something amplifier designers and audiophiles have long suspected: That music's peak SPL requirements can extend in bursts to 300mS and beyond, with a 20% duty cycle, ie. lengthy transient bursts *can* recur every 1 to 2 seconds.

Ramifications

Clearly, IHF 'burst power' ratings are too short to be useful. They don't provide the data to distinguish good amplifiers from mediocre specimens. The better modern amplifier designs have followed Mitchell's recommendations [57]. Others at least quote burst power over more realistic intervals than IHF 202.

Another US body, the FTC, insisted that an amplifier's advertised power rating must be the average ('rms') power with a continuous sine wave, into a resistive load. This ruling exists to prevent the public being mislead by 'watts peak' (twice the average or 'rms' power) or similarly inflationary claims. For a long time, all reputable pro-audio and the better high-end domestic power amplifiers have been routinely designed to meet this kind of standard, and have been tested accordingly (e.g. by *Stereophile* magazine in the USA). Given a suitable speaker, amplifiers of this calibre maintain their rated power on all kinds of signals: continuous or erratic, triangular, compressed, asymmetric or just plain random, for days on end.

Amplifiers of this type that are designed principally to meet the FTC's requirements tend to have dynamic headroom approaching 0dB. Their stiff power supplies result in an output which very nearly doubles with every halving of load impedance, until the output protection cuts in.

Power rating wars

A possible argument against the FTC specification is that it penalises legitimate designs which sacrifice high continuous power for effective burst power. As a result pro users have perhaps been paying for unnecessarily heavyweight amplifiers. And why? Is it solely to sustain the rated power under unrealistic conditions, namely with continuous high level waveforms, surely the stuff of industry? Surely, granted that pro-audio amplifiers are restricted to music and speech, the only steady tones are slates, occasional violins, or avant-garde synth players, none of them requiring anything like 0dBvr from an amplifier that's adequately rated? Or is there some other reason?

4.7.1 Class G

> *Note*: **The following text employs the Anglo-Japanese class nomenclature. In the USA, the meaning of class designations G and H is commonly *reversed*.**

In 1977, Hitachi introduced a range of domestic amplifiers and receivers labelled *Dynaharmony*, which aimed to overcome the problems of clipping during loud passages, by radical and well considered means. Back then, the professional's curt reply would have been to buy an amplifier with a high enough continuous power rating in the first place. But Hitachi saw that they could offer substantial added headroom against occasional loud passages without the domestically prohibitive cost, weight and size involved in 'true' high power unit, as used hitherto solely by professionals.

Auto headroom

Recognising that the 50 to 70% of the power that's needed to avoid clipping on most (but not all) full-range signals with (note) wide dynamic range *is only called for 2 to 10% of the time*, Hitachi had produced an amplifier (their HCA-8300) which could extend its headroom to accommodate loud passages. Looking at Figure 4.51, the middle transistors TR.1-2 comprise a conventional 100w into 8 ohm class B or A-B amplifier stage, operating on +/–40v (fed via diodes D1,D2). Provided the instantaneous signal voltage is below 40v, the outermost, cascaded transistors (TR3,4) are cut off. But once the signal swing approaches +/– 40v, they are brought into action. Being energised from +/–95v rails, the amplifier's headroom is now raised accordingly, to around 60v rms or 500 watts $\Rightarrow$ 8Ω *for just for as long as it's required*. At the same time, fast diodes D1,2 switch out or *commutate* the lower, +/–40v rails, then back on again when the peak signal level falls below 40v.

Figure 4.51

Hitachi Dynaharmony, Class G principle.

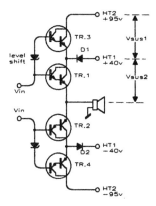

The outcome is cooler running, than if TR.3,4 with their +/–95v rails, were passing current all the time. In effect, switching is being used to improve efficiency within an analogue amplifier.

Dis-interest

In domestic realms, Dynaharmony and class G is just a distant memory. Amongst the critics, high transient power capability was not the answer. One leading UK reviewer didn't rate the HCA-8300's sonic quality. However, the class G topology (soon afterwards taken up by other Japanese consumer-grade hi-fi giants) wasn't held directly responsible. In particular, Hitachi overcame the most immediate objection: they succeeded in attenuating switching spikes that might be expected to be generated by the diodes during the commutation (rail switching). At the time, Martin Colloms argued that class G was a needless complication, as the "...ear is surprisingly insensitive to momentary and well behaved clipping, hence Hitachi have little guarantee their amplifier will sound better." [58]

The Dynaharmony and its ilk was not made for long. The best recordings used in the home in 1977 were still all vinyl, and excepting native Japanese music and audiophile releases, were precisely lacking in the dynamics and spikiness of live percussion and rock, that class G was ideal at supporting. The average domestic user in the UK or Japan must have been concerned about not having good enough sound isolation from the neighbours, let alone speakers able to handle potential 500w peaks. Now whereas most drive units can handle 3 to 10 times their rated power for the periods of typical music transients in their band, most domestic HF drivers, being rated at just 3 to 5w rms for continuous signals, are *already* making use of this safety factor. On this basis, domestic users must have suffered zapped hf drivers. *They were in a sense let down by lack of co-operation into the 'increased dynamic headroom', by speaker makers* – and this just five years before digital audio.

Technologically at least, class G was doomed in the domestic field owing to the absence of drive units which had excursion and thermal capabilities, commensurate with the transient power on offer. Without these, sonic quality was inevitably marred by thermal compression, cone and suspension break up, or sudden silence – if not by worry. Yet those companies making speakers for concerts and recording studios had already long pushed for the development of drive units that would set them free of such limitations. Whatever clean amplifier power existed, speakers with enough power handling had somehow to be made! By 1980, ATC, JBL, Gauss and others were making bass-drivers that could keep up with 63v rms (500w $\Rightarrow$ 8Ω). But recognition and acceptance of class G was slow to dawn.

Carver's cube

In 1981, Bob Carver, having left Phase Linear to set up his own Carver Corporation, unveiled a domestic amplifier which confounded the establishment. Model M400A (better known as the 'Carver Cube') weighed under 10lbs and measured just 6.8ins³, yet claimed 500 watts output. When driven with music signals, the claim was found to be true. Although relying most of all on a new and radical power supply for its exceptional power-to-weight ratio, the Cube's designer hadn't forgotten what he'd learnt nine years earlier. The M400A's output stage contributed to economy of heatsinking, by employing a subtle variation on, and refinement of, Hitachi's class G topology. Anyone who has perused Carver's literature should be forgiven for thinking otherwise, for as one reviewer put it, "...Carver has a penchant for creating names for...circuits which don't always reflect exactly what the circuits actually do..." *.

Figure 4.52

Carver's M400A 'Cube' had a 'commutator control', which Carver used to justify the description 'Class H', but the topology really is 'Adaptive class G.'
Courtesy Carver Corp.

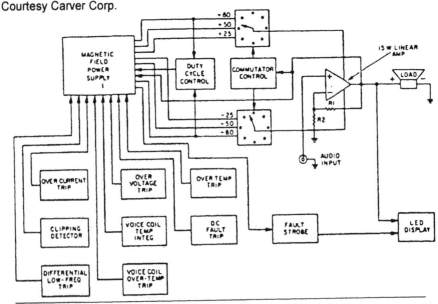

On reading this, Bob Carver exclaimed "If I invent it, I get to name it anything I want. Of course, it's only fair if my circuits actually do what I claim they do."

In this instance, the brochure explains how a '15w linear amplifier' achieves the headroom of a 500w amplifier by being 'commutated' between a series of ascending rail voltages shown being connected through selector switches, driven from the 'commutator control' (Figure 4.52). Although dubbed 'Class H' by Carver (probably the start of the confusion between US and European nomenclature) the M400A's output stage is no more than a cleverly disguised description of (or a 'new way of looking at') what Hitachi had already named class G, given that the commutated 'switches' are output devices connected as followers (i.e. with unity voltage gain).

The new, hidden part is the signal commutation control – which is a fast 'window' comparator, directing signal to the appropriate tiers, according to the signal's amplitude. In working at a small signal level, this aids cleaner, faster switching than Hitachi's principal method of solely diode-OR'ing the supplies. The latter is simpler, but relies on the less than perfectly clean and less than blindingly fast switching, of a diode that has to have a high current and high voltage rating, hence not a Schatty type.

Carver's class-G circuitry has long since been refined and repackaged for pro-audio users in various models. Figure 4.53 shows a fragment from the output stage designed for pro touring PA in late 80's. Variants of the series, class G scheme have subsequently appeared in Crest Audio's models *6001*, *7001*, *8001* and giant *9001* and *10001*; and earlier, in Yamaha's mammoth 5u high model *5002M*, now long obsolete. If anything, the history of class G demonstrates how the Japanese giants have tinkered with audio technology, while the smaller US companies 'got stuck in'.

Figure 4.53

Carver Class G OPS - simplified.

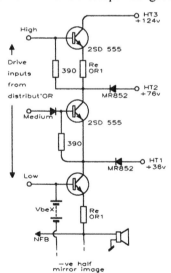

4.7.2 Class H

Quilter

Also in the first half of the 80's, but on the US west coast, a manufacturer called *QSC*, founded by Patrick Quilter, had established a line of amplifiers called Series I. Their engineering had blended some of the elegant ideas employed around the same time by Crown (i.e. a CE output topology executed in BJTs, and with mica-less output device mounting), to provide a simplified circuit with workable protection and effective yet lightweight heat exchangers. In 1984, QSC introduced their *Series III*, which retained the ground referenced output, marrying this with the first class *H* output stage (Figure 4.54). This scheme differs in having the transistors untiered, i.e. in parallel. Diode-OR'ong (series diodes in line with one or more rails) is/are needed on the lower rails to prevent inter-rail short circuits as the upper tier(s) is/are cumulatively brought into conduction. In exchange for much simpler driving (no beanstalks), the transistors switching the higher rails have to handle the full voltage alone. About this time Japanese researchers at Trio-Kenwood were working on a similar scheme [59], but this didn't get so far as QSC's implementation.

Figure 4.54

QSC series III, simplified Class H OPS.

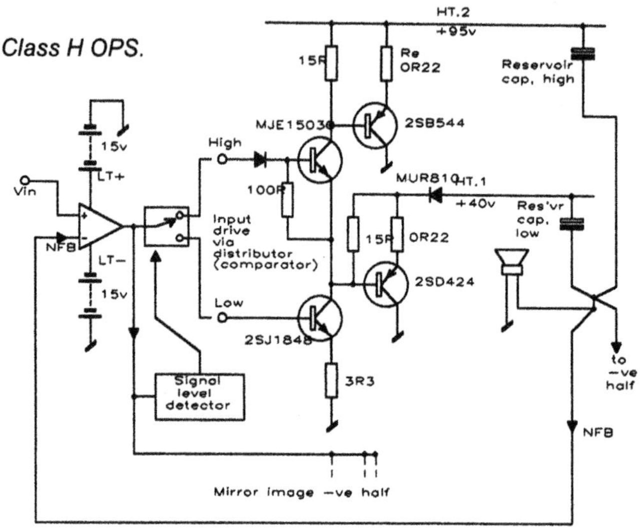

Class A/H

In 1988, Stan Gould, a co-founder of BSS Audio (cover picture, upper), then based in London, took class H still further. In BSS's model *EPC-780*, shunt-mode commutation involves *progressive* ('non-switched') drive transfer between three supply rails, instead of hard switching. As if this wasn't enough, the output stage is configured as an *asymmetric bridge* (Figure 4.55) [60], with one half operating in class A. The output of the low voltage class A amplifier is connected to the ground (0v); the bridge's HT1,2,3 rails and their common centre-tap midpoint are all floating. This most innovative circuit may be viewed as a hybrid of Super classA, NSB, *and* class H, *and* bridging! Efficiency is 70% at full power. At average program levels, up to around 20w (13v rms) the signal experiences class A operation with a relatively low dissipation. In these conditions, the amplifier is operating as a half-bridge on +/–27v rails (HT1). Since 1989, the topology has seen use in thousands of concerts worldwide, as the EPC amplifiers are part of the Turbosound's *Flashlight* and *Floodlight* PA systems.

Figure 4.55

Asymmetric Grounded Bridge in Class A/H.

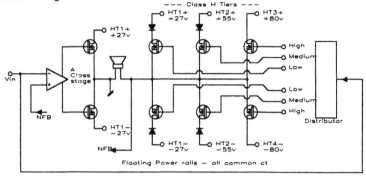

4.7.3 G and H, the comparison

As with all 'smart' solutions there will be setbacks, lurking.

Waste analysis

At the outset, class G seems to save on heat exchanger capacity. If the high voltage tiers are energised only 5-10% of the time, the average waste heat must be reduced, almost in proportion. Why ? Because waste heat is the power dissipated in the output devices. In turn, dissipation is the product of the voltage *across* the active output devices (V_{sus}), and the current (I) passing through them, ie.

$$P = V_{sus} \cdot I$$

The latter (I) depends on the speaker(s) connected, so it lies beyond the amplifier designer's control. Which leaves the HT voltage to be sustained, V_{sus} When a class B or A-B amplifier is driven with programme, V_{sus} typically averages 80-95% of the unloaded rail voltage. But with class G, the output devices are only being asked to sustain the higher rail voltages (V_{sus1} in Figure 4.51) that provide the desired headroom, for only 5 to 10% of the time. This is presumes the signal has the PMR and dispersed characteristics of reasonably uncompressed music.

Diversity loss

The occasions when the PMR of music program becomes dangerously close to that of a continuous sine wave are threefold. Two are well known:

(i) when the signal is severely clipped (overdriven) or

(ii) when the signal is highly compressed (by processing).

One other is less obvious. When an active crossover is used to split the audio spectrum into 2, 3 or even 5 bands, dispersion is likely to be reduced; there may be less happening, but what does could come all at once. *Also, more than anywhere, bass is where compression* (which reduces PMR) *is most heavily used create desirable sounds.*

It follows that to be satisfactory, a power amplifier handing actively-split bass, low bass and sub-bass in a PA rig musn't rely overmuch on burst power offered by classes G and H. The bottom end of an actively crossed-over monitoring or PA system is one place where a fairly continuous power capability still counts, at least with some kinds of music, e.g. high compressed bass guitar, or drone organ, synthesiser, didjeridoo, or any material processed with sub-sonic synthesis or pitch lowering, e.g. 'Jungle' bass.

In the original series mode 'G' scheme, developed by Hitachi and in the variants by Carver, the transistor(s) between each tier handle a fraction of the total rail voltage. For example, Tr.3 in Figure 4.53 sees no more than:

$$(HT3 - HT2) = (124v - 76v) = 48 \text{ volts}$$

Conversely, the series connection means *each* transistor potentially needs to handle the stack's maximum rated current output. This is no bad thing, since the reactive load presented by most loudspeakers can demand maximum current at almost *any* output voltage (not just at maximum output voltage swing, as with a resistive load). At the same time, for three tiers of rail voltage, three times as many output transistors are called for, compared to a comparably rated class A-B output stage. Fortunately, as with bridging, the relaxed voltage rating facilitates the economic sourcing of transistors with high current and high SOA ratings.

Tier count

Although extra dissipation arises from the series connection, it's offset by the fact that music and speech programme average 10 to 30% of maximum output power (section 2.4.4), unless hard clipped. Some of Carver's PM series amplifiers use three or four supply rail tiers to gain further, second order improvements in efficiency. There remain two snags. First, the output devices have to be mutually insulated. This means the transistors *either* suffer higher average temperatures owing to the thermal resistance of mica insulating washers, *or* else must live on separate heatsinks. Carver (today the professional division) adopts the former. But if the latter option were chosen, full-scale heatsinks would be required for each tier, to be sure each could safely handle worst case dissipation, i.e. an unfortunate combination of capacitative loading and a signal with a small PMR. Second, the switching spike residual is finite. If an amplifier had crossover distortion spikes as large, it might be laughed out of court. But class G commutation is not such a problem in practice, even to critical ears. No doubt this arises because the glitching occurs well away from the sensitive, central zero-crossing point (where the great amount of musical detail is focused). Then again, in Bob Carver's own heroic words "If they had been audible, I would have engineered the glitches away somehow".

Class H review

In QSC's Series III output stage topology (Figure 4.54), there are just two tiers of output devices. Notice how these are shunt connected. Because the OPS topology is CE, with floating rails, the outputs appear on the rails and are summed via the reservoir capacitors. *In consequence, there are no large scale spikes in the output; the period of transition is more like class AB, moving from A into B.* To support this,

144

QSC quote a near worst case distortion contribution of 0.03% at 20kHz at 6dBvr. Though it's not immediately apparent, the upper tier of output devices normally sustain less than the full rail voltage, and with music programme, the transistors in each tier share the power dissipation almost equally between them.

As with the series scheme, for *n* rails, *n* times as many output devices are required. But with the shunt connection, there's a 'soft fail' mode; the amplifier can continue working with reduced efficiency if one or more transistors expire (assuming the power supplies blows them into an open-circuit state). The shunt connection also reduces saturation losses. It follows that a class H amplifier will run cooler than one based on series topology, *when handling some actively split LF program*, or muzak (definition: program with PMR below 10dB) as well as whenever the abuse of hard drive deep into clip occurs.

Other benefits

For a given amount of feedback, class H goes on to exhibit a better, lower, open-loop output impedance under large signal conditions, where it matters most. This helps to keep interface intermodulation effects caused by the back-EMFs made by loudspeaker components, at bay. Finally, given QSC's clever *Series III* topology, shunt connection allows all the transistors of common polarity between different tiers, to be directly mounted on a single, common heatsink. The upshot is that a given volume (not to mention weight) of aluminium alloy is kept 'filled with heat' most of the time, dispersion notwithstanding. Here lies another commonly *unmeasured* and *uncited* dimension of amplifier efficiency: the percentage utilisation of net thermal capacity.

Real efficiency

The efficiency of QSC's Series III amplifiers (seen in Figure 4.56), below the tinted curve for standard class B and AB output stages) shows off the performance of pure class H at its best. By contrast, the *class G scheme is progressively less efficient for*

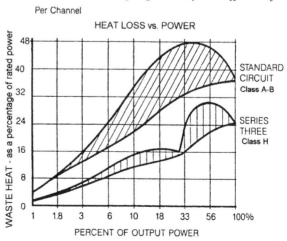

Figure 4.56

HEAT LOSS vs. POWER

Upper curve of each envelope shows sine wave losses.

Lower curve of each envelope shows losses for typical program having same average power.

Courtesy QSC

Per Channel

HEAT LOSS vs. POWER

STANDARD CIRCUIT
Class A-B

SERIES THREE
Class H

WASTE HEAT - as a percentage of rated power

PERCENT OF OUTPUT POWER

145

signal levels averaging >5dBvr, i.e. above 30% of full power. If the signal is compressed, clipped or bassy enough for this kind of condition to be prevalent, then the difference is a significant one: At 56% of full power (2.7dBvr), Carver quotes 65.7% whereas QSC's curve is indicating 76% with sine waves, and 82% with music drive. But with substantially uncompressed music and speech, i.e. purer live sound, the difference recedes.

One question remains. Did the Japanese really invent [61] class G? The answer is no. The late Peter Baxandall kindly unearthed a reference [62] to a circuit described as a 'Quasi linear' amplifier. It appears in a 1964 American textbook, and anticipates Dynaharmony in every significant detail.

4.8 Beyond analogue

In spite of their differences, all the amplifier types we've looked at so far are analogue: they are concerned with linear power conversion, where the output devices (which may be active at different times) act as a modulated resistance. Take any ordinary class B or A-B power amplifier, drive it into hard clip with a clean square wave, and it's likely to run cooler than when it's driven with music – for the same rms power into a resistive test load. So what?

Loss review

Dissipation in all linear amplifiers' output stages varies according to the signal's instantaneous voltage (expressed as a % of the supply rail) and the time spent in regions of high or low dissipation. Thus practical efficiency varies with drive level, waveform shape and duty cycle. In class A, standing current masks these effects, whereas for classes A–B, G and H, dissipation reaches a minima when the signal voltage is close to its maximum, near either rail. At this point, dissipation in both halves of a class B, G or H output stage is close to nil – always presuming the wiring and output transistors have low resistance, hence negligible I^2R losses.

For any split rail, push-pull amplifier (i.e. with conjugate pairs, running off +/– supply rails), the 0v (quiescent) point is the opposite minimum, only here, dissipation is somewhat higher, depending on the standing current employed. During the waveform's transition to all other voltages (lying between HT+ and 0v; and 0v to HT–), dissipation increases to a lesser or greater degree. *It's this area that both sine waves and music waveforms spend most of their time traversing.* Imagine though, that music were conveniently composed of perfectly neat and oversized square waves, meaning the middle areas were transversed extremely quickly, i.e. the output devices acted purely as switches. Dissipation in them would then shrink and we might look forward to output stages that approach 100% efficiency.

Some key sonic benefits are:

(i) no crossover distortion, as the zero, central point would be 'mapped out' of significance.

(ii) power device linearity is unimportant and will not affect sonics; specifications reduce to those needed for fast, clean switching alone.

(iii) global negative feedback can *in theory* be typically 5x to 10x faster than the 1 to 2μS of most ordinary, linear amplifiers.

4.8.1 Class D

While the origination and later harnessing of class G owes much to engineers in the USA, the development of class D has been a largely British affair (the same could be said for class B, and error correction, with European input). While class D principles have been cited as far back as 1947 [63], it is regarded as having been invented in the '50s in the UK by Dr. A. H Reeves, father of Pulse Code Modulation.

From 1960

Although a practical design is cited in 1960 [64], and the Anglo-Dutch transistor maker Mullard also published a design, the world's first *commercial* unit was a DIY module, designed by Gordon Edge and engineered by Jim Westwood at Sinclair Radionics, Cambridge. His X10 amplifier was launched in the UK in fall of 1964 (Figure 4.57). It was temperamental, and when it did work, the output power was barely 2½ watts. Sinclair's Mk.2, the X20, easily produced 20 watts in 1966, but

Figure 4.57

Advertisement in Practical Wireless, December 1964.

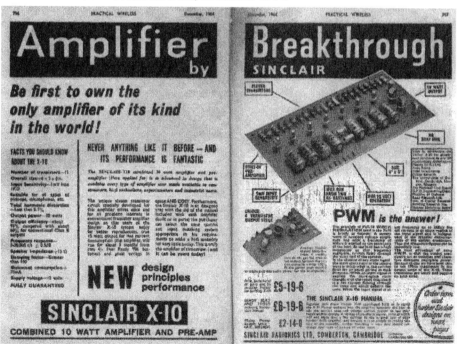

suffered from the inconsistencies and limitations of the germanium transistors available at the time [29]. With the explosion of so many other possibilities resulting from transistor developments, Sinclair went off in other directions, ironically including class A-B amplifiers. The lack of persistence was understandable. Also in 1966, the distinguished expatriate engineer, Norman Crowhurst (ex-chief engineer at Tannoy, and 50's inventor of bi- and tri-amping) patented an improved scheme [65], but the makers he approached were disinterested.

80's revival

Twenty years had to pass before class D techniques again received serious attention from professional power amplifier manufacturers. In the interim, in 1974, Infinity Systems had produced a 125w/ch PWM amplifier, developing in 1976 into *Swamp* ('switch amp'), a 300w$\Rightarrow$8Ω/ch model. In practice, this had to have far larger heatsinks than theory predicted, and reliability in the silicon BJT output stage varied unpredictably and mysteriously. Sonics were also chequered. By 1978 Infinity abandoned class D, but simultaneously Sony introduced their TA-N88. This class D unit employed the first generation of power MOSFETs. Even the power supply was a switching type. Then like many Japanese innovations, it was withdrawn after a couple of years. From the early 80's, switching amplifiers were subsequently employed for brute-force 'low-end' car systems, but that is outside our scope.

It was the rapid developments in vertical MOSFETs (DMOS) between 1979 and 1985 set the scene for some relatively successful professional amplifiers. By being able to switch with these, at up to 500kHz with tolerable losses, the poor hf and variable quality of earlier designs would be overcome.

In this period, a third Englishman, Brian Attwood had a consultancy, PWM Systems. In 1983 he published a paper [66] which got the ball rolling again, as his ideas were duly embodied in a range of high power professional audio amplifiers. These were Peavey's 'DECA' (Digital Energy Conversion Amplifiers), conceived in 1986. About the same time, Harrison (ex- of HH in the UK) launched their own range, employing similar circuitry [67] to achieve a claimed 2x 600w $\Rightarrow$ 4Ω/ch, in a 2u enclosure of below conventional weight.

PWM is not PCM

Some class D power amps have being sold under a misleading title: 'digital'. There is *some* truth in this but it's coincidental that the classificatory 'D' letter (originated nearly half of a century ago) appears to stand for **D**igital, as the Pulse Width Modulation (PWM) techniques employed under the lid of all class D amplifiers so far are *un*related to CD's and DAT's Pulse Code Modulation (PCM). But see the subsequent section. Instead, the classic embodiment of class D is more of a hybrid affair between a switching power supply and an FM radio transceiver.

Figure 4.58

Class D, Generic PWM after Attwood.

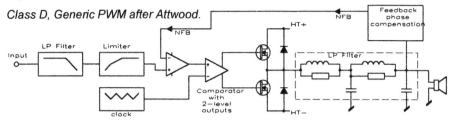

Classic technique

Looking into Figure 4.58, the class D recipe begins with a precise triangle wave clock with a frequency that's much higher than audio, typically 500kHz, and preferably shared between two channels, for mutual synchronisation. The incoming audio is low-pass filtered (as in any well designed amplifier) and its amplitude is limited or at least *bounded*. The signal is then summed with the global feedback. There need not be any, but so far, all practical PWM amplifiers use it. The combined signal is applied to a comparator, and compared with the clock. The difference between the two signals emerges from the comparator as a square wave with a *duty-cycle* that's proportional to the signal's *amplitude*, at a 'sample rate' that's much higher than audio. In turn the square wave switches a pair of output devices, placed as usual between + and –ve rails. When signal is nil, the triangle wave's exactly 50/50 duty cycle leaves the output devices switching equally to maintain 0v.

MOSFETs are a keynote. The switching speeds and losses of comparably rugged BJTs are inadequate and too high respectively. At the output, a low pass filter (LPF) 'reconstructs' the original signal, at the same time removing the worst of the sampling RF 'hash'. With the amplifier's output devices reduced to on-off switches, efficiency should approach 100%. There was just one little snag – with device physics. The MOSFETs with low 'On' resistance had high capacitance (C_{DS}). With high switching speeds, this capacitance draws substantial current, so the amplifier's dissipation would be high even when it's idling. But if C_{DS} is reduced, R_{on} increases, which in turn would make the amplifier inefficient when driven hard.

Scrutiny

While simple on paper, *the calibre of engineering design needed to produce a PWM amplifier that doesn't radiate EMI, and is measurably and audibly on a par with equivalent 'analogue' amplifiers, is truly formidable.* For example, only limited amounts of global negative feedback can be applied to 'kiss it better'. Why ? For analogue feedback, it's necessary to 'sample' the actual, reconstructed output. So the feedback must be derived after the output low pass filter. The accompanying phase shift typically limits feedback to just 20dB at high frequencies. There is also, for 500kHz, an effective 2μS lag. Added phase compensation (as shown in the figure) enables a greater amount of global feedback to be usefully employed, but not so much as a linear amplifier is often able to employ.

It follows that the PWM amplifier's open-loop linearity has to be good to secure respectable harmonic and intermod distortion figures at hf. Linearity depends on the comparator's incisiveness, the accuracy of the triangle wave, and the *precise matching* of each MOSFET's switching speed at high frequencies.

Efficiency

Makers of class D amplifiers routinely claim efficiencies of 90% and over, irrespective of output level. Amplifier efficiencies of this order need careful definition, taking care for example, to exclude the power used by auxiliary parts, such as LEDs and relays! High efficiency demands good output inductors. At 500kHz, or just 10kHz, not much skin or proximity effect is needed for there to be high copper losses, which wastes energy in heating. Efficiency is aided by use of suitable 'efficiency diodes' to recycle RF energy to the supplies. A tapped inductor may also be used.

It is informative to check class D efficiency figures across the audio bandwidth. At the state of development attained by Peavey in 1989 in their pioneering DECA series, comfortable operation up to 40kHz was at the 'leading edge'. But with first and second generation models, ascending frequency (or for music, increasing hf spectral content, or 'spectral shift', typically caused by tape rewind or other inaudible ultrasonic signals, or clipping of line stages) causes switching losses to multiply rapidly, and full drive levels above 6 to 10kHz, even only mildly sustained, could cause OPS flameout ! In such cases, an increased steepness of low-pass input filtration above 20kHz is a tempting 'solution', but really only a palliative, since clipping can generate high frequencies from low ones.

In another of the pioneering, professional class D amplifiers of the late 80s, made in the UK by Harrison, hf response was arranged to rolloff dynamically for protection as 0dBr was approached. It might be called *'level and survival-conscious'* hf filtration. The effects of the dynamic protection might be benign but could not be described as favouring dynamic accuracy.

Noise and EMI

With an RF ($\geq$500kHz) squarewave let loose amongst sensitive circuitry, there's plenty of scope for wideband radiation, modulation – and the nub of the problem, demodulation in audio places. Inside a PWM amplifier, any ingress of RF to the comparators' inputs is particularly devastating, as it will likely create a positive feedback loop at RF. The thought of irradiating people, or RF in a studio via the speaker, mains or input cabling is even worse. To have any chance of meeting EMC specs or even FCC, the output reconstruction filter must in practice be followed by filtering that can be forceful at mopping up residual, mostly high RF, that has escaped with the reconstructed signal. Meanwhile, overall filter design is not closely optimisable, as the amplifier's load impedance is almost completely undefined in most cases. Speaker makers would not have this difficulty; with a built-in PWM amplifier, even the speaker leads can be 100% defined.

Loudspeakers are largely protected from being harmed, or attempting to transduce any escaping RF (alias switching noise) by the low pass effect of their voice coil inductance. Any speaker cables other than non-inductive types will radiate residual RF hash. In Peavey's DECA (Digital Energy Conversion Amplifier) series, representing the most refined class D technology in 1989, the maker was able to cite conformance to FCC regulations for '*A class B computing device, part 15(j)*' – with a pk-pk output ripple of <10mV. The SNR of class D amplifiers also needs careful assessment, particularly for studio monitoring, as the spectral content will be quite unlike linear amplifiers, and *noise modulation* is likely, where *background noise is intensified in the presence of the music.*

Development

A recurring practical problem in PWM schemes is to do with preventing *cross-conduction* (alias *shoot through* or *common-mode conduction*), which causes a *current-spike*. Here, one half is still 'turning-off' while the opposite half turns on. Even if the devices were perfect, at the speeds involved, the inductance in the PCB track is enough to pervert the instantaneous switching that purist audio users would prefer, for plausible sonics and detail. A modern solution [68] is to 'soft-switch', so the switching path is flushed-out, while *transient resonance* is supposed to keep the audio going.

The dual of cross-conduction is the back-EMF kick-back from the output inductor, causing the output to shoot to one or other rail. As well as testing device voltage ratings, this too diminishes efficiency.

Round-up

Peavey's DECA series was re-engineered and renamed in 1993. Peavey now use a patented form of class D, called *Phase Modulation Control*. Instead of simple PWM or duty cycle variation of a square wave, the conjugate output switches operate at the same frequency, but with a variable phase difference. A present day example of this approach delivers 750w/ch while weighing 12 lbs.

In 1992, Motorola published an application note [69] showing how their own parts could be used in an application circuit, which claimed 92% efficiency with 0.08% THD at 10Hz rising to 0.31% at 10kHz, for 30v into 8Ω, alias 112 watts, with a mere 120kHz rate. Refinements include voltage balancing MOSFETs, load current sensing, current compensation feedback, and a duty cycle limiter to prevent clipping.

Production of complete PWM amplifiers for hi-end car and even domestic systems has been re-commenced by Infinity systems, with design by Mack Turner. Such amplifiers are the complement to homes that are wind or solar powered by 12 or 24v DC. Also in the US, Acoustical Supply are manufacturing OEM PWM amplifier modules and PWM-powered sub-woofers rated at up to 400 watts, that are expected to see widespread use in other products – although EMI conformity is unmentioned. As is usual with PWM, the power stage is remarkably small, e.g. at the cited nominal 95% efficiency, 400 watts will see the two output devices dissipating just 10 watts each.

Doubts

Other than meeting EMC requirements, what is never mentioned about analogue PWM, except perhaps in some hard hitting review reports, is the difficulty in any of commercial implementations of getting a %THD+N under realistic conditions that's below about 1% in the top 10dB of voltage swing – and not just above 5kHz. Loudspeaker damping in the form of low enough output impedance may also be questionable.

4.8.2 'Digital' amplification

Today, audio is increasingly derived from digital sources. Digital first came from reel-to-reel tape in the late '70s, joined in 1982 by Compact Discs, then R-DAT, and more recently, a broadening range of sources including radio broadcasting; and ISDN and other plumbing connected to an 'Information Highway'. Audio is also increasingly processed, and more recently, mixed in 'the digital domain'.

The parts of the audio path that have yet to be acceptably *digitalised* are loudspeakers, microphones and mic amps, and power amplifiers. A wholly digital amplifier is not much use without a digital loudspeaker. There are none of either on the horizon – yet. The nearest there is to a digital amplifier is a PWM amplifier, with a digital input.

The digital audio's PCM (Pulse Code Modulation) is translatable into PWM (Pulse Width Modulation). The amplifier is then almost all digital, bar the output reconstruction filter, which might be seen as a passive DAC. But unless we add a proper DAC just for control, there is no chance of global feedback. Still, with sufficiently high sample (or clock) rates, and noise shaping (e.g. with 32x oversampling), there should be no need for analogue feedback, or heavyweight filtrative parts. While elegant on paper, the ideal readily pushes clock speeds into Giga-Hertz (1GHz = 1000MHz), and the fringes of what's possible in digital electronics. The approach of a team led by Dr. Mark Sandler, at Kings College, London [70] has been to use noise shaping to reduce clock speed to a more practical 125MHz. This allows an 8 bit word to be encoded as an integer multiple of 8nS. Success has proved elusive so far and research continues as we go to press. Meanwhile, Harris is the first semiconductor maker to offer its series of 'HIP408X' MOSFET driver ICs, originally designed for disk head positioners, for audio use in class D power amplifiers. Accepting a digital input, these ICs operate at a more modest 1MHz clock rate. The OPS is bridged, to get more useful swing from Harris' own state-of-the-art 80v rated power MOSFETs. Apex (in the USA) meanwhile, have produced similary rated but almost *completely* integrated PWM amplifiers, as SA50 and 51.

The practical reality (so far) is that whether 1MHz or GHz, protection against adverse loading is a moot point, while EMI radiation control demands can readily eat up the anticipated parts and design cost savings. In future, surface-mount power parts are expected to greatly ease present EMI problems. Another problem has been sonics. But with an increasing awareness of digital jitter, and how it can particularly (and counter-intuitively) alter *low* frequency rhythmic accuracy, the things that need fixing, that give low jitter, have become clear [71]. It still remains *difficult* however. It's difficult enough getting low jitter in a CD player. But when sampling at 1MHz (the high sample rate avoids the need for PWM having sample and hold devices, most of which can't even manage 44kHz very well), and passing maybe tens of amperes and swinging from + to −100 volts at a rate of 100,000v/µS, then jitter-generating phenomena such as *ground bounce*, are particularly hard to engineer out. Wasn't the whole idea of class D simplicity, light weight, and cheapness...?

In 1997, and despite much hard work, the only kind of 'Digital' power amplifier which is likely be described as high performance in high-end domestic terms, and that is being made commercially, is the type where some (any) kind of analogue power amplifier is fronted by an integral DAC. This kind of amplifier may also accept analogue signals, but otherwise, it gives the appearance of being a digital amplifier, *and may be passed off as such.*

Conclusion It has been said since 1960 that (words to this effect) "Once the potential shortcomings of class D have been overcome to everyone's satisfaction, PWM amplification will be all there is for anything over a 100 watts or so". But it hasn't happened yet.

4.9 Class summary

Type Key Features

A

Pros: Ease of low distortion, both static and dynamic

Cons: High hardware cost, high weight, high electricity cost, high heat output, power capability practically limited to about 100 watts.

B, A-B, A-B$_1$, A-B$_2$, QPP.

Pros: Simple circuitry, low hardware cost, medium electricity cost. With good design and application to ameliorative tools, almost as low a distortion as class A, is possible.

Cons: The reference condition.

Super A, π-mode, Dynamic and Plateau biased-A.

Pros: Potentially lowered dynamic distortion over ordinary B and A-B types. Much of the class A advantages at some fraction of the cost.

Cons: Slightly more complex circuitry than A-B. Topologically more limited. Higher electricity cost and heat output if quiescent periods dominate, but still much lower than class A types.

C   Radio Frequency amplifiers only, not for audio.

D

Pros: High efficiency, lightweight, potentially low hardware cost,

Cons: Complex circuitry, sonic performance may be poor, particularly above 5kHz, EMI emission likely.

E, F: Category undocumented, not for audio.

G, H

Pros: Both permit high transient power capability. usefully higher efficiency than class A-B under medium to hard-drive conditions.

Cons: Slightly more complex circuitry than A-B. More PSU parts and hardware and higher build cost. Added dynamic distortions possible (but well containable in practice) due to multiple-supply switching and finite part matching. In inexpert designs, multiple power supplies might be improperly apportioned, leading to wasted hardware and unsuitability with some programme.

S   See section 4.9.3. Not really a class.

per watt output:	⇐Lowest		– non-linear scale–		Highest⇒
Waste heat	D	G, H	B, A-B	SA	A
Weight	D	G, H	B, A-B	SA	A
Parts Cost	D	G, H	B, A-B,	SA	A
Distortion	A	A/H, SA	B, A-B	G, H	D

Notes: A-B includes NSB (non-switching class B).
'SA ' (special A) includes Super class A and sliding A and π-mode A-B.

The following table concludes, with a listing of power amplifiers established in professional sound, that go beyond ordinary class B and A-B operation.

Class examples – includes classic models

Notice! US Designation for class G and H are exactly the reverse of British and Japanese nomenclature used here.

Maker	Model	OPSD	Class
ATMC, UK - see MHA.			
BSS, UK	EPC-780	DMOS	A/H
	EPC760	DMOS	A/H
Carver, US	*PM1.5	BJT	G
	*PM2.0	BJT	G
	*PM175	BJT	G
	*PM350	BJT	G
	PM 700	BJT	G

	PM 950	BJT	G
	PM & PT 1400	BJT	G
	PM & PT 1800	BJT	G
	PM & PT 2400	BJT	G
C-Audio, UK	RA 4001	LMOS	H
	XR-3801	LMOS	H
	ST-1000	LMOS	H
Crest, US	7001	BJT	G
	8001	BJT	G
Hill, UK (MHA)		* DX series BJT	SA
HIT, UK	*DSA series	DMOS	D
Peavey, US	AMR/PMA	BJT	G
	VX series	BJT	G
	*DECA 424	DMOS	D
	*DECA 1200	DMOS	D
	DPC-1000	DMOS	D
QSC, US	*Series III	BJT	H
	MX Series	BJT	H
	EX Series	BJT	G
	MXa Series	BJT	G
	CX Series	BJT	G
	PowerLight series	BJT	G
Soundtech		BJT	G
Tytech, Korea		BJT	G
Yamaha, Japan	*PC5002M	BJT	G

OPSD = OutPut Stage Device type. DMOS = vertical. LMOS = lateral.

* indicates classic models believed to be no longer made.

4.10 Introducing modes of control

So far, throughout this chapter, the details of negative feedback (NFB), or whatever else is used to achieve stable performance, has been omitted from circuit drawings and discussed only where of particular concern.

While *parallel-mode* and *bridge-mode* are in everyday use different *modes* of operation, '*Mode*' is used here in a higher sense, to distinguish between a third, fundamental feature, beyond class and topology, that is often taken for granted. This concerns:

(i) *which* dimension of the signal is aiming to be *controlled* at the output – V or I ?

(ii) *what* dimension is being *sensed* at the output – volts or I ?

(iii) *how* it is sensed – as volts, or amperes; in shunt or series ?

(iv) *how far* the control loop(s) wrap(s) – global, or part, or local ?

4.10.1 Negative feedback modes

Nearly all conventional high NFB amplifiers operate in what could be called *'voltage-voltage-voltage shunt global'* mode, i.e. they act as voltage sources, with voltage sensing and shunt-mode voltage feedback, connected globally. Figure 4.59 shows the circuit arrangement for conventional, global NFB, as employed in the vast majority of solid-state audio amplifiers that have existed, period. The feedback is said to be shunt or *shunt-derived*. This same configuration is called non-inverting, since at midband frequencies, the signal polarity at the output is the same as that at the input (irrespective of phase shift). Also shown are optional RF compensation networks ('comp'); the low-arm high-pass (hpf) capacitor that can so greatly affect sonic quality, and if fitted, enforces 100% feedback at DC, with the upper (UAR) and lower arm resistors (LAR) defining the AC (signal) feedback ratio. In most of the preceding circuits in this chapter, this form of network is what would normally be connected between the two points marked 'NFB', and also ground, in most of the preceding circuits in this chapter.

Figure 4.59

Universal 'Voltage Mode' shunt (derived) Feedback.

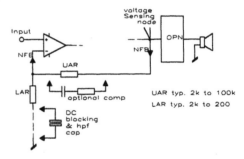

Figure 4.60 shows the more purist *series* mode of NFB, alias inverting, as commonly used for summers and mixers, but far less often for power stages. One reason is plain habit. Another may be experience of RF instability. For unlike the shunt mode, the output has nominally opposite polarity to the input, so it is better proofed against interaction with input signals – except at some high frequency. When the phase become within 90° of being the same, stability becomes an issue, as positive feedback takes over. When due care is taken to keep input and output signals separated, this scheme has the advantage of not developing any common-mode non-linearity, that would cause CM distortion on medium to large input signals – see chapter 3.

Figure 4.60

Series-derived Voltage Feedback.

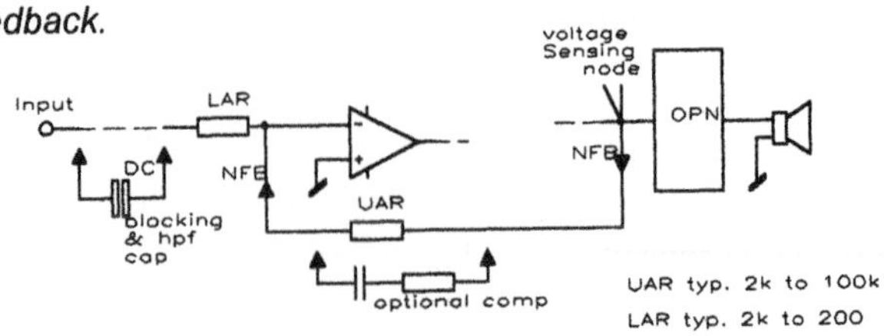

Notice that in *either series or shunt mode*, the feedback (NFB) is *always* returned to the –ve (inverting) input, thus it is the input connection that has moved. This inverting terminal is not an absolute property of the input device, but always relative to the output polarity. With the CE OPS topologies, the amplifiers' output is always inverted, hence the positive (+) non-inverting input of an op-amp used to receive the feedback often becomes the effective –ve input. Where in a circuit global feedback appears to be applied to a positive (+) op-amp input, prior inversion is usually the answer.

What the most common form of 'feedback' (alias NFB) achieves, when well adjusted, and whether series or shunt-derived, is well known:

* Input impedance is high. Little current is drawn from the signal source.

* Output impedance is very low, so the signal presented to the speaker at the output behaves almost like a pure voltage source, i.e. voltage does not sag or flinch no matter how much current is demanded.

* Non-linearity is smoothed out, hence % distortion is reduced. As discussed in the section on class B and A-B biasing, this presumes all signal path semiconductors are alive in forward active mode throughout the cycle.

* DC balance at the output is assured and principal DC conditions are stabilised at switch on. Only *DC feedback* is needed to achieve this, but making the AC and DC feedback proportions different requires unnatural HP (filtrative) capacitors. This may degrade sonics for other reasons, as discussed elsewhere.

Limitations

It is also well known that global NFB, as practically employed, has limitations. In particular, the amount of it you can usefully apply in practice reduces with increasing frequency. Thus its beneficial effects are gradually or eventually lost with increasing frequency, often significantly within the audio band. Trying to increase the feedback causes RF instability, what other electronic engineers call *Nyquist's instability*. Nelson Pass [25] describes NFB as adding a kind of 'distortion in reverse' to the input signal. This can only happen after the event. The lag can be very, very short (in tens of nanoseconds, the region of 0.00000003 of a second, say), but for music signals, where there is little continuity, NFB is condemned to be *always* trying to catch up with what has just happened. It is in this dimension, namely time, in which global NFB is most likely to cause errors while appearing pristine for repetitive signals. This is particularly so over the surprisingly short periods of time that are important to musical detail.

NFB is able to loose control transiently, for a variety of reasons. When this happens, performance momentarily degrades to the open-loop response, which is how it was before feedback was applied. In amplifiers employing high NFB, especially those with very high open-loop gain (a result of having cascaded two or more voltage gain stages) the open-loop condition is commonly a highly non-linear state, hence momentary loss of NFB control will inevitably cause a large change in measured and sonic quality. Then a form of electronic indigestion may occur, when an amplifier is paralysed, sticking to one supply rail, or RF oscillating, for several tens or even hundreds of milliseconds after a transient overload – before control is regained.

NFB alternatives

Although their total, cumulative physical manifestation is small in percent alongside a majority that numbers in millions, there are many other schemes beyond global NFB.

Lesser loop

Many high-end domestic power amplifiers claim not to employ global NFB. But they still employ modern circuit topologies which require some 'wrap around' feedback correction to work at all. Their solution is to arrange the principal feedback loop to bypass the output stage (Figure 4.61). Using the more fully descriptive nomenclature, this would be termed *Voltage-voltage-voltage shunt partial* mode, or possibly *Voltage-voltage/current-voltage series local* mode. In this case, what will become the output voltage may be sensed as a current, but 'read' as voltage. The output stage may be lightly coupled, either at all frequencies using a resistor (UAR2), or just at DC, using an RCR network. *This latter connection is also useful to investigate the open-loop behaviour of all amplifiers employing global and partly-global NFB,* as without DC feedback, the mostly high gain circuitry would be locked-up by the slightest input mismatch of dc offset, creating a dangerous DC fault.

Figure 4.61

*Partial Global Feedback
(Shunt-derived voltage).*

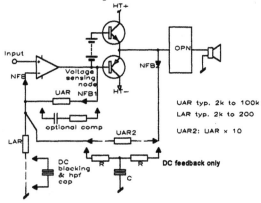

UAR typ. 2k to 100k

LAR typ. 2k to 200

UAR2: UAR × 10

DC feedback only

'Zero' feedback

In amplifiers which operate purely in class A, or NSB, but not in B or A-B, global feedback may be abandoned altogether. This is plausible with circuits that are suited to *open-loop working* because they do not require global feedback to maintain near ideal DC conditions of bias and output midpoint centering. Usually, such circuits are simple, with minimal gain build-up. 'Zero' really means no overall feedback. It is usually practically necessary in production, and is also no less advantageous to measured linearity and sonics, to add local feedback where it isn't already used, and to increase the amounts. Generally, local feedback is achieved at signal frequencies by unbypassed resistors in line with the emitter (e.g. R_e in the many previous schematics) or source terminals, of BJTs and MOS- or J-FETs respectively. So-called 'local' (but really global) feedback also occurs when gain blocks comprising more than one active device are created. The connections creating the local feedback are akin to those in Figures 4.61 and 4.62, only the loop is shorter, and the same resistors may double at other tasks. One problem with solely local feedback comes in production, as the various parameters of individual amplifiers that are important to audio quality, such as hum levels, are likely to measure quite variably, mainly as a result of the wide tolerances of many active device properties.

Intermediacy

Global NFB suits mass manufacture's presumption of consistency, because it smoothes out tolerance variations. With its widespread use, we take the consistency of electronics for granted. In some audio power amplifiers, ZFB is not claimed. Instead, a larger than usual degree of local feedback has been employed to achieve high open-loop linearity, and thus acceptable sound with no overall feedback. A *small* amount of overall feedback is then applied to smooth-out tolerance variations in the unit overall (Figure 4.62).

Figure 4.62

Combined NFB: Local & Global.

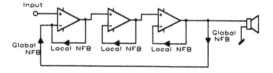

Nested NFB

In the early 80's, Professor Cherry revived interest in multiple 'nested' negative feedback loops. Here, there are successive the global loops. If set up properly, non-linearity can be greatly reduced, but the likely compromises to transient accuracy, and harmonic phase relations, and hence sonic quality, are uncited. The slight additional complexity of such circuitry, and added fiddle-factor when there are already enough variables in power amplifier manufacturing and hf stabilisation may explain why amplifiers employing nested NFB are scarce.

4-wire sensing

Figure 4.63 shows a connection method using 4 wires, that is well known to readers familiar with high current regulated DC power supplies. Here, global NFB is used to full advantage, by extending the sensing node *and ground reference* out to the loudspeaker's terminals. This has the effect of including the speaker cable in the NFB loop, thereby forcing the amplifier to use its loop-gain to compensate for signal errors or losses caused by the cable. Thus, the feedback works to perfect the signal at the speaker terminals, rather than at the amp's output. In practice, this scheme is little used, although it was offered by Deltec, a UK domestic maker, in a past model, and in the late 70s and early 80s there were some brave attempts by Japanese consumer-grade product makers to introduce it or similar techniques to

Figure 4.63

*4 wire NFB connection.
A scheme like this was used
by professional loudspeaker
maker UREI in the USA, in
their 810 series monitor
speakers, between the late
'70s and early '80s.*

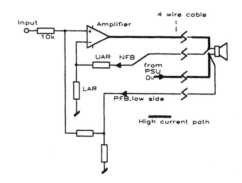

the general public. One difficulty is that most conventional speaker connectors are only 1, 2 or 3 pole. Another is that with the feedback 'brought out' there's considerable scope for wiring errors which could defeat the purpose and/or blow up the amplifier. Third, the absence of suitable speaker cables, comprising small, twisted and ideally shielded sense wires which do not carry appreciable current, so can be quite thin, running alongside a speaker cable, itself optimised for driving speakers, *without interaction or contamination between them*. Over several yards (metres), the current levels and the frequency range required for high performance, may be a tall order. The other option, of running the sense wires down a separate, spaced, and ideally orthogonal route, is a logistical nightmare in all but the simplest system. The exception is in powered speakers, where the speaker wiring is fixed, short and installed by the maker. Ironically of course, this is where the speaker wiring is most likely as short as it can be, and thus where the compensation on offer is least needed!

Current types

An increasing number of the IC op-amps designed in the past seven years have employed so called '*current-mode*' feedback [84]. This has real benefits when high gain is required, with high global NFB. A power amplifier using this technique has been described [72]. Sonic reports are encouraging but the reasons are debatable. As pointed out by the late Peter Baxanadall, the description is careless talk. Its feedback is more like *Voltage-voltage-current series global* mode, or something similar.

It is rare but possible to control the output voltage while sensing output current (*voltage-current-voltage*, again rather loosely styled 'transconductance amplification'); or a combination of voltage or current [73]; or more usual combinations. The control mode description might include the desired or implied mathematical operations, e.g. *voltage*, geometric-mean-*voltage-voltage*, or *voltage*, vectorsum-*voltage-voltage* (etc). It is also rare yet possible to drive a loudspeaker with current [74, 75], while using a sensing coil to feed back a corrective voltage proportional to the voice-coil velocity. This control mode might be described as: *Current-velocity-voltage shunt global*.

Current limits

These and related current-involved configurations are considered 'too esoteric' and neglected by the vast majority of power amplifier makers. Firstly, they require specially adapted drive units, cabling, connections and considerations that would make the amplifier unsuitable for general sale, while likely requiring the maker to be a speaker manufacturer as well. Second, current-driving a modern loudspeaker drive-unit, long refined on the basis of being driven by a near pure voltage source, is an unexplored art and not automatically any panacea; there will be losses in long-established good areas of amplifier and speaker performance alike, such as flat frequency response. Thirdly, learning to make non-voltage-source power amplifiers as unconditionally stable, foolproof and reliable as the better conventional audio power amplifiers, has only just begun.

Damping control

Figure 4.64 shows the gist of simplistic schemes to control loudspeaker damping (assuming ordinary global NFB referred to the amplifier's output terminals), by varying the amount of voltage-sensing NFB (a), and/or by introducing varying amounts of positive feedback (b), or else by introducing current-sensing feedback (c). These were in vogue in the 1950s, with valve amplifiers. Twenty years later, one of Crown's industrial amplifier designs was one of the few solid state amplifiers to offer the facility as a tweaker knob. Over the years, other methods have been suggested to improve or selectively modify drive-unit damping behaviour, such as selective negative impedance driving [76] achieved by adding a measure of positive feedback. In conjunction with a class G OPS, Philips used similar *negative impedance driving* in their DSS940 DSP active speaker. This causes the drive-unit's Q to vary with level, which may make tidy measurements, but is almost certainly sonically unnatural.

Figure 4.64

Variable Damping Controls

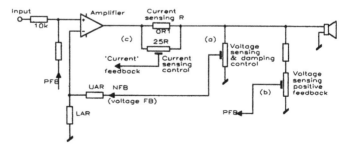

Further study

Some of the more common of the many possible simple 'feedback modes' are identified and analysed in electronics textbooks by way of generic 'black box' feedback connections, e.g. [8, 23, 77].

4.10.2 Other Error Correction Modes

Error takeoff

In the '*Short Circuits*' column of the January 1973 issue of Wireless World, A.M. Sandman (UK), having made proposals for class B amplifier improvements in 1971, presented a simple circuit showing how to use a low cost IC op-amp to build a low-quality amplifier, then use a second op-amp to correct the errors (Figure 4.65). *This did not involve NFB and was in addition to the global NFB already used* by the low quality amplifier. Instead, Sandman had noted that the error signal is approximately equal to the residue at the inverting input of the first op-amp. This is true for the series NFB configuration used. The error signal is then simply amplified (with a non-inverting amplifier note), and applied in antiphase by a second power stage, which makes a full-bridge. So here we have a bridge topology, IC op-amp-based, with NFB and one kind of *error feedforward*. Sandman titled his circuit 'Reducing distortion by error add-on'.

Figure 4.65

Distortion Reduction - by Error Add-On.

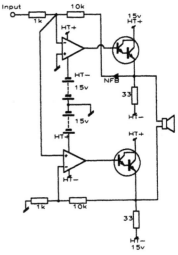

In the February 1973 issue, Bollen [78] followed with 'Distortion reducer'. In this article, a similar, rather more convoluted scheme was proposed (Figure 4.66). First the input signal is subtracted from an attenuated version of the output. This should leave just distortion and noise. Then feed this back into the input stage in anti-phase, to cancel the distortion – and ideally the noise. In practice, the IC op-amps of the day added some substantial noise and also distortion of their own. In turn the

Figure 4.66

Active Error Feedback (after MacDonald & Bollen).

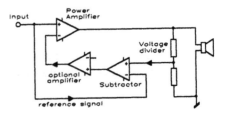

circuit was limited to attaining a nonetheless creditable reduction of non-linearity and hum by 15 to 20dB or so. Subsequently J.Ross Macdonald drew attention to his 1955 patent and paper [79], where he had described much the same system, but using valves (as you then did) and calling it 'Active error feedback'. Others call this 'multiple feedback'.

In the October 1974 issue of Wireless World, Sandman then presented a more in depth article, showing the 'distortion add-on' circuit providing what he now termed 'error takeoff' and also 'error feedforward'. He also showed an iterative, parallel scheme, where the main signal and error canceling signal are passively summed at the output, *n* times *ad nauseam*. He revealed that Howard Black, the inventor of NFB (with 5 patents filed from 1928 to 1932) had patented the latter earlier, in 1925, but it was too far ahead of its time. Sandman now felt that the time for putting *feedfoward error correction* (FFEC) to work had come. The gist was to measure the error at the output of the amplifier, amplify it, then subtract it from the signal. According to which way you see it, FFEC has a lot in common with current-dumping.

Sandman later called this 'class S' [80]. But it is really a mode of control, and again, quite unlike any of the feed*back* schemes previously discussed. It has seen commercial use in several Japanese domestic amplifiers (e.g. Sansui [81] and Technics), but nowhere else high profile. As a measure of what is possible in the realm of improving conventional specifications, in the Technics SE-A100 (Figure 4.67), which employs Sandman's 'FFEC', Hood [27] cites 0.0002% THD @ 1kHz, and a reasonably wide bandwidth, up to 150kHz, for a nominal 100w⇒8Ω. As in current-dumping, a small, clean amplifier 'A', with its own NFB is used to 'fill-in' the relatively small deficiencies of the larger-current-handling, high gain amplifier, 'B'. Each operates in the same class as its respective name. One way to look at the operation of the circuit is that the high gain error amplifier (B) makes the speakers impedance appear to be higher than it really is. But low distortion, including high cancellation of the crossover errors in amp B, requires a precise resistance match, as suggested by some of the ohmic values. And, as soon as the speaker is driven, or even the room temperature varies, nearly exact cancellation is bound to be lost.

Figure 4.67

Error Feed-Forward (after Sandman).

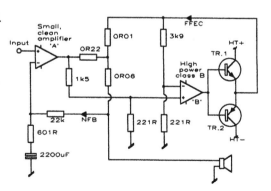

Error-correction

Work on the general topic of what is nowadays called *error-correction*, continues with further contributions from Dr. Sandman and particularly, Prof. Hawksford, both at British Universities. Having developed 'current-steering' amplifier circuits with significant benefits over conventional topologies, error-correction is now being developed [82], including schemes that need no ground reference [83].

4.11 Conclusions

After ploughing through this chapter, the heaviest in this book, the reader will have some feel for the way that only a fraction of the possibilities have been explored or widely exploited. Many an 'ideal' scheme has been tried by Japanese domestic hi-fi makers, for example, but few of the smart solutions have been presented, in a way that would bring out their sonic strong points, to listeners already attuned to the nuances perceived with simpler signal paths.

You may also have glimpsed the possible permutations (and chaotic possibilities) of topologies, classes and modes. The futility of power amplifier reviewers and pundits trying to link anything less than a full and accurate run down of all the circuit elements, class(es) and control mode(s), as barely outlined in the preceding 29,000 words, with any particular sonic attribute(s), should now be clearly visible. If a particular amp should have great bass slam, or liquid midrange imaging, it should be seen as sloppy practice to simply (and naively) associate this with some bulleted technical detail the maker's advertising house has decided to trumpet, for even if it is not some relatively minor topological or component 'technology', the context is bad while the *whole picture* includes so many other facets. In future, quality makers should *ascribe all elements* – or none at all.

References

1   Geddes, Keith, & Gordon Bussey, *The Setmakers,* BREMA, 1991. ISBN 0-9517042-0-6

2   Sziklai, *Proc.IRE, vol.41*, p.717, June 1953.

3   Self, D.R.G, *Distortion in power amplifiers, parts 1-8* , Electronics World, Aug 1993 to Mar 1994 and subsequent correspondence.

4   Otala, Matti, & Jorma Lammasniemi, *Intermodulation at the amplifier-loudspeaker interface, part 2,* Wireless World, Dec 1980.

5   Rosenberg, Harvey, *The Search for Musical Ecstasy, Book 1 - In the home,* 1993, Image Marketing Group, USA, ISBN 1-884250-01-7.

6   (Nameless), *Trans IRE, CT-4,* p.207, Fig.6, Sept 1957.

7   Brociner, Victor, & Recklinghausen, Daniel.R.Von, *Interrelation of speaker and amplifier design,* JAES, April 1964.

8   Cherry, Prof. Edward.M, & Hooper.D.E, *Amplifying Devices and Low-Pass Amplifier Design,* Wiley, 1968. LCCCN 67-29933.

9   Dinsdale, Jack, and Tobey, Richard, *Transistor Audio Amplifiers*, Wireless World, vol.67, Nov 1961.

10 Duncan, Ben, *Simulated attack on slew rates*, EW+WW, April 1995.

11 Shaw, *Quasi-complementary output stage modification*, Wireless World, June 1969.

12 Baxandall, Peter. J, *Symmetry in class B (letter)*, Wireless World, Sept 1969 and Aug 1970.

13 Linsley-Hood, John, *The Art of Linear Electronics*, Butterworth-Heinemann, 1993.

14 Bailey, Arthur.R, *30 watt high fidelity amplifier*, Wireless World, May 1968.

15 Self, Douglas, *Distortion off the rails*, EW+WW, March 1995.

16 Duncan, Ben, *Evaluating audio op-amps, part 1*, Studio Sound, July 1990.

17 Jones, Morgan, *Valve Amplifiers*, Butterworth-Heinemann, 1995. Also excerpted in EW+WW, 1995-1996.

18 Glogolja, *Biasing for the output stage of a power amplifier - the V_{BE} multiplier*, AN-6297, RCA, reprinted May 1974.

19 Hawksford, Malcolm, J, *Optimization of the amplified diode bias circuit for audio amplifiers*, JAES, Jan/Feb 1984.

20 (Nameless), *Low distortion class B output (staff report on QUAD 303 design)*, Wireless World, April 1968.

21 Bongiorno, Jim, *Ampzilla III*, The Audio Amateur, 4/1984.

22 Linsley-Hood, John, *A direct-coupled high quality stereo power amplifier, Part 1*, Hi-Fi News, Nov 1972.

23 Feucht, Dennis.L, *Handbook of Electronic Design*, Academic Press, 1990. ISBN 0-12-254240-1.

24 (Nameless), *Audio amplifier designs using IGBTs, MOSFETs and BJTs*, report by John Linsley-Hood, Toshiba Electronics (UK) Ltd, 1990.

25 Pass, Nelson, *Linearity, slew rates, damping, Stasis and ...*, Hi-Fi News, Sept 1983.

26 Linsley-Hood, John, *Symmetry in audio amplifier circuitry*, EW+WW, Jan 1985.

27 Linsley-Hood, John, *Solid-state audio power, Parts 1,2 & 3*, EW+WW, Oct, & Nov 1989, & Jan 1990.

28 Otala, M & Lohstroh, J, *An audio power amplifier for ultimate quality requirements*, IEEE transactions on audio and electroacoustics, Vol. Au-21, No.6, Dec 1973. Also presented, AES 44th convention, 1973.

29 Dale, Rodney, *The Sinclair Story*, Duckworth, 1985.

30 Giles, Martin (ed), *Audio/Radio Handbook, section 4*, Nat Semi, 1980.

31 Duncan, Ben, *Evaluating chip power*, Studio Sound, Sept 1995.

32 Losmandy, B.J, *Operational amplifier applications*, JAES, Jan 1969.

33 Alexander, Mark, *Boost op-amp output power with complementary power MOSFETs*, Siliconix Inc, AN83-5; also p.6-125 of *MOSPOWER applications handbook*, Siliconix, 1984. ISBN 0-9305190-00

34 Williams, Jim, *Power gain stages for monolithic amplifiers*, Linear Technology, AN-18, March 1986.

35 Olsson, Bengt, *Better audio from non-complements?*, EW+WW, Dec 94.

36 Duncan, Ben, *Ultra high-power amplifiers: The rationale, Proc.IOA, Vol.15*, part 7, 1993.

37 Taylor, *Proc.IRE*, p.444-5, March 1959.

38 Tharma, Poothathamby, *Transistor Audio Amplifiers*, Iliffe/Butterworth, 1971.

39 Brady, R.M, *The floating bridge*, Wireless World, 1980. See also letters, Jan 1981.

40 Duncan, Ben, *PSU regulation boosts audio performance*, EW+WW, Oct 1992.

41 (Nameless), *A 4.5w sliding bias amp, using an OC16*, Mullard technical communications, Vol.4, No.31, July 1958.

42 Worcester, *IRE audio transactions, Vol Au-7*, p14, Jan & Feb 1959.

43 Sano, N, T.Hayashi & H.Ogawa, *A high efficiency class A audio amplifier, preprint 1382*, 61st AES convention, Nov 1978.

44 Linsley-Hood, John, *Class A-B amplifiers*, letter, Wireless World, Aug 1970.

45 Blomley, Peter, *New approach to class B amplifier design*, Wireless World, Feb and March 1971.

46 Hartsuiker, Hans, *50w Blomley*, EW+WW, June 1992.

47 Walker, Peter.J, *Current dumping audio amplifiers*, Wireless World, Dec 1975.

48 Olson, Bengt, *Current dumping audio amplifier (letter)*, Wireless World, July 1978.

49 McLoughlin, M, *Current dumping review*, Wireless World, Sept 1983.

50 Visch, Nico.M, *A novel class B output*, Wireless World, April 1975.

51 Tanaka, S, *New biasing circuit for Class B operation*, J.AES, March 1981.

52 Margan, Erik, *Crossover distortion in class B amplifiers*, EW+WW, July 1987.

53 Gevel, van de, Marcel, *Audio power with a new loop*, EW+WW, Feb 1996.

54 Sato, T, K. Higashiyama & H.Jiko, *Amplifier transient crossover distortion resulting from temperature change of output transistors*, 72nd AES convention, Oct 1982.

55 Chater, William.T, *A 40w all-MOSFET power amp*, The Audio Amateur, Q2, 1988.

56 Carver, Robert, *Designing a 700 watt amplifier*, Electronics Today International, Nov 1972.

57 Mitchell, Peter, *A musically appropriate dynamic headroom test for power amplifiers*, AES preprint, 83rd convention, 1987.

58 Colloms, Martin, *review of Hitachi Dynaharmony HMA 8300*, HFN/RR, Oct 1977.

59 Funada, Saburo & Akiya, Henry, *A study of high efficiency audio power amplifiers using a voltage switching method*, J.AES Oct 1984.

60 Gould, Stan, *High power/2U*, Studio Sound, Nov 1988.

61 Fox, Barry, *Nothing new in amp design*, HFN/RR, June & July 1981.

62 SECC, *Handbook of basic transistor circuits and measurements*, vol.7, John Wiley, 1966.

63 Fitch, *Vol 94, No.13, Pt.IIIA*, J.IEE, 1947.

64 Ettigen & Cooper, *Proc.IEE, Vol.106-B-18*, 3092e, April 1960.

65 Crowhurst, Norman, *Updating pulse code modulation*, TAA, Q1, 1987.

66 Attwood, B, *Design parameters important for optimisation of very high fidelity PWM audio amplifiers*, AES, convention preprint, 1983.

67 Harrison, Mike, *The high power digital audio power amplifier*, Pro-Audio yearbook, 1987 (UK).

68 Poon, N.K, and Lau, W.H, *A novel cross over delay PWM amplifier, preprint 3969-G5*, 98th AES convention, Feb 1995.

69 (Nameless), *Switching audio amplifier uses power MOSFETs*, EW+WW, July 1992.

70 Sandler, Mark, B, and J.M.Goldburg, *Noise shaping and pulse width modulation for an all-digital audio power amplifier*, JAES, Vol.39, June, 1991.

71 Fourré, Rémy, *Jitter and the digital interface*, Stereophile, Oct 1993.

72 Alexander, Mark, *A current-feedback audio power amplifier, Preprint 2902-D5*, 88th AES, March 1990. Also available from Analog Devices as AN-211.

73 Hibbert, *Loudspeaker feedback circuit*, Wireless World, Aug 1976.

74 Mills, P.G.L and M.O.J. Hawksford, *Distortion reduction in moving coil loudspeaker systems using current-drive technology*, JAES Vol 37, March 1989.

75 Mills, P.G.L and M.O.J. Hawksford, *Transconductance power amplifier systems for current-driven speakers*, JAES Vol.37, Oct 1989.

76 Werner, R.E, & R.M. Carrell, of RCA, *Application of negative impedance amplifiers to loudspeakers systems*, 9th AES convention, Oct 1957.

77 Grey, Paul.R, & Robert.G. Meyer, *Analog Integrated Circuits*, Wiley, 2nd ed. 1984. ISBN 0-471-81454-7.

78 Bollen, D, *Distortion reducer*, Wireless World, Feb 1973.

79 Macdonald, J.Ross, *Active error feedback*, Proc.IRE 43,808, 1955.

80 Sandman, A.M, *Class S, a novel approach to amplifier distortion*, Wireless World, Sept 1982.

81 Takaashi, Sasumu, & Susumu Tanaka, *Design and construction of a feedforward error-correction amplifier*, JAES, Vol.29, Jan/Feb 1981.

82 Hawksford, Malcolm O, *Quad input current mode asymmetric cell (CMAC) with error-correction applications*, Proc.IEE, CDS, vol.143, Feb 1996.

83 Hawksford, Malcolm O, *Differential current derived feedback in error-correcting audio amplifier applications*, Proc.IEE, CDS, vol.141, No.3, June 1994.

84 (Nameless), *AD844 data sheet*, Analog Devices.

Further reading

85 Borbely, Erno, *A 60w MOSFET power amplifier,* The Audio Amateur, Q2, 1982.

86 Borbely, Erno, *Third generation MOSFETs: Pt,1, the Servo 100,* TAA, Q1, 1984.

87 Carruthers, J, J.H.Evans, J.Kinsler & P.Williams, *Power Amplifiers efficiencies, matching and 'rms' power,* Wireless World, June 1973.

88 Gevel, van de, Marcel, *Audio power with a new loop,* EW+WW, Feb 1996.

89 Goldberg, J, and Sandler, Mark, *Comparison of PWM modulation techniques for digital power amplifiers,* Proc.IOA, Nov 1990.

90 Hawksford, M.J, *Power amplifier output stage design incorporating error feedback correction with current-dumping enhancement,* AES 74th convention, Oct 1983.

91 Hawksford, M.O.J, *Towards a generalisation of error correction amplifiers,* Proc. IOA, Nov 1991.

92 King, Bascom. H, *Switched-on Amps: Power with a Pulse,* Audio (US), 1995.

93 Miloslavsky, Y, *Wideband audio power amplifiers,* Wireless World, June 1980.

94 Roddam, Thomas, *Transistor Amplifiers for Audio Frequencies,* Iliffe, 1964.

95 Sandman, A, *Low crossover distortion class B amplifier,* Wireless World, July 1971.

96 Sandman, A, *Reducing crossover distortion,* Wireless World, Oct 1974.

97 Tremaine, Howard.M, *Audio Cyclopedia,* Howard Sams, 2nd ed. 1969. ISBN 0-672-20675-7.

98 Vanderkooy, J. & Lipshitz, Stanley.L, *Current dumping does it really work?,* Wireless World, June 1978.

99 Williams, Peter, *Voltage Following,* Wireless World, Sept 1968.

Note: Hawsford, Malcolm O.J., M.J. and M.O.J. are all the same author.

Features of the power stage

5.1 Overview

This section considers the more *nitty-gritty* details of the power stages introduced in chapter 4.

The *power stage* is what sets power amplifiers apart from the more lightweight electronics in the remainder of the audio path. Today, the historically gross physical differences of size and particularly weight may have been ameliorated, but the fact remains that whether solid-state or valve, power stages built to drive practical loudspeakers always involve the delivery of higher voltages, far higher currents, and also high temperatures.

In transistor amplifiers, these quantities can range up to and over 300 Volts, 100 Amperes and 150 degrees Celsius (°c). In valve equipment, such temperatures are unremarkable and voltages are even higher, typically 450v and up to 1500 volts. But currents before the final output terminals are much lower.

The design of high performance power stages involves other considerations not encountered in the remainder of audio electronics – such as the *safe operating areas* of the active parts, adequacy of current sharing, management of hot air, guarding against potentially expensive and catastrophic damage, and commoning ground currents that could be 160dB (100 million times) apart in magnitude.

5.1.1 Operating with high voltages

The voltages in *solid-state* audio power amplifiers are not high enough to require special insulation procedures. But even without considering the AC mains connections, the voltages employed can still be lethal in the higher swing units. Nowadays, the safety line is drawn lower than it was, since potentials over just 45v DC or AC peak rms are widely considered potentially hazardous to life – particularly to young humans and pets. In Western Europe, the LVD (Low Voltage Directive) makes it compulsory for all equipment (from 1997, and including all audio amplifiers) placed on the market, to meet basic safety standards if a voltage above about 45v above ground or between two accessible parts, is present. This thus applies to all amplifiers offering above some $120w \Rightarrow 8\Omega$ or $240w \Rightarrow 4\Omega$ (either per channel, or if bridgeable, when bridged). Above these power capabilities, users must be guarded from readily coming into contact with both the internal circuitry, and the output signal. From now on, this is one of the considerations that anyone creating a new power amplifier will have to contemplate from the outset.

5.1.2 Operating with high currents

High currents, above 1 Ampere, greatly increase the risk of heating, to the point of destroying insulation, the generation of toxic fumes, and fire, since

$$\textbf{Watts (heating work done)} = \textbf{I}^2 \times \textbf{R}$$

So 10 Amperes potentially results in one hundred times as much heating as 1 Ampere – for a given circuit resistance. For this reason alone, conductor gauges and the dissipative (wattage) ratings of resistors (*both of which are scarcely ever given a first thought in small signal and digital audio electronics*) must be adequately scaled. Appropriate overcurrent protection is of particular importance for amplifiers mounted in (or adjacent to) wood, cardboard or plastic enclosing materials. Most professional and many high-end domestic power amplifiers are mounted in all-metal enclosures which would contain a substantial internal fire. But many that are fan cooled or open-vented would not detain the no-less lethal and damaging smoke.

Loss of sonic quality through intermittency is reason enough for requiring amplifiers with high current capability to be built with all-soldered joints, or with physically stable, vibration-proofed, preferably latching, and gas-tight connectors. The combination of a high current path and screw connections is inadvisable in any amplifiers used in a strong bass soundfield. Whenever high currents do attempt to pass through loose connections (the likelihood of a nut undoing itself seemingly increases by the square of the current passing!), arcing ensues and ozone is produced. The ozone then produces free radicals from materials it comes into contact with. This readily degrades and embrittles most materials with elastic properties, including protective coatings on components and conductors.

At high frequencies, much above 1kHz, high currents may behave counter-intuitively. First, at *any current level*, locally circulating eddy currents cause *skin effect*. As implied, the effect is to force the current to flow primarily near the surface. With the conductor's apparent size reduced, resistance increases gently with increasing frequency – acting as if the wire were a *transmission line inductor*. Second, with high current levels much above 50 Amperes, and where send and return conductors are close and parallel (as they should be for low EMI), *proximity effect* causes a further constriction in the conductor gauge, as the current flow is forced to the side closest to its opposite conductor. *With conductors in high performance audio amplifiers potentially handling significant currents at up to 10kHz and above, it follows that great care must be taken in the design of conductors and their routing, beyond what is passable for small signal audio.*

5.2 Power devices

Power devices are active components rated to handle high currents, voltages and the consequent power dissipation (VxI). They are required in audio power amplifiers because many good loudspeakers are highly inefficient, because amplifiers with high sonic quality are inefficient, and because of the acoustic power levels required for some kinds of music performance, for a few listeners in a well appointed sitting-room – let alone in front of up to 250,000 people. The line between small signal and power devices is commonly defined on the basis of power dissipative ability, or current handling. 1 watt and 1 Ampere are common thresholds. The line is also a fluctuating one, since with BJTs in particular, it may be defined as the point where subtle performance parameters, particularly f_T and switching speed, degrade appreciably.

5.2.1 Bipolar Junction Transistors (BJT)

These are the original, 'ordinary' transistors. They remain the dominant active devices throughout the input, process and drive stages of the vast majority of solid-state audio power amplifiers. This dominance was total from 1956 to 1976, as it included the output stage. Twenty years on, BJTs remain the more common OPS device in low budget amplifiers, but in the UK and Europe particularly, BJTs are used in the output stages of not more than about 50% of professional and high-end domestic solid-state amplifiers (see Appendix 2).

The BJT's history and some of its developing characteristics are briefly outlined in chapter 4. Other keynotes of modern silicon BJTs are:

Non-linearities

Beta (β, H_{FE} raw current gain) varies widely. It varies typically +/–50% with unselected parts at a given temperature and current. Beta varies a further, similarly large percentage over the span of operating current in an output stage, e.g. 20mA to 20 Amperes. A drastic reduction occurs at relatively high currents (on a log scale), typically above a few Amperes. Figure 5.1 (overleaf) shows how these changes may be extracted from the output characteristic curve supplied by the maker.

Beta also varies significantly with temperature, reducing to a fraction of its room temperature value when hot, precisely when the transistor is working hardest. In circuits employing global NFB, reduced Beta in the output BJTs shifts the workload to the drivers. Their Beta may decrease in turn, and the buck passes down the line. The overall result of these variations in beta is increased non-linearity, and ultimately, current clipping.

Figure 5.1

Common Emitter mode Output Characteristics.

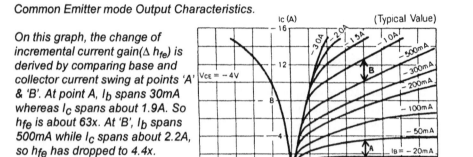

*On this graph, the change of incremental current gain($\Delta\ h_{fe}$) is derived by comparing base and collector current swing at points 'A' & 'B'. At point A, I_b spans 30mA whereas I_c spans about 1.9A. So h_{fe} is about 63x. At 'B', I_b spans 500mA while I_c spans about 2.2A, so h_{fe} has dropped to 4.4x.
Copyright Sanken.*

Excepting micro-amplifiers operating on single cells, low V_{ce} arises when a transistor is driven hard on, within 1dB or so of clipping. A BJT's ability to sustain a given level of collector current (Ic) is reduced at low V_{ce}. This is one symptom of *Early Effect* (discovered by Mr. Early at Bell Labs in the 1950s) which is another cause of non-linearity, particularly in BJTs handling high voltage swings, i.e. the output, driver and VAS BJTs in most conventional amplifiers, unless with cascode 'shielding'. Physically it involves modulation of the thickness of the effective base, the central layer 'where it all happens', inside any BJT. For audio, it amounts to an output impedance that is dependent on the signal voltage. The *Early voltage* is not a physical property but a notional measure of the effect, based on the projected-back origin of a characteristic curve [1]. The best BJTs exhibit twenty to forty times less Early effect than mundane types, but the data supplied by semiconductor makers remains so meagre that only one BJT maker in the world openly publishes figures.

Voltage ratings

BJTs have a forest of different safe voltage ratings. For conventional audio amplifiers, the voltage that can be safely withstood is generally the lower of the various ratings cited, under one of the following aliases:

$$V_{CEO(max)} \cong V_{(BR)ceo} \cong BV_{CEO} \cong V_{ceo} \text{ or } V_{ce} \text{ max}$$

In this book, V_{ceo} is used, with 'max' being understood. Even this rating is only safe provided that any back EMFs caused by inductances in the speaker circuit are prevented from exceeding the supply voltage, e.g. by use of Baker clamp diodes (Fashionably but incorrectly described by semiconductor makers as '*anti-parallel*' - which at best means 'in series').

Like beta, the V_{ceo} rating falls off at high temperatures. In output stages, and under hard-drive conditions, it may be typically 75% of its rating. If the applied voltage were to be further derated for reliability, then the highest voltage available in complementary BJTs, where Vceo = 230, would restrict power capability to about 45v rms, irrespective of the loading. This is explains why the greater complexities of bridged OPS or cascaded output devices are seen as fundamental requirements by some designers, despite the fact that many designers of conventional BJT-based output stages developing over 45v rms 'get away' with working the BJTs under greater voltage stress.

Power limits

Of particular importance in power output stages, instantaneous 'power' dissipation (really 'energy handling') is the product of 'real' (resistive) current flow at the collector (Ic) and the sustained voltage, Vce across the main terminals. The highest individual device power ratings of BJTs made for audio power have long leveled out at about 250 watts. But this description (once thrust at purchasers in advertisements, as in '2400 watts total dissipation' - for a 200w/ch amplifier) is particularly misleading or 'optimistic' when the output stage employs BJTs.

Figure 5.2 shows the SOA (Safe Operating Area) of a medium power transistor in an IC power amplifier. At low voltages, below 25 to 5 volts for most BJTs (here 21 volts), steady dissipation and current handling are limited by the current rating of the internal bond wires. This is the region marked 'Current Limited, DC'. A '150 watt' rated part for example, implies a capability of 60 Amperes at 2.5 volts, but the bond wires are likely limited to handling less, maybe 20 Amperes. In this case, we have 2A. If the current is discontinuous, with pulse periods in the audio midrange and above, peak values 2 to 3 times higher can be managed, here to 5A, marked 'Pulsed Limits'.

A little documented facet of BJTs is that the base's ohmic resistance increases at high drive currents. This can help protect from runaway.

Figure 5.2

Typical Safe Operating Area curves for a medium BJT (Bipolar Junction Transistor), rated at 70v and 2A. The safe limits for bass frequencies are shown by the DC line, except that the 25°c case temperature to which the line refers isn't maintainable for very long. Temperature derating means all the curves pull-inwards (towards the bottom left) with increasing temperature. Second breakdown starts at 44v. Notice how the pulse-rated SOA and 2nd breakdown boundaries are otherwise lifted at 5mS while the latter vanishes at 1mS. BJT's second breakdown may therefore be described as a low frequency problem - exactly where BJTs are else best suited. Copyright Burr-Brown.

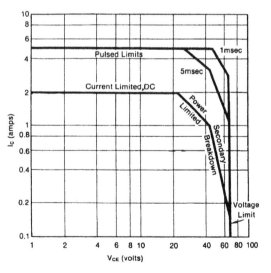

1st breakdown

At voltages above 5 to 25v, BJT power ratings enter the region of being 'thermally-limited' or 'power-limited' (they mean the same). Here, the SOA curve slopes characteristically at 45°. *Only in this region, and with their cases kept at room temperature, do BJTs display their rated power.*

In and above this region, ratings vary more according to the pulse duration. For example, at and above 20 volts, the current rating for a 100µS pulse is typically five to a hundred times greater than for DC, while a 5mS rating is up to ten times higher. This form of derating, shown at 1mS and 5mS in the graph, *is unsafe for audio* application, except when an amplifier is dedicated to mid or high frequencies (anything above 500Hz, say), and is fed from an integral active crossover and will *never* have DC offset co-existent with a shorted output! Excepting this dedicated active drive, operation up to 0dBr down to 20Hz or lower with continuous signals is expected from any other high performance audio power amplifier, effectively denying any significant uprating over the DC values of the SOA curve.

2nd breakdown

Past the power limited zone, at between 30 to 80 volts, the BJT's *second-breakdown* region is entered. Here, the power rating is considerably reduced. In a typical 250w rated BJT, at 100v (Vce), allowable Ic is just 1A, so the transistor's dissipation limit has reduced to:

$$(^{100}/_{250}) = 40\%$$

of what is was, e.g. a 250 watt rated part is reduced to 100 watts. Here, the approximate 40 watts rating implied at the centre of the power limited zone ($\cong 1.35A \times 30v$), is reduced at 60v to some 0.25A, i.e. 15w, or a similar reduction, to just under 40% of the maximum capability. Finally, the curve, which has grown steeper, becomes a vertical line – the voltage limit, usually the V_{ceo} condition.

Thermal limiting

Even this power dissipation presumes there is *thermal headroom*. If the BJT is heated above 25°c by external causes (let alone internally by biasing), the dissipation (wattage) rating left for audio reduces linearly, typically to 50% at around 105°c case temperature. In effect, the current handling is derated at a given sustained voltage. Either this, or reduced voltage is necessary, because less heat can flow out of the junction when the outside is also hot. Where else can it flow ?

Scope

Secondary breakdown doesn't apply to small signal transistors. It comes into play as the transistor die (or wafer) is enlarged to handle higher current. The maximum collector current ratings about which it becomes significant vary from 150mA up to 3 Amperes, depending on type and internal *geometry.*

Second breakdown occurs when it is possible to have different parts of the wafer at differing temperatures, some forming local hotspots. It happens only at relatively high voltages, as the Early Effect (see above) comes into play and reduces the effective base width. This in turn increases the variability of the current distribution, as the thinned base is less uniformly active. This sets off the formation of localised hotspots, and consequent *thermal avalanche breakdown*. ZZHUTT-*BANG!!* This can occur at mean power levels far below the power that the transistor can dissipate at low values of collector voltage [2].

Sharing

Paralleled connections are shown in section 4.1.9 (Figure 4.14). Paralleling of BJTs so current and power dissipation are shared-out equally, is the obvious solution to their dissipative and SOA limitations, and any non-linearity caused by high current operation. But sharing depends on matched V_{be} and H_{FE}. Unless correctly sized individual R_e 'ballast' resistors are used in line with each emitter (these also increase non-linearity and losses), and also ideally R_b ballast resistors are inserted in line with each base, then sharing will not occur (or else be very accurate) at some high current. Due to the BJT's positive tempco, thermal runaway will then occur, with consequent destruction. At the very least, one or two BJTs will work harder than the others, so the work is not shared, performance is not enhanced, MTBF is reduced and the other BJTs are a wasted effort. It follows that BJTs require precisely-sized extra parts and/or very careful selection, together with careful checking during production testing, to ensure that a high level of sharing occurs.

Leakage

Leakage current is naturally highest close to the BJT's voltage rating. It is commonly measured under the worst connective condition, with the base terminal floating ('open'). But the figure this gives, called I_{cbo}, is not the principal current of concern. Rather, it is I_{ceo}, the resulting main terminal current. This is I_{cbo} multiplied by the low current beta. For OPS (power) BJTs at room temperature it's typically 100μA. But at realistic worst case junction temperatures, say 150°c, leakage can rise to as much as 10mA. In a conventional power stage with global NFB, should negative feedback lose control, or in a power stage with nil global feedback, leakage currents ranging into milliamps can upset DC balance, and voltage and current sharing. Large scale destruction may then ensue.

HF limitations

The speed of 'power' BJTs has risen from barely 5kHz in 1958 to above 40MHz in the best modern examples, almost exclusively made in Japan. The upper curve 'β' in Figure 5.3 shows how the transition frequency (F_T) specification relates to the effective frequency response of a BJT. Beta begins to reduce at frequencies that are $(^{F_T}/_{Beta})$. For example, if Beta is 75, a BJT with F_T = 1MHz experiences a reducing Beta from above (1MHz/75) = 13kHz. This amounts to hf non-linearity. The lower curve (α or alpha) applies to BJTs in common-base mode only, e.g. cascode devices. It shows how in this configuration, Beta remains independent of frequency up to F_T, typically Beta times higher than for BJTs in the other configurations (CE and CC), thus hf linearity is enhanced.

Figure 5.3

Power bandwidth of BJTs in CE and CC stages depends on Beta and H_{FE} (upper curve, β). Paradoxically, β of higher-β BJTs begin reducing at a lower frequency in these modes. Notice how BJTs operating in cascode exhibit a wider power bandwidth that is independent of Beta (lower curve, α).

From 'Electronic Components' by Daniel L Metzger, (Prentice-Hall)

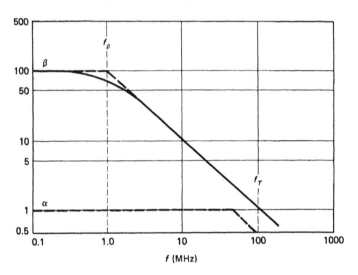

Selection

Makers of the more high performance power amplifiers routinely test all incoming BJTs. The following characteristics may be tested by plugging BJTs into a curve tracer or appropriate test circuit:

* V_{be} for matching both paralleled parts and also conjugate pairs at a given median operating current (e.g. $Ic = 1A$) and junction temperature (e.g. 80°c).

* H_{FE} changes over span of operating currents and voltages, hence linearity. All devices may be usable but not randomly. Batching at constant operating current (while being careful to keep absolute junction temperature the same) by E-series is performed by the more dedicated makers.

* Noise. Peak and rms noise *vs.* frequency may be plotted and devices with excess noise rejected.

* Voltage ratings (V_{ceo}) can be quickly established and parts batched using a high-speed breakdown tester. It is not unusual to get a yield above the maker's rating. Lower voltage rated parts may be employed in lower-swing models.

* Leakage current, when immersed in an oven or in hot oil

* Second-breakdown rating - a tricky test [3].

* Sonics. Individual BJTs may be listened to, typically in a simple circuit (e.g. class A, 1 watt$\Rightarrow$8Ω) fed by a high resolution music source.

The result of these tests is frequently the discovery of way-out parts that would otherwise compromise reliability, RF stability or sonics later in the production or use of an amplifier, and cost either the maker or someone else a great deal of money and grief. However, the savings have to be paid by everyone, and up front. That's because if as many as 5% of the BJTs pass all the tests, then a part that costs £3 in quantity can end up costing in effect £60, to the maker. The purchaser pays even more for the same part. This is (or *should be*) one difference between high end amplifiers costing about £10/watt and other units costing £1/watt or less. It is also part of the true cost of quality engineering in output stages with BJTs.

5.2.2 MOSFETs (enhancement-mode power FETs)

Perspective

Power MOSFETs are alternatives to Bipolar Transistors (BJTs), with particular advantages in high performance audio amplifier's output stages. MOSFETs are Field Effect transistors that (unlike J-FETs, and depletion-type MOSFETs) have been developed so that they can be made to handle high power, i.e. combinations of high currents and voltages. Low power versions of these 'power' MOSFETs are also 'available for work' alongside J-FETs and BJTs in small signal circuits and positions. In the family tree of FETs (Figure 5.4) the power MOSFETs appear as n- and p-channel enhancement MOSFETs.

History

The first MOSFETs (it's superfluous to say 'transistor', as that's the 'T') to be designed for power and used in a manufactured audio amplifier appeared in domestic Hi-Fi amplifiers made by Yamaha, in the autumn of 1975, appearing in the UK

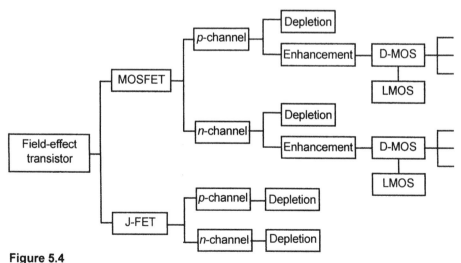

Figure 5.4

Family tree of FETs.

177

in the spring of 1976. They came in complementary pairs, but had a high turn-on threshold voltage of between 7 and 25 volts. The history of power FETs is complicated by many false starts and confusing nomenclature (there have been many names and geometries) but the devices used by Yamaha are believed to have been the commercial exposition of the geometry developed in Japan in 1974 by Nishizawa (one of the fathers of FETs, whose research dated back to 1950), at Tokuko University.

Two tribes

The first practical MOSFETs – *of a completely different type* – were almost simultaneously launched in the USA in October 1975 by *Siliconix* [23]. These 'MOSPOWER' parts or 'VMOSFETs' used a V-groove construction, which had also been pioneered by Nishizawa in Japan, back in 1969. This type subsequently developed into 'Vertical DMOS', which is 'Double diffused MOS', and later DMOS-II, amongst other names. In this book, these parts will be referred to as DMOS, D-MOSFETs – or switching MOSFETs, as they were optimised particularly for switching (*cf.* linear amplification). Thus on-resistance (R_{on}) was low. This species includes the majority of MOSFETs made today. IR's HEXFET is one of several geometric variants. Initially at least, the available parts were primarily *n*-channel (equivalent of an *npn* BJT); complementary parts had (and still have) limitations.

LMOS

A year later, in 1977, Hitachi announced a planar type of DMOS that was practical for audio power output stages. These 'lateral' MOSFETs (LATMOS or LMOS) were readily made into *closely matched* complements (n and p-ch). They had low gate-drain capacitance, and the drain current where their gate threshold voltage's temperature coefficient was zero, was much lower than the other types, at around 100mA, rather than 3A or so.

In 1978, the first professional power amplifiers to use Hitachi's lateral MOSFETs were launched in the UK and sold worldwide by HH Electronic for musicians and PA systems. HH's amplifier range was called the *V-series* - though V-FETs did not come into it (they might have when the name was earlier conceived). Ashley in the USA is believed to be the second or equal-first professional amplifier maker to manufacture with L-MOS, but by and large, US makers have been loath to give up BJTs. Between 1977 and 1993, Hitachi was the sole world source of lateral MOSFETs, and British and some European power amp makers, particularly those supplying professional users, were the main market. Since 1992, a new family of lateral power MOSFETs, including equivalent parts and replacements for withdrawn Hitachi parts, have been designed in the UK and are dual-sourced by UK makers, Profusion and Semelab. *It is important to recognise that lateral MOSFETs are made expressly for audio – probably the only power transistors to be so favoured.*

Advantages

The advantages of lateral MOSFETs in particular, and power MOSFETs in general, over BJTs, are as follows.

Figure 5.5

Showing absence of cross-conduction in MOSFETs compared to BJTs, as measured with circa 1978 devices.
Copyright Hitachi

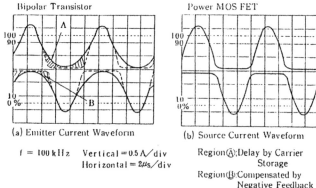

(a) Emitter Current Waveform

(b) Source Current Waveform

f = 100 kHz   Vertical = 0.5 A/div
Horizontal = 2μs/div

Region Ⓐ:Delay by Carrier Storage
Region Ⓑ:Compensated by Negative Feedback

1. Wider bandwidth

In operation, MOSFETs have no minority charge-carriers clogging them up; they can switch several Amperes in tens of nanoseconds. This is 30 to 100 times faster than equivalently rated BJTs. Audio power amplifiers with MOSFET output stages can readily have *full power* bandwidths that extend a decade or more above audible frequencies. The MOSFETs' higher speed means hf compensation is less fussy and generally much lighter. By comparison, few BJT output stages have significantly above 15kHz to 20kHz full power bandwidth into their rated load. This is as a result of the measures needed for stability and to prevent transistor destruction through common-mode conduction, i.e. through 'sticking', due to charge-carrier 'storage' (read: sluggardness). Figure 5.5 from Hitachi illustrates the cleaner current waveform at 100kHz in a class B amplifier, with the original lateral MOSFETs being compared to the best BJTs that were available for the same task, ten years ago.

2. More rugged

MOSFETs exhibit no 'second-breakdown' region, only a limit on their dissipation. Thus the 'power limited' region in Figure 5.2 would continue until it reaches the voltage limit. For this reason, advertised MOSFET power ratings are much closer to practically usable and safe power ratings. Given adequate heatsinking and thermal protection, the SOA (Safe Operating Area) curve is not a survival issue – at least with lateral types.

Compared to BJTs, if amplifiers made with MOSFETs should go RF unstable, they aren't nearly so prone to speaker voice-coil frying oscillations at ultrasonic frequencies. MOSFETs can certainly oscillate at higher RF frequencies. Self-oscillation in lateral types is rarely self-destructive (as in BJTs and often, D-MOSFETs), most commonly occurs between 1MHz to 10MHz, and does not usually harm loudspeakers, even tweeters. 'Parasitic' or self-oscillation in D-MOSFETs typically occurs at still higher frequencies of 10 to 100MHz, also does not harm loudspeakers, but can be self-destructive. With these switching types, knowledgeable layout is essential for stability.

In practice, the reliability of properly configured lateral MOSFETs in sustaining real world abuse, notably overheating and driven output shorts, and without overt protection, has been vastly higher than that of comparably rated BJT output stages even with the usual extensive protection.

3. Easier drive

MOSFETs' high input impedance at most audio frequencies lightens the task of the driver stage – at least for continuous signals and for frequencies below 10kHz. For pulses and higher frequencies, the input capacitance of MOSFETs requires high current capability – but only momentarily. The long-term rms current demanded by a MOSFET gate with most music (whether full range or just hf) is fortunately low. With many power amplifier drive stages arranged to provide only 8 to 30mA to MOSFETs, dissipative stress is not an issue with 100v rails and class A operation of the driver stage. But the question of repetitive transient error then arises (i.e. 'loss of edge'), as there is no instantaneous extra current available to drive the MOSFETs' capacitance. It follows that MOSFETs are still relatively easy to drive, but some circuit design subtlety is required for handling live music or wide range recordings without loss of realism.

4. Easy paralleling

MOSFETs' transition from positive, through zero, into negative temperature coefficient of gate threshold voltage ultimately forces balanced sharing. In lateral MOSFETs, the transition takes place at such low currents (below 200mA, compared to typically 2 Amperes and above for D-MOSFETs) that these types can be simply paralleled without any ballast resistors, or even selection, particularly when the FETs are mutually well thermally coupled. A prime example are the double die MOSFETs, e.g. ECF20-XXX, BUZ-XXXD (see Appendix 3.)

Because D-MOS 'switching' MOSFETs have a temperature coefficient that changes from positive and thus suicidal (like BJTs), to being negative and thus self-protecting, at a current that will cause substantial dissipation (2A @ 100v = 200 watts), the paralleling of switching MOSFETs requires individual V_{gs} and even conductance matching at a high percentage of rated drain current, and ideally also ballast resistors in line with each source leg. This rather deflates the D-MOS's great benefit over L-MOS, of lower cost due to large volume industrial use.

5. Simple biasing

Class A-B biasing with lateral MOSFETs is relatively uncritical. Depending on the design, between 25 and 50mA per device (or dice) is adequate, even though 80 to 100mA has been recommended to coincide with the drain channel's zero temperature coefficient point. There is usually some un-nullable crossover residue yet sonics are perceived as benign or absent by many critical listeners. *In part this is because the bias setting is not very critical, nor thermally sensitive, and thus it does not suffer continuous, signal-dependent under-and over-biasing, as with BJTs.* By contrast, the crossover distortion that is heard on most bipolar amplifiers is not principally the static figure but a higher, transient level of crossover distortion occurring after loud peaks, unrevealed by static tests with steady-state signals.

Biasing of D-MOSFETs is more fussy than lateral types as the turn-on is harder, and it requires some temperature tracking. But it is still a relatively minor matter compared to biasing BJTs.

Figure 5.6

Grandson of Lin test-circuit used to demonstrate performance of Toshiba's BJTs, MOSFETs and IGBTs. Courtesy of John Linsley-Hood/Toshiba.

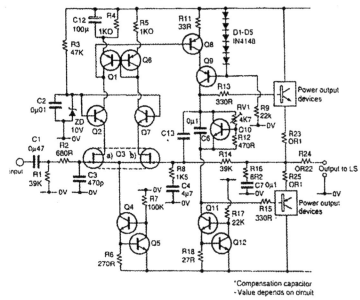

'Compensation capacitor
- Value depends on circuit

6. Ease of application

Lateral MOSFETs slip readily into most existing Class A-B circuit architectures, usually with simplifications. Figure 5.6 shows a modern circuit following the 'Sons of Lin' topology [section 4.3.6] where the OPS devices may be Sziklai connected BJTs, or D-MOSFETs [4,5].

MOSFETs can also be used by themselves in simpler circuits than those that have been satisfactory with BJT output stages. The world's simplest high performance audio power amplifier is constructed with one D-MOSFET [6]. An amplifier built with a single BJT, valve or even L-MOSFET would not perform nearly so well.

Unlike most BJTs (RCA's classic 1C15 and 1C16 are examples of exceptions), all MOSFETs have an *intrinsic* flyback or 'Baker clamp' diode within them, from drain to source. These 'parasitic' diodes can do the job of the reverse biased fast diodes connected from output to each supply rail, that provide essential protection against back–EMFs in knowledgeably-protected BJT output stages.

7. Ease of repair

When lateral MOSFETs expire, they usually fail 'soft', where the gate shorts to the source. At worst, the main channel shorts out. Either way, they cause little collateral damage. The common chain-of-destruction, back to pre-driver stages when direct-coupled BJT output stages fail, is rare with MOSFET output stages.

Disadvantages

These are few and relatively minor.

1. Extra dissipation

The bias currents needed for individual MOSFETs leads to significant heat dissipation on standby when a number of MOSFETs are paralleled for higher transconductance, hence lower distortion into a given load and increased current capability. In a 1kW/ch power amp, the difference is of the order of 50 watts. A 60w domestic mains lightbulb gives an idea of the amount of waste heat. At the same time, this is a necessary consequence of operation in class A-B; you are, after all, getting the sonic benefit of class A most of the time (see section 4.6).

2. On-resistance (R_{on})

Lateral MOSFETs have a higher 'on' resistance (R_{on}) than either BJTs or switching MOSFETs, which have resistances of 0.5Ω ranging down to just few milliohms. Alas, the very low resistance D-MOSFETs are intrinsically limited to low voltage ratings, allowing only 25v of swing, say. Conversely, the R_{on} rises with voltage rating.

At room temperature, a lateral MOSFET, when turned hard on, measures between 0.5Ω (for double die), up to 4Ω. In conventional power amps, MOSFETs in parallel reduce the open-loop output resistance (=1/gm) pro-rata, while global negative feedback reduces it again by a factor of at least ten, and up to a thousand times, particularly at low frequencies (<500Hz), where it matters most.

Losses brought on by MOSFETs' R_{on} scale in the opposite direction to static bias, being highest when an output stage with say just one or two pairs of single die MOSFETs is driven hard, into a low or 'difficult' speaker impedance. In high power designs, multiple paralleled MOSFETs can reduce the I²R dissipative losses to relative insignificance. On this basis, and assuming equal bias levels, the conversion efficiency of MOSFETs working in class A–B is not much less than that of BJT output stages.

3. Transconductance

Related to R_{on}, MOSFETs' transconductance (gm) is lower than BJTs by about an order of magnitude. L-MOS have the lowest gm of 1.0. The outcome is that the gate drive signal has to support 1 volt per Ampere of output current, on top of the signal voltage, for one pair of L-MOS (single die) MOSFETs, reducing pro-rata with more pairs. This voltage adds to the threshold, which is fortunately low (<1v) for L-MOS.

D-MOSFETs' gm is typically 4 to 8. Thus a lower voltage is added to the drive signal for a given current (although the initial threshold voltage begins higher at about 4v), and fewer devices are required in parallel for a given current capability.

4. The parasitic diode

In high power, very high slew-rate amplifiers, the intrinsic diode may be destroyed by its inability to handle the back–EMF current, destroying at least a pair of MOSFETs in its wake. Faster, discrete diodes may have to be added after all.

5. Threshold voltage

The gate threshold voltage (V_{gs}) of all MOSFETs is higher than the V_{be} threshold of BJTs, while the difference in the abruptness of the 'turn-on' voltage between LMOS and DMOS, reflecting the gm differences, is akin to that between Germanium and Silicon BJTs (see section 4.3). Lateral MOSFETs are biased on between typically 200mV and 1v, whereas D-MOSFETs require 1 to 4 volts. Yet for the reasons explained in section 3 above, signal handling headroom can end-up similar, as the DMOS' higher V_{gs} is compensated directly by the higher gm. The fact remains however that MOSFETs are unsuited to tightly-engineered amplifiers on low volt supplies, where power capability figures are everything and where a few volts of extra rail voltage cannot be afforded or justified. The sharper turn-on of D-MOS means that for reliable application in parallel, D-MOSFETs must be selected in tight *E-series* bands of V_{gs}.

5.2.3 Insulated Gate Bipolar Transistors (IGBT)

The IGBT is a relatively new part, coming into large scale production in the past decade, and a cross between a MOSFET and a BJT. It has the low drive power demands of a MOSFET, and the low saturation voltage of a BJT. But this latter feature is not so important for driving music signals, unless IGBTs are employed as switches in HF power supplies, or in a Class D OPS. IGBTs with high voltage (up to 1kV) and high individual current ratings (e.g. 50A) are readily available, but there are not many complements. Like BJTs, the first generation of IGBTs had parasitic elements that can cause latchup and catastrophic failure, under conditions that MOSFETs are immune to. So far IGBTs are mainly used as power switches in medium frequency SMPS, where conditions can be better defined. John Linsley-Hood's work on behalf of Toshiba showed how the basic linearity (hence % harmonic distortion) of Toshiba's own IGBTs 'dropped into' the output stage position of the modern Lin circuit (Figure 5.7) with the necessary degeneration resistors, was little lower than Toshiba's own superb BJTs, and both were up to five times *less* linear (at least at 1kHz) than Toshiba's own D-MOSFETs. Few comparisons between semiconductor technology have been more free from vested interest.

Figure 5.7

Output stages used in Linsley-Hood's test circuit for Toshiba. These are the 'Power output devices' in figure 5.6.

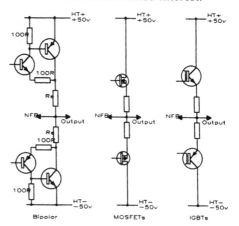

5.3 Recognising large signals

Introduction

When signal levels in any amplifier stage pass a certain threshold, the signal moves from being considered *small signal* to being a *large signal*. There is no hard definition as to when this occurs. Rather, the changeover to the large signal condition is gradual, as it is defined by several independent factors. All affect linearity. Examples of conditions that occur only with large signals are:

(i) slew limiting

(ii) inflexions, usually slope reductions, in key transistor action parameters, eg. transconductance, in mA/v

(iii) significant signal and/or temperature modulation of transistor parameters, e.g. *Early Effect*.

Thresholds

Signal magnitudes that will cause these effects are spread widely, from above (>)1 to 30v and >3mA to 3A, depending on circuit topology, design finesse, and active device type.

Progression

The following description considers how aspects of large signal behaviour can apply *throughout* conventional audio power amplifiers made with BJTs, while becoming most acute as the output stage is approached.

In modern, semiconductor-based power amplifiers, the input stage may have to be designed for a high slew limit, as the output stage swing requires this. Often, the input stage also has to sustain HT supply rails equal to or even higher than the output 'power' stage. In amplifiers with high enough rail voltages, this combination may demand consideration of the safe operating area of the input stage transistors, if BJT. In these senses, it is a 'large signal stage', even while the signal being passed on is not a voltage swing at all, but just a small current with a peak value of which is rarely above +/−5mA.

In common, Lin-based amplifiers, the next stage is a voltage amplifier, or 'VAS' (Figure 5.8). Here, the small current input swing (I_3, say at most +/− 2mA) is converted into the full voltage swing, which might be +/−100v, and equal to or greater than the final stage's swing. Now the Early Effect and C_{ob} become of concern. As a 'voltage' (*cf* 'power') stage, maximum current delivery into the next stage is typically restricted to below 50mA. Even so, the SOA (Safe Operating Area) of the semiconductors may again need evaluating at this point.

In the driver stage, if operating at above about +/− 25v, and driving BJTs, SOA may be a significant issue as output currents are 'wider open' (being dependent on two wide variables, namely Beta x loading), and may range up to several amperes, with high junction temperatures to match. BJT parameter inflexions over this current

Figure 5.8

Voltage and Current Progression.

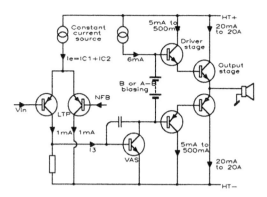

range must now be evaluated, as well as considering C_{ob} and VAF again, *vis-à-vis* the peak voltage swing.

Finally, in the output stage, the full output current is handled. Here the SOA problem reaches a crisis, as one transistor is rarely adequate, and multiple, parallel-connected transistors (or valves) are almost universally required to share the load current and dissipation resulting from the maximum power delivery.

5.3.1 The slew limit

Introduction

Slew limiting or 'slewing' is a highly unpleasant kind of 'large signal' distortion, resulting from a catastrophic loss-of-control at high frequencies[8,9]. It mainly (but not exclusively) afflicts any analogue signal processing stage that employs global feedback, whatever the active devices. As a large-signal effect, it is of particular concern in power amps. Sadly, the subject is rife with misunderstanding and misleading information, some of it disseminated by makers who should know better.

Terminology

The confusion begins with words. In general, '*slew*' may be taken as an abbreviation for 'limiting the rate of change'. In this book, '*Slew limit*' is the rate-of-change of voltage (volts per second) ceiling. '*Slewing*' and '*Slew limiting*' describes the behaviour of signals that attempt to change faster than the slew limit ceiling. The ambiguous term 'Slew rate' is best avoided where possible. Sometimes it means the slew limit, but at other times it has commonly been used to describe the rate of change (in volts per sec) of a signal that may or may not cause slewing, yet is being discussed in the context that it might.

Identification

Slewing on a test sinewave is seen on a 'scope, at the output of amplifiers (usually those with global feedback) as a sloping but straight 'cutting of the corner', when the rate-of-change of the input signal exceeds the rate at which the feedback (from the output) can rise to correct the error. A similar effect can be seen when an

Features of the power stage

Figure 5.9

Output current clipping (lower waveform) may not be noticed on program, as with conventional voltage monitoring (upper waveform) it is seen as a section 'bitten' out of the side of the waveform, not as the familiar flattening in the waveform.
From: Baxandall, P.J, Chapter 2, Audio Engineer's reference book, Ed. Talbot-Smith, Focal Press.

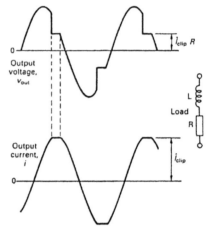

amplifier's output current clips into an inductive load (Figure 5.9). The feedback loop's restriction is dominantly caused in turn (at least in many conventional topologies) by a limit on the stage current I_e, that is available to charge an internal compensating or *lag compensation* capacitor, C_c:

Slew limit = $(^Ie/C_c)$ volts per second.

e.g. for I_e = 1mA and C_c = 47pF, SRL = 21v/µS.

Usually, this capacitor is in the Miller integrator position around the second stage – or wherever the VAS is located, through which the signal is converted from current back to voltage (Figure 5.10) [10,11,12].

Figure 5.10

Slew Limiting in the Lin Topology.

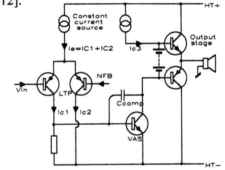

Effects

Slewing or 'slew-limiting' causes loss of control in the global negative feedback loop that encloses most power amplifiers. By itself, slewing causes unpleasant distortion that ceases as quickly as the effect causing the slewing lasts. In practice though, the upset amplifier circuitry may not recover immediately, due for example, to charge accumulation and upset DC operating conditions. Moreover, while the NFB is upset, the amplifier looses control of the speaker. The speaker's back EMF then appears enlarged at the amplifier's output port, which can only worsen the situation. When the amplifier's NFB control is regained, it takes a significant fraction of a second with a large cone driver, to regain electrical control of the damping.

Even the speaker cable can add substantially to delay recovery, if wrongly constructed (see Chapter 9).

Musical causes

Signals that can cause slewing, depending on an amplifier's slew capability and HF filtering, are first of all, in the music itself. These comprise full and high level (above −10dBr) high treble music signals, usually above 5kHz and up to 100kHz, and usually from heavy percussion and wideband electronic keyboards (particularly analogue synthesizers, but also samplers with above-16 bit capture).

Slew limiting caused by instruments' ultrasonic content is most likely to be experienced when live sound is captured, and particularly when EQ is in use. On mixing consoles, the author has yet to meet an hf shelving equaliser, which does not also boost ultrasonic frequencies, when set to boost lower treble frequencies by a lesser amount. The hidden boost can range up to +20dB and can counter the effects of portal filtering and transducer and system hf roll-offs. If the slew limit is low enough, then there will even be audible distortion with breathy, close-mic'd female voices, electric guitars, and numerous other instruments with high level partials above just 5kHz. In vinyl replay systems, whether used at home or publicly by DJs, fast-edged dust clicks will cause momentary slew limiting which should not be audible in itself, but would be if the amplifier takes over a few milliseconds to recover.

External causes

Second, interfering signals that might otherwise possibly remain inaudible, either ultrasonic or RF, can cause slewing. As the frequency of a signal applied to the input of a power amplifier's drive stage increases, less and less amplitude is needed to overdrive the amp into slew. These external causes of slewing may include:

1. RF pickup, most often from a nearby transmitter, e.g. mobile phone, taxi, other mobile radio users, local broadcast station or relay mast; or unsuppressed, noisy electrical equipment, etc. Another cause is local, particularly EHT power-lines (up to 400kV *Supergrid* in UK) arcing in wet weather, or due to damaged insulators [13].

2. Badly designed or wrongly connected source equipment oscillating at RF. Even if the RF-generating stage is not in the signal path, the RF is often able to leak through, to surreptitiously infest the audio.

3. Ultrasonic leakage of analogue tape being fast wound while playing.

Prevention

Slew limiting can be prevented by:

(i) proper design so that *all* conceivable legitimate music signals can be handled at 0dBr; and

(ii) proper portal RF filtration, so that signals above 100 to 200kHz are robustly excised.

For any given degree of RF filtration, a more-than-adequate slew limit can distance people from the ear-fatiguing distortion that can begin when an amplifier's slew limit is approached by a factor of two, or even ten. *At the point where slewing is visible on a sinewave on a scope, %THD is already about 1% and the damage has already been done.* This may happen rarely in some systems and situations, and in others not at all. But no one would remove the rear fog lamps from their car on the grounds they had never been run into!

Safe ratings

Equipment slew limit ratings needed to avoid slew limiting were established in the USA by Walt Jung two decades ago [7,8,9]. Recognising that unpleasant distortion begins before hard slewing, Jung's recommendation is an allowance of 0.5v/µS ('volts per microsecond') *per peak output volt* (V_O pk), based on an 80kHz power bandwidth. In other words, 0dBr ultrasonic signals are planned for, up to 80kHz. This condition provides a safety factor for most reproduced sound, by a factor of four for 16-bit CD and R-DAT; and at least five for VHF, FM broadcasts with 15kHz (or so) bandwidth.

Slew Limits for Power Amplifiers

Minimum figures to meet Jung's and Pass' criterion for audio quality:

nominal Power into: 16Ω	8Ω	4Ω	nominal rms output voltage swing:	80kHz PBW *Minimum* Slew Limit:	200kHz PBW *Minimum* Slew Limit:
8w	16w	32w	11v	8v/µS	22.5v/µS
16w	32w	64w	16v	11v/µS	31v/µS
32w	64w	125w	22.5v	16v/µS	45v/µS
64w	125w	250w	32v	23v/µS	65v/µS
125w	250w	500w	45v	32v/µS	90v/µS
250w	500w	1kW	63v	45v/µS	126v/µS
500w	1000w	2kW	89v	64v/µS	179v/µS
750w	1.5kW	3kW	109.5v	77v/µS	216v/µS
1kW	2kW	4kW	126v	89v/µS	249v/µS
2kW	4kW	8kW	178v	125v/µS	350v/µS

This table reinforces the often neglected point that *unless a maximum peak voltage swing is defined, a slew rate limit in v/µS is fairly meaningless.* 10v/µS may be 'fast' for a mic input stage, but is inadequate for all but the lowest power deliveries into speakers.

Nelson Pass, a leading light for twenty years in the design of high-end domestic amplifiers in the USA, goes one step further, with 1.4v/µS per peak output volt. This corresponds to a full power bandwidth of 200kHz, giving some added safety margin (+8.9dB) for live sound reinforcement, as well as for future digital record-ings with wider bandwidths commensurate with 20+ bits. If adopted, the minimum slew limits expressed in the fifth column of the preceding table would be multiplied by a factor of 2.8 times, so the 32v/μS for a 250w⇒8Ω amplifier becomes 90v/μS.

Increasing

An amplifier's slew limit may be raised by at least three means, of varying utility and practicality:

(i) At the design stage (usually), the power stage's input stage transconductance can be lowered by breaking apart the LTP's joined emitters and adding degen-eration resistance in line with each emitter (Figure 5.11) (or by increasing exist-ing *degen* resistor values), while the input stage current(s) are raised. This won't be feasible at the outset if the long tailed pair is a monolithic dual with a commoned emitter leadout.

Figure 5.11

Increasing Slew Limit - with constant gm degeneration.

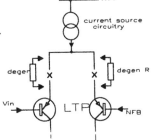

The effect of degeneration (alias *local feedback*) is to increase the linear range of the input pair, irrespective of global feedback. The effect of the increased current is that more is available to charge a given size of lag compensation capacitor. By this technique, the slew limit may be increased several fold. Mak-ing the slewing symmetrical (so it is +/–100v/μS rather than +107 and –58v/μS, say) is another matter. For quality audio this is highly desirable, unless there is easily more than enough. But symmetry may not be realisable with some topologies, notably the Lin [10,14].

ii) By reducing the lag compensation capacitor. This is usually set at the design stage, usually in concert with any degeneration. The effective compensating 'Miller integrator' capacitance is *not just the explicit capacitor indicated in the circuit* schematic, but also comprises the collector-base capacitance (C_{ob}) of the VAS transistor, if BJT – or the equivalent inter-electrode capacitance where the VAS is some other device (eg. J-FET, BJT-JFET, etc). Semiconductor device capacitances have wide tolerances, typically +/–30% or more. They cannot be relied upon for compensation across a population. In addition, device capaci-tances also vary with the electrode voltages (Figure 5.12).

This variation, which may be driven by the signal in some cases, is numerically far greater. For this reason, it is usual to make the added compensation cap's value large enough to swamp out the worst of the variation, even if a smaller

Figure 5.12

A BJT's C_{ob} (parasitic capacitance) variation with sustained voltage. After Sanken.

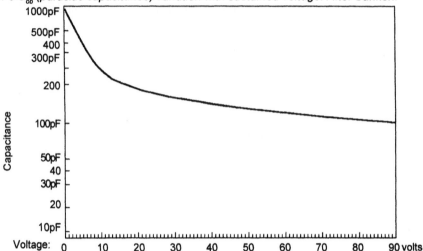

value permits stability. For example, if the VAS transistor's C_{ob} is 12pF, and an approximate 20% uncertainty in the total lag capacitor value is acceptable, then a 47pF cap with 1% tolerance would be used, giving a total lag value between (at worst) 46pF and 60pF. The slew limit would then vary accordingly. Even higher values, maybe 150% to 200% higher may be called for by cautious production engineers, 'just to be sure'. It follows that in many individual amplifiers, the slew limit may be tweaked upwards by as much as four fold, by judiciously decreasing the lag compensation capacitor. It is easiest to do this by installing a suitably rated trim-cap which spans down from the existing compensation value. At some setting the onset of RF instability or oscillation may or will be observed and the final setting will be backed off at least 20% from this point. The caveat with reducing the compensation is that there might still be a risk of then having the amplifier become RF unstable under some untested operating condition, e.g. a cold start.

iii) By using a 2 channel power amp in mono bridge mode. This doubles the slew limit, as the voltage swing capabilities of the two channels add. As this is at the cost of turning two channels into one, it is not a economic solution, but a possible pragmatic solution in the field to achieving a cleaner, less gritty and harsh treble sound from an amp that has audible slew problems. The input drive level would be backed-off by –6dB, to keep the output level it as it was.

iv) By increasing amplifier gain. This can allow lower values of compensation capacitance (ii, above) for a given stability margin.

v) By replacing the BJT input pair with FETs (J-FETs or small-signal MOSFETs). Their lower transconductance has the same effect as the added degeneration resistors, R_e (section i, above).

5.4 RF stability

'Instability' is electronic engineers' name for an amplifier that oscillates or otherwise produces signals of its own accord. In conventional, mainly direct-coupled (certainly transformerless) transistor amplifiers, irrespective of how much NFB there is, oscillation is usually restricted to frequencies above audio, typically 30kHz to 300kHz in BJT output stages; and up to ten times higher in the far faster lateral MOSFET output stages, and often ten times higher frequencies again, when the output stage employs D-MOSFETs.

LF instability ('motorboating') is a pathology limited to amplifiers that employ cascaded RC and/or transformer coupling, and suffer from high impedance and/or unclean power rails.

When a BJT power amplifier becomes an RF oscillator, due to whatever combination of bad design, unfortunate loading or unlucky part tolerances, it commonly destroys the HF drive unit. *This may not happen so directly as is imagined.* Above 20kHz, all conventional moving coil drive units have rapidly rising (series inductive) impedance. This means that the power they are able to dissipate reduces rapidly when driven above 20kHz. The power that an amplifier can deliver at RF may also be limited by increases in its own output (source) impedance. That's because the oscillation implies that the feedback is at least partly positive, which counteracts the global NFB's output impedance reducing effect.

Still, fatal overdriving does not need much power delivery with many domestic HF drivers being rated for just 5 watts of a continuous signal, while already handling up to 50 'watts' of occasional program transients. Indeed, RF oscillation is often a continuous or at least repetitive condition, and the foremost occasion for RF oscillation causing burnout is when audio program is also present. If the RF is just a few volts, the amplifier will be driven sooner into clip, by 1dB or more. If the RF is much higher than 5v, the average amplifier being used towards its limits will be driven continuously into clip. Either situation is enough to burn-out even ruggedly-rated HF drive-units.

5.4.1 Power stage, critical layout requirements

The physical layout or *topography* of parts and hence wiring, can have a far greater effect on signal purity and proper operation in power amplifiers, than in other audio equipment. Poor layout causes inter-stage *feedthrough* or *backtalk*, which cause various kinds of distortion, and even RF oscillation.

Caused by instantaneous signal *voltage difference*, such backtalk or feedthrough may readily be well guarded against – if not by spacing, then with conventional 'electrostatic' shielding, e.g. shielded wires and any, even very thin, metal bulkhead. The scope for voltage differences in adjacent circuitry is in any event relatively limited – few power amplifiers have voltage gains much above 100x, and with most, gain is nearer to x20 to x30.

Where power amplifiers differ from nearly all other audio equipment is that peak output currents range into tens of amperes. With a high input impedance, the input current, for tens of output amperes, may be as little as 1µA. *So current gain can be over a million times.* Without a well designed physical layout, there is far greater scope for the induction of hostile signals by this route, notwithstanding that incoming signal voltages may be at a comparatively high line level. For example, when current differences of a million-fold exist within a few inches (millimetres even!) of one another, the interfering magnetic field induced, or set up by the larger current is far, far harder to shield against, than the electric field caused by a large voltage. To be at all effective, only thick steel or far better, thinner but much more expensive *mu-metal* will do. These are brute-force techniques, that are not often seen in practice. Instead, *induced current* backtalk is lessened in well engineered designs by three low cost, elegant means.

First, by keeping the high current conductors as far as possible from sensitive parts of the driver circuitry – as magnetic field intensity falls off rapidly with distance.

Second, by minimising the area and length in particular, of sensitive nodes. Generally, these are parts of the circuitry where the signal is in the form of current. In nearly all cases, these are the inverting input of the driver stage (assuming conventional global NFB) and the input differential pair's collector nodes. The same points are no less sensitive when an IC op-amp is employed instead of discrete circuitry – but at least they then start out highly miniaturized, thus compact. The designer need then only keep the -ve input's nodal tracking compact.

Third, the radiated field is reduced further by keeping the area of high current loops to a minimum. This may be achieved by arranging PCB tracks so the current return is parallel and close, and by twisting high-current conductors, into send/return pairs. A good sign this has been done properly is to look for *pairs* of twisted wires. *A third wire is almost certainly out of place unless it is twisted in the opposite direction.* Even better than twisting is to use audiograde cables which employ various proprietary techniques to maximise mutual inductance and retard skin-effect. Such low inductance cables are self-shielding. PCB tracks cannot be twisted but can be arranged back-to-back for least series inductance. Altogether, these steps can attenuate the voltages or currents caused by the induced magnetic field(s) by up to 60dB, or 1000 times, or more.

5.4.2 Critical nodes

A *node* is a point where different currents either originate and split-out from, or converge at. *Every junction between more than two parts in an electronic circuit is a node.* In power amplifiers, the critical nodes are those where dirty and clean, and/ or small and large signals come together. Why critical ?

Interaction between high and low level signal currents can occur independently of magnetic or electrostatic induction – by superimposition. This happens when 'common' connections to which both high level, possibly 'dirty', and small (and possibly clean) signal connections have significant common length and hence resistance.

Figure 5.13

0v Noding

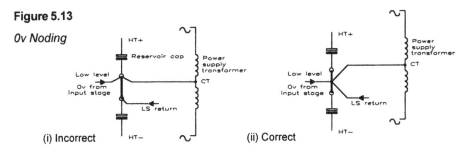

(i) Incorrect   (ii) Correct

This may be called *bad noding*. If 1 ampere flows in a 1" long connection with 1 milliohm of resistance, there will be (1.0A x 1.0 mΩ) volts, i.e. 1mV, across the connection. If a low level signal shares this, 1mV of noise in series may be rather large by comparison. If it is anything more than 100,000th of (or −100dB below) the small signal, sonic degradation is possible. 1mV is therefore on the threshold of affecting even a 60v rms *large* signal, let alone all the other, mainly far smaller signal voltages, that are at large in a power amplifier. In turn, considerable non-linearity in audio power amplifiers may be attributed to layout. One investigator writes "having examined ... (class B) power amplifiers I feel that this effect is the most widespread cause of unnecessary distortion" [15]. Figure 5.13 shows a classic example of superimposition, where the busbar 'wiring' to one of the reservoir capacitors, which is carrying a 1 volt peak of 100Hz (or 120Hz) ripple with a high harmonic content, is also shared by the 0v reference for the driver stage. The result is that the driver stage is riding on top of ripple, and buzz pollutes the signal - no matter how much shielding and other remedies are used. Another classic cause is the same asymmetry in the connection of the NFB takeoff, which in a class B or A-B circuit must be at the exact point where the half wave currents coalesce. Or, even better, take off may be at the output socket.

Good noding practice involves hierarchical 'star earthing' or 'star-point earthing'.

5.5 V&I limits on output, the context

Figure 5.14 follows Figure 5.2 but shows the SOA for one transistor, or one half of a push-pull output stage, replotted with *linear* voltage and current scales, along with additional curves. The common consideration is dissipation or power delivery, hence current and voltage relations, hence the term 'V-I' or 'V&I': V for voltage, I less obviously, for current.

The effect on the linear x linear scales is to re-map the BJT's power limit line (here nominally 160 watts) as a hyperbolic curve. The second breakdown region appears as a further, prouder hyperbolic-like segment. The voltage and current demanded by resistive loads are mapped linearly i.e. as straight lines. The first of these, on the bottom left, is that for an 8 ohm resistor, then a 5.5 ohm resistor; notice how lower resistance appears steeper, using the hill analogy. Both curves have their origin at 25v as this is the maximum output swing in this example. The maximum voltage

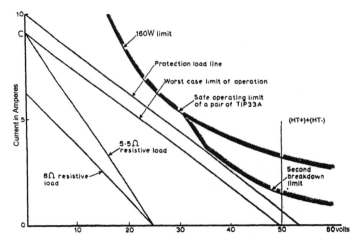

Figure 5.14

Load Limits

The upper hyperbolic segment represents 160 watts, the limit for the output transistors being used. The inner, lower hyperbolic segment shows the secondary breakdown limit coming in about 30 volts, reducing the allowable current. These limits shrink inwards with higher temperatures. The next curve is the V&I protection's load-line, at about 45°. Beneath it, the hypothetical limits of worst case operation are plotted. Finally, the load lines for two resistive load conditions.

From Hardcastle & Lane, High power amplifier, Wireless World, Oct 1970.

seen by the transistors is double this, 50v, or HT+ plus HT−. Notice the large, safe distance between the transistor SOA and what the nominal speaker loads would demand – at least for 25v peak swing. If the transistor is MOS, the second breakdown limit line is absent, giving added dissipation margin at high voltages.

Figure 5.15 shows why any margin we can get is going to be needed when driving many real speakers. In this graph, the resistive part is 5.5 ohms in all cases, while the reactive part is cited individually in +/− j ohms, for inductance and capacitance respectively. The first surprise, perhaps, is that reactive loads describe ellipses, giving rise to more marked, non-monotonic changes in dissipation, compared to that of the diagonal in the centre, which is the basic 5.6 ohm resistive load-line. In particular, each load has two V-I values below its maximum, one benign, the other here shown at a tangent to the limit line. Notice also how it is counter intuitively the high

Figure 5.15

Speakers, unless they are ribbons or have conjugate networks, are reactive not resistive loads, and appear as ellipses when plotted as shown here. The ellipses may be mapped onto the previous figure. The thick line is resistive.
From Hardcastle & Lane, High power amplifier, Wireless World, Oct 1970.

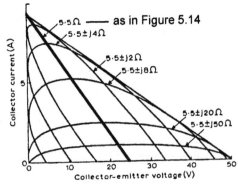

194

impedance reactances (+/–50j Ω) that probe the high V_{ce} regions, where BJTs are most vulnerable to 2nd breakdown !

Figures 5.16b-d (overleaf) use Micro-CAP IV simulation to demonstrate some unexpected facets of instantaneous voltages, current and dissipation in one output transistor (it could be BJT or MOS), driving a 15", nominally 8Ω bass drive-unit in a sealed box. The voltage swing is quite high at just under 100v peak. This would amount to about 563w into an 8Ω resistor, assuming 95v peak. The level is set at just ($\cong$ –0.1dB) below clip since this is the worst condition for dissipation into *reactive* loads. Note that $\cong$ –3.5dBr (see section 7.3.2) is the worst case for output stage dissipation when driving a pure *resistive* load.

Fig.16b shows the conditions at 30Hz, the resonant frequency. Current in the monitored transistor is relatively low, peaking at about 2 Amperes. Instantaneous power dissipation is peaking at about 200 watts, although this increases off-scale to about 240 watts on the third cycle. The notional power delivered by the transistor to the speaker in this condition would be about 100 watts (95v pk x 0.707 x 2A pk x 0.707). Thus the transistors are instantaneously dissipating 2.4 times more power, than the average power that is being delivered. Fortunately, what is being delivered is well below the rated maximum power.

In fig.16c the frequency has been increased to the resistive zone at 435Hz. Here the current is peaking at over 15 Amperes, while instantaneous power dissipation is at least 400 watts. The power delivery under this condition would be about (95v x 0.707 x 16A x 0.707) = 800 watts. So now the instantaneous dissipation, while much higher, is about half the average power that is being delivered.

In fig.16d, the frequency is moved to 600Hz, where the speaker is starting to appear slightly inductive, past the resistive zone. The peak current is about the same as before at about 16 Amperes, but the instantaneous dissipation is now about 600 watts. Average power delivery is about 800 watts as before, so again, more power is being delivered, but the dissipation is creeping up.

In his last work [16] Baxandall cites how with a highly reactive (i.e. capacitive or inductive speaker loading) causing an exceptional 90° phase shift (in current, relative to voltage), transistor dissipation is near to its highest, as high current is having to be sustained with substantial Vce. The dissipation in each half of the OPS is then raised to:

i)   At clip, an instantaneous maximum of 2.6 times the average maximum rated delivery into the same value of purely resistive load; or

ii) 4 times the instantaneous dissipation into the same value of purely resistive load; or

iii) instantaneously about 15 times the *average* dissipation in that half of the OPS when driving the same value of purely resistive load.

All loads are of the same value. It follows that *heatsink temperature is no measure of power transistor stress*, due to potential 15 fold disparity between peak and average power dissipation.

Features of the power stage

The impedance behaviour of a simple, single drive-unit speaker, and the resulting instantaneous voltage, current and power dissipation in one output transistor, in a class A-B Output Stage, at different frequencies.

Figure 5.16a

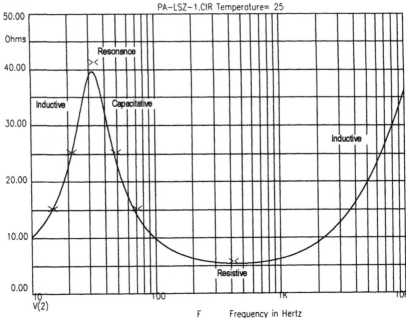

Figure 5.16b

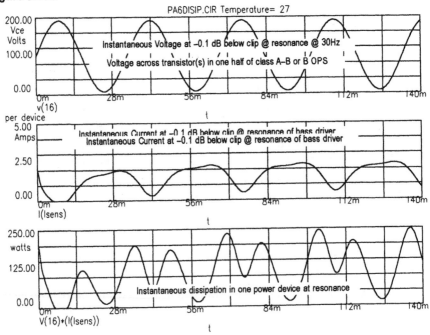

Figure 5.16c

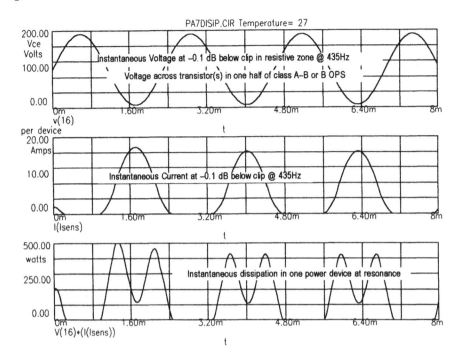

PA7DISIP.CIR Temperature= 27

Instantaneous Voltage at –0.1 dB below clip in resistive zone @ 435Hz

Voltage across transistor(s) in one half of class A–B or B OPS

Instantaneous Current at –0.1 dB below clip @ 435Hz

Instantaneous dissipation in one power device at resonance

Figure 5.16d

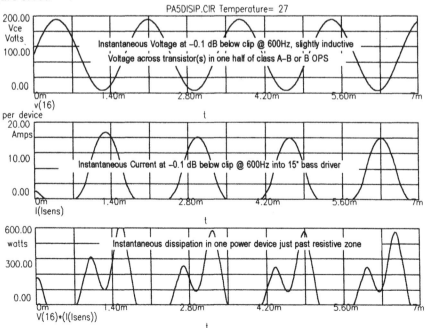

PA5DISIP.CIR Temperature= 27

Instantaneous Voltage at –0.1 dB below clip @ 600Hz, slightly inductive

Voltage across transistor(s) in one half of class A–B or B OPS

Instantaneous Current at –0.1 dB below clip @ 600Hz into 15" bass driver

Instantaneous dissipation in one power device just past resistive zone

These figures assume that the maximum dissipation is divided between the two halves, and most textbooks glibly state this. However, program asymmetry means that the worst case instantaneous dissipation may be even higher. For a road worthy BJT power amplifier that will not 'crap out', a 90%/10% division should be assumed, i.e. *the rating of each side should be doubled.* This allows up to 19dB of instantaneous asymmetry in the programme.

The upshot is that with an *unprotected* output stage, and a stiff power supply, BJTs that will withstand driving any loudspeaker, cable and crossover combination will need to be rated at an average that is (x2.6 for 90° reactive loading, x2 for asymmetry) = 5.2 times the amplifiers' rated power, i.e. for a nominal 100 watts of maximum potential delivery, parts rated at 520w *per side* should be used. This amounts to about eight 250 watt BJTs inside a stereo 100 watt amplifier. To employ any fewer transistors, or survive connection to loads into which power delivery is above the rated capability, the amplifier must be protected.

5.5.1 V-I output capability

The map of an amplifier's 'V-I' or 'V & I' output capability or limits, is rarely shown. One US maker that did once publish the information, later withheld it. The map of V & I limits is not just the sum of all the SOA curves of the transistors in the two halves (or one half if single-sided), but also shows the effects of any output protection circuitry or systems, as well as power supply limitations, against excess current. What constitutes 'excess' current varies widely with its duration, and also depends greatly on the instantaneous voltage across the power devices - not forgetting that finite dissipation handling in them (V x I) is them is the foremost reason for V & I limits. It is important to recognise that when this voltage is high, the voltage across the speaker is low, and *vice-versa.*

If there is no protection, or the protection has gaps, a V&I graph shows, by implication, either theoretical boundaries, or else boundaries beyond which the amplifier will no longer exist. This is particularly relevant to amplifiers with BJT output stages. The V-I 'plot' also supplies graphic evidence of ability to drive and satisfy the demands of specific loudspeakers. A speakers' own V-I 'demand' may be quite well defined by cumulative V&I plotting with music signals [17] , but again, this is rarely seen. As with electron shells, there is a 'probability space'.

For conventional class B and related push-pull amplifiers, there is an important distinction to be made between V&I plots like those shown in this chapter, which are fully symmetrical (positive and negative current and voltage), and *output-referred*; and others that are like SOA plots (Figure 5.14), and asymmetric, showing limits in one half of the output stage only, i.e. in just two of the four 'quadrants'.

In the former type illustrated in this section, the voltage being shown is the output voltage that will be applied to the speaker. This is valuable enough to informed users. It's also of value to designers, but only for a given model, since the graph cannot be very easily rescaled for other supply conditions [18]. In the latter kind, the current and voltage is that respectively in and across the transistors in only one half of the output stage. This presentation is of most use at the design stage and for analysis.

Figure 5.17

In this, one of very few V&I capability maps ever published for any audio amplifier with a BJT output stage, the central area is legal in all cases. The outer areas, towards the top left and bottom right, are 'out-of-bounds' except for brief pulses. The two lowermost, unlabelled pulsed curves are the 'mirror image' of the uppermost ones. But only the lesser curve marked 'DC' would be safe for (or available from) the output devices under all conditions with full-range music reproduction. *The amplifier being considered is thereby limited to delivering 2 Amperes and 30 volts, both peak values.* Copyright Burr Brown Inc., reproduced with permission.

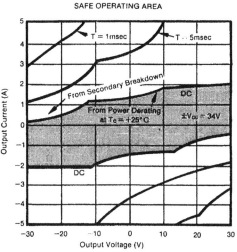

Figure 5.17 shows the V-I limits map for an IC audio power amplifier, not intended for high performance use but nonetheless, representative enough for illustration. As in the SOA curve of an individual transistor (Figure 5.2) we see the limits of simultaneous output voltage and current that the amplifier's output is able to traverse, according to duration. Notice the areas marked 'from power derating' and 'from secondary breakdown'. As this amplifier has no V-I limiting protection circuitry, these directly correspond with the broadly like-named regions on the transistor SOA graph (Figure 5.2). One difference is that 'DC' is labeling the whole bound line; the right hand section it is lying against is the same as the bond-wire 'current-limited' zone in Figure 5.2.

In comparison to the SOA graph of an individual power device notice how the limits are comprised of both straight and curved regions. The curved (strictly: hyperbolic) regions always appear when a given power (VxI) is plotted on a linear x linear V-I scale.

With any unprotected and under-rated amplifier, the limits might be easily transgressed by an unfortunate choice of loudspeaker and music, let alone abuse. With a protected amplifier, the same V-I plot would represent the limits. If an attempt is made to exceed them, by connecting the wrong kind of loudspeaker, distortion will result. Whether the limits are the unfenced borders of vaporization, or physical restrictions, has to be inferred from a knowledge of whether the amplifier has effective V-I protection (even if it claims it hasn't any). This particular amplifier has none and is restricted to at best an instantaneous, hence peak current of 2 amperes.

In concert with the maximum 30 volts peak, the amplifier's capability is:

$(VxI) = (0.7 x 2 x 0.7 x 30) = (1.414 x 21) = 30$ watts (average power),

and this maximum power is revealed as being developed into:

$$V^2/W = (\{ 21 \text{ x } 21 \} / 30) \cong {}^{454}/30 \cong 15 \text{ ohms.}$$

5.5.2 V-I output limiting (adverse load protection)

Most audio power amplifier output stages require some form of protection. A list of 'illegal', stressful load conditions appears in section 8.6.2. The complexity of the protection can vary widely. BJT output stages commonly require more involved and explicit protection. And the higher the power capability, the more likely it is that a complicated but watertight protection system is justified. Here 'complexity' might mean 20 components, but they are most likely small parts, usually amounting to far less than 1% of a high power-rated amplifier's parts cost.

Comparisons

Before transistors, valve power amplifiers survived with only loose protection from 'bad' loads (excepting open circuit conditions) by their output transformers and by fuses. How did this work ? First, as the load impedance reduces below the intended value, the power deliverable through an output transformer falls off rapidly. In effect, transformer-coupled valve amplifiers act like a rather limp voltage source. Second, valves are more able than transistors to withstand temporary dissipation beyond their ratings for minutes at a time, without failure. This gives time for fuses to react (see section 5.5.4). Third, with typically only some 20dB of global NFB, the output stage is not driven so hard when the feedback tries to act against the shortfall, as heavy output loading prevents the proper output voltage being developed.

Class setting

Output transformers continued to afford some protection to the first transistor amplifiers, as did class A operation. Here, the output transistors' maximum current is set and already flowing while quiescent; a short circuit or low load should not be able to increase the total draw, at least in single ended circuits. In fact, hard-driving and feedback overdrive should back-bias the OPS into class B, so OPS dissipation actually reduces. Where not, a pair of series connected diodes between the bases and common point of the drive or OPS transistors, or else a bias regulator (see section 4.6.3) are all that should be needed to limit overdrive. Having (in theory at least) *no great need for output protection is a small but significant advantage for Class A amplification* in terms of both engineering and sonics. In the following section, the protection that's necessary for class B and A-B amplifiers will be considered. To the extent that most class G and H amplifiers are also one of the B species at low signal levels, they are included.

Fuse rejection

When BJT output stages became class B and OTL in the early '60s, leaving just fuses, thermal switches and relays for protection, it was commonly observed that ordinary 'quick blow' fuses were more often being protected by the transistors.

Power switching relays and external thermal sensors were plainly never going to act fast enough, and while high-speed fuses were developed to protect critical semiconductors, they were too costly or esoteric for widespread consideration at least until the mid 1980s.

Active protection

Since the dawn of transistors, electronic protection faster than any fuse has been on offer, with the possibility of auto-reset, 'keep trying to reset' and other 'intelligent' functions. This is the natural corollary, the reason why transistors need *fast-acting* protection. But a change of mindset had to occur to adopt existing schemes made to protect transistors in other apparatus for audio use, since protection device which arbitrarily shut the amplifier on and off with little 'intelligence' could create havoc in the context of listening to music. For example, an early, crude protection scheme was to connect a current-sensing resistor (Rs) in line with the speaker, as shown in Figure 5.18. If the current through, hence voltage across this, exceeds the diodes' threshold voltage (typically 0.25v or 0.65v), the diodes conduct heavily. This lowers the gain abruptly, which ideally saves the transistors, but also draws attention to itself. It might even cause RF oscillation, adding 'frizz'. Now the search for sonically transparent protection had begun.

Figure 5.18

Early RCA protection scheme.

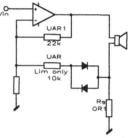

The next generation of transistor protection was more subtle, as circuitry which limited current alone developed into 'Load line-limiting' or Load line protection' alias 'V-I limiting' or 'V-I protection'; they are all the same. The following figures 19 to 26 cover the evolution.

Figure 5.19

Output Current Limiter.

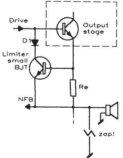

In Figure 5.19, the emitter degeneration resistor R_e in the output stage develops a voltage across it in proportion to the output current in this half. The output current is clamped (or limited) at the current where this exceeds the current required to turn on the limiter BJT (a small signal part) hard enough to divert any further drive.

What is rarely mentioned in this context is that frustrated global NFB will force the preceding stage to deliver all the drive current that it is capable of.

In Figure 5.20 the effect of this circuit relative to the safe operating bound (including 2nd breakdown, which is the curved line on the right) is shown by the horizontal line A-B. Clearly, a horizontal, purely current-limit is not going to allow a very big speaker-load ellipse to develop while keeping it clear of the SOA boundary. Of course, the maximum size of the ellipse broadly represents the deliverable SPL.

Figure 5.20

Bounds set by Load-Line Limiters Vs. *SOA*

The thick curve is the safe operating bound set by the transistor's maximum dissipation. The 2nd breakdown area is not shown, but obtrudes adjacent to F,G, as seen in Figure 5.14. A-B shows the bounds of a simple current limiter as fitted to many power ICs. This would prevent unsafe high currents at low voltages; yet not protect against unsafe lower currents at higher voltages. Current limiting bounds are useful when appended to the V-I bound E-F, as in F-G; or to C-D, as in D-DD. The sharp corners are idealised. V-I limiters, essential in ruggedised BJT amplifiers, may employ up to 5 differences of 'slope', some of them time or frequency dependent.

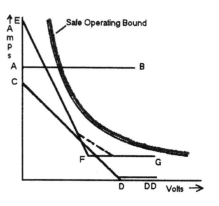

Returning to Figure 5.19, the diode D1, usually germanium for a low forward voltage drop (often the venerable OA47, now over 35 years old) is included in most circuits to prevent the limiter BJT performing in *inverse mode*. This is likely to occur during normal operation every time the output swings (for this half) negative. Considering the class B or AB bias current that will still flow through Re, the base is slightly biased on relative to the emitter, *while the drive has flipped polarity.* Thus the BJT conducts, even though the collector is more negative than the base. The diode prevents this, and the unwanted limiting that would occur.

Here and in all the following circuits, there are three points to bear in mind:

(i)   the output stage, shown as a single transistor for simplicity, will likely be a Darlington, Sziklai or triple in practice. It should not invert the signal, however, or have voltage gain.

(ii) The positive half circuit is shown. Mostly the negative half's mirror-image begins at the dashed lines.

(iii) As implied, and as seen in chapter 4, the incoming drive signal in Lin-based topologies (for which all the illustrated load line limiters were developed) is the same on both sides, and is a large signal voltage, as the output stage has only current gain.

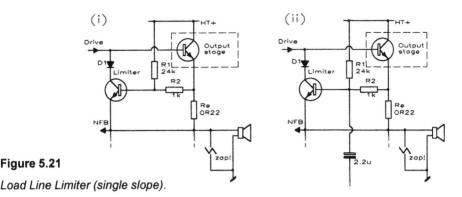

Figure 5.21

Load Line Limiter (single slope).

In Figure 5.21 (i), R1 has been added, so when the V_{ce} of the output stage is high, proportional current in R1 increasingly contributes to the base-emitter current. Then at maximum Vce, little or no current will be needed to limit the output. Values are typical but vary widely with the rail voltage and R_e value. Looking back at Figure 5.20, the result is the resistive-like load-line C-D. Moving to (ii), the Crimson Electrik topology (see 4.3.4) had this basic, single-slope kind of VI limiting. The sonic difference is a small electrolytic which coupled the limiter BJTs' bases, delayed limiter action for some a few mS, and probably more important, filtered hf supply noise – a sonic factor which is never mentioned in this context.

2 slope limit

The common shortcoming with this kind of V&I limiter circuit is that it doesn't follow the SOA curve very well. If set *back* from the SOA, as with the line C-D, the V-I limiting is perceived to over-react, clamping output down to a few watts when strong bass is played from (say) a vented cabinet. The alternative is to set the limit too close; then as temperature rises, an increasing portion will pass beyond the SOA bound, and eventually, a particular piece of music crossed with a particular speaker load-line will cause this unprotected spot to be traversed, and *Fizzle zzt!*

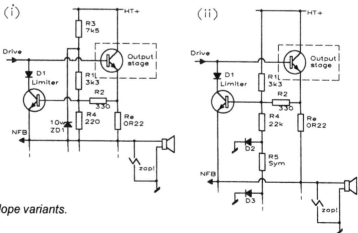

Figure 5.22

Improved multi-slope variants.

203

Higher slopes

In Figure 5.22,(i), R4 is added optionally to control the voltage and current sensing ratios independent of R_e, the value of which is set by far more important other agenda. More important, R1 is split. The upper half is renamed R3 and a zener inserted to clamp the voltage at a level near halfway. Separate from the issue tackled by D1 earlier, this prevents the voltage being sensed, from increasing from the rail (HT+) to ground voltage, and on towards the total rail voltage (2 x HT+), as the output swings negative. Without this, and even with D1, the limiter on each non-conducting half will be stealing drive current, especially if the OPS is class A-B biased.

The zener clamping has the effect of allowing the limit current at HT+ to extend outwards. The line D-DD in Figure 5.20 shows this. This modification gives a second slope to the V-I protection limit line. In Figure 5.22 (ii), R4 is taken across to its opposite neighbour via a common resistor, R_{sym}, and this point is also grounded on each side via diodes D2,3. This generates a more complex characteristic variously described in the literature as having dual or triple slopes, or even semi-elliptical, like the reactive load itself. What is *most* needed though, is a load line that closely matches the SOA bound's outline.

Figure 5.23

3 slope V–I limiting.

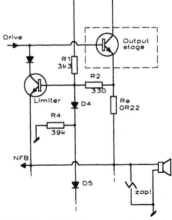

The line E-F-G in Figure 5.20 comes closest to this. The third slope produced by the circuit in Figure 5.23 comes even closer. Notice how the main difference is that the diodes and resistor have swopped places. This arrangement (seen in amplifiers made by BGW, for example) works rather like the zener earlier. Looking again at Figure 5.20, it produces a segment approximating the dashed 45° portion over F, leading, into the section F-G. Assuming competent design, this 3-slope arrangement, signified by the inter-diodes D4,5, gives the most power delivery from a given set of BJTs with the greatest safety.

Mid-Floyd

Figure 5.24 shows a practical limiter used on one of the first of the high power, high performance amplifiers that was made for high-end domestic use, but was also adopted by Pink Floyd and others as the sound was so much better than the indus-

Figure 5.24

Phase Linear 700 V-I scheme.

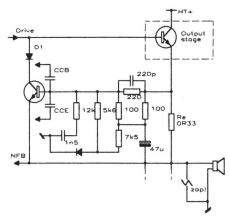

trial power amplifiers they had been having to use (back in 1972-3) to get enough power. Over 15 years, the Phase Linear 700 was used for the many bands who hired systems from Britannia Row Productions (see Appendix 1).

Turning to the circuit, CCB and CCE show two locations for compensation capacitors where parasitic HF oscillation caused by positive feedback at RF, and other instabilities respectively, may be slugged – in this or any of the preceding limiter circuits. Notice the voltage sensing is not from the rails, but from 0v (a cleaner place that gives the inverse information) through the diode and resistor array, to give the dashed section down to G again, at high voltages. Note also the relatively elaborate control of the limiter's 'attack and release' response with the 47µF, 1.5nF and 220pF capacitors.

5 dimensions

US maker Crown (Amcron) had patented their own scheme as early as 1967. Here, the V-I limit turns back into a current-only limiter at low bass frequencies. This would be like the load line towards E in Figure 5.20 turning hard left before it reaches point E. But only at low frequencies. So a V-I load limit-line's slope changes may be frequency conscious, as well as varying with temperature, and five spacial dimensions would be needed to 'see' what is going on dynamically.

The next circuit illustrates a comparatively recent development used in designs by UK professional makers, after Tim Isaac. In Figure 5.25, improved protection fitting is attained by splitting the voltage sensing, with R6 diverting around the zener clamp. The zener-clamped voltage is then controlled by a second small signal transistor, TR.2. This turns on hard when the output swings more than a volt into the opposite polarity, effectively shunting the zener with R7. Other than providing enhanced short circuit protection near zero volt output, this circuit provides up to five slopes, including the 'anti-flyback' simple current limit line.

Shared sensing

A practical detail common to all the preceding V&I limiter circuits is the value of R_e. As the output stage in the C-Audio *TR-850* amplifier in which this scheme is

Figure 5.25

Multi-Slope Isaac's V-I limiter.

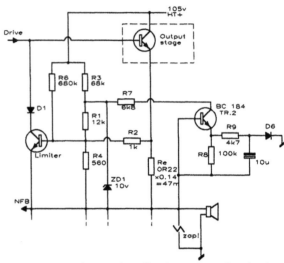

used, comprises seven paralleled power transistors, the effective value of R_e is about 47mΩ, if each leg resistor is 0R33.

In some designs with multiple transistors, perfect sharing is rather rashly presumed and so current is sensed in just one BJT per side. Figure 5.26 shows how multiple sensing is commonly arranged for rugged professional BJT-based amplifiers like the TR-850, where nothing is left to chance. R2 is split in *n* parts each 1/*n* of the originally computed value. This summing system averages the readings. In more sophisticated designs, a 'highest wins' readout would be used, but by now our protection circuitry is getting complex.

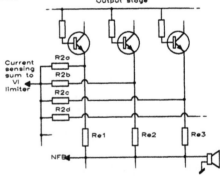

Figure 5.26

Current sampling

Some formulae for calculating the values of V-I limiter resistors are cited in the references [2, 24 and 25].

V–I computing

In Crown's PSA-2 (pro) and SA-2 (domestic) amplifiers of 1978/9, the grounded output topology (see 4.5.1) simplified the provision of analogue computing circuitry to monitor the output voltages, currents and even compute the junction temperature. The analogue output limited the drive-stage current in the normal way, much as in the previous circuits. In theory by being able to follow and fit against the curved limit line more closely – and this had to involve automatically expanding or contracting the limit as the computed junction temperature varied – safety can be more certain while the highest power delivery into speakers was assured. Ironically, the maker's use of sluggard but rugged hometaxial BJTs in a their low voltage bridge configuration rather made it the amplifier least in need of protection ! In 1984 Crown's launched the first of their Macrotech series with 'ODEP', alias 'Output Device Emulation Protection', a development of the PSA computer with improved thermal response modeling and consideration of the recent programme history. A raw current limiter continues to operate at low bass and down to DC.

Alternatives

The re-introduction of CE-type (grounded emitter/source) output stages a decade ago triggered the development of alternative protection schemes – one of which QSC patented in 1983. Also, with BJTs operating in class G or H, the reactive load stress near zero output volts was greatly reduced, as the low-rail transistor is sustaining a relatively low voltage. Stress on the higher tier parts could be similarly reduced in class G, allowing them to operate with simpler, less sonically obtrusive V&I limiting.

MOSFETs

To the advocates of MOSFET amplifiers, Crown's sophisticated schemes, with more circuitry performing protection than the already large amount involved in amplifying with a bridge, appeared obsolescent the moment they were introduced.

With lateral MOSFETs, only a handful of passive parts are needed for thermally limiting self-protection to occur across the entire range of adverse conditions. Figure 5.27 (i) shows the simplest scheme. Each zener clamps the drive to the gates in one half. The zener voltage simply sets a maximum output current, shared between the MOSFETs. The series diodes prevent reverse conduction on opposite half-cycles in most circuits, where, as in the Lin topology and others, the drive for the two halves is common at signal frequencies.

Figure 5.27

MOSFET Over-current Protection.

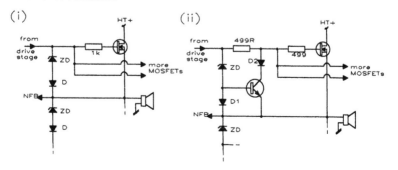

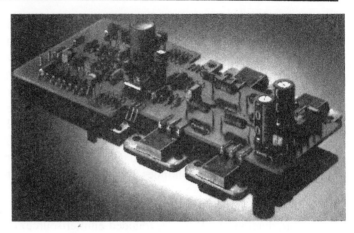

Figure 5.28

Hitachi's original plastic-cased 'H-PAK' MOSFETs required hardly any components for foolproof protection, while giving high sonic performance. This is a 200w rated D-MOSFET amplifier card from B&W's 'Active One' speaker system of 1985, showing overall simplicity. Note open bias trimmer, centre left, that should really be sealed from contamination; strings of seriesed bias diodes; and right centre, four R_e of 0R22, giving each D-MOSFET a little local feedback. © B&W Ltd

With double-die L-MOS and also D-MOS types, the clamping may not be abrupt enough to prevent destruction under all conditions. This is due to the zener's slope resistance. The problem is exacerbated by a drive stage with high current capability, and by using low current rated zener-diodes, which may otherwise be desirable to restrict the distortion caused by zener's non-linear capacitance *Vs.* applied voltage. Fig.27(ii) shows a scheme that clamps harder (against high drive current), and folds-back. Another, no more complicated type uses the BJT to buffer and 'stiffen' the zener voltage. A third scheme achieves stiffness with high current zeners used non-invasively.

These simple schemes and their variants essentially control current. L-MOSFETs survive all kinds of abuse without multi-slope load-line limiters because (i) they withstand considerably higher junction temperatures than a silicon BJT, for which 200°c and often 150°c is the absolute maximum. (ii) the current that can be passed reduces at high junction temperatures, due to increased channel resistance, and (iii) the integral protection diodes quickly hard-clamp the gate drive (provided the drive current is limited) when the junction temperature exceeds published ratings.

IC catches-up

Most monolithic IC power amplifiers have operated from low enough maximum voltages for secondary breakdown not to be an issue. Some ICs have nil protection; others have simple current limiting. Information on more sophisticated schemes is scarce.

In 1993 National Semiconductor launched the first of a series of IC power amplifiers with their own SOA protection. Clumsily named 'SP-i-Ke' – 'Self Peak instantaneous **Kelvin** – (sic), the reality is no more than another, reasonably thoroughly executed SOA computer [19], arriving at the party about twenty years late. The miracle is that it is inside a chip that costs less than fish and chips.

V-I limitations

Although the V-I limiter in all its forms is still in widespread and continuing use in amplifiers with BJT outputs today, it is rare in the best sounding, high performance amplifiers. It is easy to disconnect a V-I limiter and many listeners have noted just how much load-line limiter circuits degrade sonics even when they are not overtly at work. Causes include:

(i) Current clipping, an occurrence with particular speakers (Figure 5.9) and program, *which is not visible with a conventional voltage-sensing scope connection.*

(ii) Negative resistance conditions. If not destructive, these result in RF oscillation(s) that may be local and not readily noticeable at the output.

(iii) Related to these, flyback pulses caused (with the 3 and more particularly, 2 slope limiters) by instantaneous shutdown of current into an inductive load after it has caused voltage spiking. The flyback pulse is heard as a popping or rasping sound.

(iv) V-I limiting is an urgent, brute force form of protection. The side can often be 'bitten out of' a few cycles of a bass sinewave. This takes the 'edge', 'slam' and 'thrust' out of music that has it. Otherwise it changes the tonal and harmonic structure. More subtle means of protection would not interfere with instantaneous values of the waveform so abruptly or crudely. See below.

(v) Non-linear positive feedback.

(vi) Supply noise injection. As V-I limiting circuitry is required to act within a few milliseconds to V_{ce} changes, to be of use, this means that anti-musical half-wave supply rail noise is inevitably injected into the signal path, through the limiter transistor, which can act as a common emitter amplifier stage.

The sonic effect of a V-I limiter may be ameliorated by delaying its action. This is risking transistor longevity if pushed too far without simultaneous derating, much beyond 10mS, say. In more sophisticated V-I load line limiter circuits, the load limit-line may be continuously adjusted against the lowest large signal frequency. Anything that backs off unnecessary V-I limiting will help recover power delivery and maximum sound level. This was the secret of one of Panasonic's 'Ramsa' PA amplifiers sounding more powerful than a certain US east-coast PA amplifier that had twice the rated continuous power.

Class D

The needs of protection in class D against adverse loads are quite different. The output reconstruction/filtering is tuned to best suit a range of speaker loading, and very low loads, or shorting, will reduce the filtering effectiveness, allowing RF to be released. Hence it may involve EMI and EMC considerations. If the amplifier can retain control, and considering the output transistors will in practice be MOSFETs, relatively simple thermal sensing of any excess-dissipation should be all that is required – in theory.

5.5.3 Mapping V-I capability

In practice, as implied in the previous section, the V-I load limit on output capability may vary with frequency and with repetition, as well as with transistor temperature. Figure 5.29 shows how Crown later published the limits of their DC300A in the mid '70s, and on the left, the equipment connections employed. Note the drawing mistakenly cites the load as being under test. The voltage across the oscilloscope's vertical (Y) input reads the current, with 0.1Ω giving a 0.1v per Ampere scaling. Looking at the composite V-I load-lines' plot, the most notable feature, the kite shape, is the DC limit for bass-frequencies. Note also that we are dealing with output voltage, the opposite of V_{ce}, so the sloping line from **X** to the three dots (• • •) is just the usual single slope limit line, seen flipped in the X plane. The various horizontal lines, keyed below the drawing, show a simple current limit bound increasing at unspecified mid and high frequencies, and still higher current limits into a low impedance, etc. Behind them, the hatching warns that the V-I limiter is operating, and that there may be added limiting, dependent on the signal. This is not the clearest way to present such complex information. The absence of the instantaneous response to voltage, alias Crown's anti-flyback feature, is not evident, for example.

The V&I plot may be seen as the 2 dimensional sum of all the large signal possibilities. A more sophisticated test-set designed to plot an audio power amplifier's dynamic and frequency conscious V-I capability, again using a 'scope to view particularly

Figure 5.29

Crown's V-I test set-up and results for their model DC-300A, back in 1974. © Crown Inc.

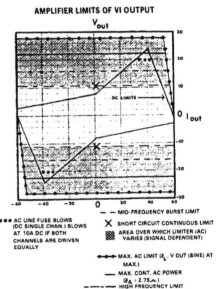

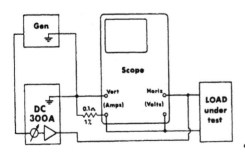

Note: Scope and amplifier grounds are not common. Vertical input reads (-) amperes vertically. If scope has an inverter, invert to read (+) A.

Figure 5.30

Baxandall's V-I test set-up

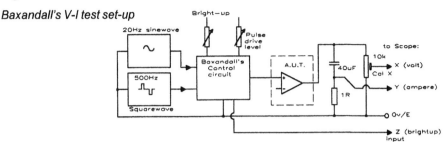

transient effects and protection circuit action, was published by Peter Baxandall [17]. Baxandall's circuit (Figure 5.30) drives the amplifier into hard but very brief clip, with 50μS pulses on the back of a 20Hz sine wave. This enables the limits into reactive loads in particular, to be fully explored without stressing or destroying the amplifier, and with relatively little dissipation in the test load either.

The value of V-I testing is fourfold:

i)   Obscure amplifier faults can be detected. Particularly momentary ill sounds.

ii) Reactive load driving capability (into difficult speakers) can be determined - with both dynamic and steady-state test signals.

iii) the V-I demands of loudspeakers can be mapped, monitored and possibly improved upon. Amplifiers with adequate capability can then be selected.

The 'disadvantage' of Baxandall's technique is that it reveals some very messy and anomalous behaviour in many V-I limiters, including parasitic oscillations!

5.5.4 Audio protection, by fuse

Fuses are the simplest form of protection. They have their uses but applied in the wrong place at the wrong time they have hindered the protection of transistors. Simplistic use also degrades sonic quality.

DC fusing

Looking at the universal fuse map (Figure 5.31,i), fuses A through C are explained in chapter 6. Fuses 1D,2D in the DC power lines, were notorious for some twenty years from the mid 60s in the first generation of transistor power amplifiers with 'split' or 'dual' supply rails (as shown). For *without DC protection, this fuse placement is potentially and readily lethal to the speakers*. The risk is ever present, because if one or the other fuse fails from aging, falls out, or is removed for inspection while the amplifier in switched on, then one or other DC supply rail will usually appear at the output. Without DC protection the speakers are at immediate risk (see section 5.7 and 5.7.1). Worse, in some of the Lin-based BJT circuits, of 'classic' 70's amplifiers, if the fuse (1D or 2D) failed or was opened on one particular rail (more often the negative one) the power stage could be silently destroyed. In turn, the DC lurked, often to destroy the next speaker that was connected.

Figure 5.31

Fuse & breaker positions in power amplifiers.

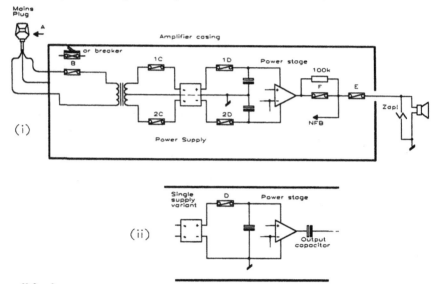

One rail fusing

The first generation of transformerless transistor power amps (say 1956-66) had a *single* HT rail (Figure 5.31 ii), first negative for amplifiers built around germanium *pnp* transistors, then positive when *npn* silicon became the dominant type. In this case, there is nothing risky to the speaker in fusing the DC supply in this case (Fuse D only), since loss of a single rail fuses can only act as a power on/off switch. Moreover, a DC fault wouldn't harm the speaker since with the single rails a DC blocking 'output' capacitor is required in line.

Output fusing

Fuse E is in line with the output. This *may* protect the speaker or the output stage, but misses the point on several counts. For example, considering output protection, the safe output current is say 1 Amp when the output is at 1 volt, and as high as 50 Amps when the instantaneous output is at clipping, at say 60 volts. Which Ampere rating do you choose ?

Tolerance

Output protection fuses have often been empirically rated to just not blow at a certain sound level, or even chosen after computation to precisely match the speakers' rms or EIA/AES power rating. However, *any precision is false unless the individual fuse is selected to high tolerance.* Without this fuses may blow at lower sound levels, or worse, fail to blow before damage occurs. But even with tight tolerance, fuses positions 'E' and 'F' cannot be regarded as anything more than a partial safety net. For example, fuses only respond to mean ('average') power. But transistors can be destroyed in smaller fractions of a second, by pulses, i.e. high instantaneous power dissipation.

Modulation

Moreover, the fusewire will experience a large change in its resistance, when the current through it, during current-heavy music passages, heats it towards melting. This has the effect of slightly compressing the output signal, with a relatively slow attack (hundreds of milliseconds) and equally slow release. This is not conducive to accurate reproduction. The compression might be masked by the drive-units' thermal compression but not if the speakers are highly rated and the fuse value is conservative.

Assuming the amplifier in Figure 5.31 is a conventional high NFB amplifier, fuse F is a better position than fuse E, being placed inside the feedback loop. This means the fuse's addition to the output source resistance, whether hot or cold, is almost wholly cancelled – at least as quickly as the loop can act. Fortunately the rate at which a feedback-controlled output can settle, in a competently designed amplifier, is at least 500 faster than a fuse's resistance can change. A high value resistor is connected across the fuse. The value is chosen to prevent any more than a few milliamperes flowing when the fuse is blown, while still maintaining the NFB loop, at least when the short-circuit and/or speaker are disconnected. This may be necessary to avoid chain destruction in some BJT designs in particular.

Heat of the moment

However much fuses are rated to protect an amplifier from abuse or damage the protection will only last so long as the exact correct replacement fuses are *always* fitted. This particularly applies to fuses that are readily accessible from panel fuseholders, and also fuses in the output positions E or F, where DIY renewal by the user is safe, and expected or invited. A practical solution that invites correct replacement (at least once!) is to have an adjacent fuseholder marked 'SPARE', containing the correct fuse. A circuit breaker is an even better solution, but rarely adopted in commercial products as a high quality type costs at least ten times more.

5.6 Clip indication – external relations

Global NFB linearises the gain across the span of signal amplitude, but 'saturation' is then abrupt, and is more aptly described as clipping. If the overdrive is more than about 6dB, it may be considered 'hard clip'. But driving just 0.5dB into clip (so the top 0.5dB of the waveform is squashed to a small fraction) can destroy hf drive-units. Hard clipping can also stress an amplifier particularly if it is already adversely loaded, and particularly if the output transistors are BJTs or IGBTs. This is covered in the next section, while clip indication and avoidance are considered briefly in chapters 3 and 9.

Clip indication was not widely introduced, even on professional power amplifiers, until the latter half of the 70s. It remains absent from most high end domestic power amplifiers, on the basis that either (i) customers never ask for it and (ii) every scheme that the particular maker has tested, detracts from the sonics; and (iii) anyone with sensitivity towards music would notice the ugly sound and quickly reduce the level. In professional use this is no less true, but there are a lot more variables. A reliable indicator then proves the point.

The first clip indicators were simple. A zener diode, resistor, LED and a few other parts were connected across the output. When the voltage exceeded the threshold of X volts set by the zener, the LED turned on. This kind of indicator not only introduces distortion (with the non-linear if tiny current-draw working against the finite output impedance of the amplifier) but the LED came on softly, with no definite threshold. And what threshold there was varied with each new batch of parts.

Next came meter ICs. The pseudo 'peak' or 'clip' indicator LEDs that are associated with them are still supplied with many low budget audio amplifiers. Quite often, the 'peak' or notional 'clip' LED is the topmost red LED in a multicolour LED bar, or the top segment in an LCD, both at the top of a peak reading bargraph that does for a 'power' level meter. Indicators of this type, in almost all instances, are arranged to turn on when the output voltage exceeds the maximum power output, expressed as a nominal signal voltage across 8 or 4 ohms, e.g. 40v peak for 100w$\Rightarrow$ 8Ω. However, the clip level is defined by the rail voltage, which can droop by varying amounts, from −5% down to −35% when loaded by speaker impedance dips, and also allowing for variations in mains supply sag. Overall, the amplifier's overload threshold is lowered, leaving the 'clip' LEDs to indicate +1 up to +5dB past the +0.5dBvr of clip that will cause damage.

Crown's patented IOC (Input-Output Comparator) scheme first appeared in 1977 on their DC300A-II. In this and the subsequent related schemes used by other makers [20], the input and output are compared with the transfer function subtracted. If input differs substantially from the output (less the normal gain and phase differences), the amplifier must be clipping, shorted, slew limiting − or otherwise sick, due to RF oscillation or a 'DC fault'. If so a LED marked 'IOC' or 'Clip' or 'Error' lights. In the Rauch *DVT* amplifiers, Figure 9.19, the True Error Detector (TED) signal was applied to a push-pull comparator, which drove a pair of back-to-back LEDs, so the error polarity (mindful of music's asymmetry) was also visible, as the flashing was then more red, or more green.

In sophisticated designs, the clip LED attack and decay is adjusted so that the shortest overload periods that produce audible distortion on the most sensitive programme are stretched using a monostable to a consistent period that's long enough to show as a consistently bright flash to the eyes. Shorter lived clipping then passes as a faster, far dimmer flash, and can be appropriately disregarded.

Today, high accuracy IOC or TED-type circuits comprising just two low cost ICs can detect the onset of clipping within +/−0.1dB *regardless of the loading, or low mains conditions*, and the other errors. Moreover, by using balanced techniques, and suitable sensing impedances, distortion caused by the sensing may be avoided.

The most sophisticated form of simultaneous level-cum-headroom metering, pioneered in BSS Audio's EPC–780 (cover picture, upper), is headroom referred. In this design, the top LED of a bargraph registers true clip. All the lower LEDs are likewise *referred down from* the instantaneous clip point, defined by the supply rail voltage, and even speaker loading. It is the *conjugate* of the normal consumer-grade 'power' or 'dB' bargraph meter, and scaled in '*dB below realclip*', is elegant and suits the large-scale touring PA systems it is part designed to be used in.

5.6.1 Overdrive behaviour – internal relations

Users do not expect their power amplifier to explode, burn out or otherwise need repair, when 'hit' by hf signals x dB into clip, or when the amplifier is checked with square-wave test signals. Yet this was a common problem with the early Lin-based power amplifiers, and remains so with carelessly designed modern amplifiers continuing to use BJT output devices and high NFB, when they are subjected to accidental large program bursts, hf spikes, or the dynamics of live music. Looking back at circuitry in section 4.1.8 for example, when driven beyond clip, the upper Darlington arrangement, assumed to be the conducting side, would be turned on harder and harder by the feedback, as it tried vainly to make the output voltage swing higher. When the output next swings negative, and thus while the lower transistor pair are beginning to pass current, the upper Darlington is still flushing-out excess charge. *Cross-conduction* then results, and if the current is high enough, the upper transistor will quickly enter second breakdown and burn out catastrophically, with the lower-side devices soon following. If BJT output transistors must be used, a good design will employ anti-saturation techniques to prevent charge accumulation in the junction capacitances. The BJTs with much higher F_T ratings and generally lower junction capacitances that have become available in the past 15 years, are a help. Early silicon BJTs with current ratings of around 10 to 20 Amperes managed 0.2 (200kHz!) to 1MHz; this crept up to 3, then to 10MHz in the early 1980s; and now, past the mid '90s, F_T can go as high as 50MHz.

5.6.2 Output stability and the output network (OPN)

Even without abuse, misuse and accident, the load seen by audio power amplifiers is most indeterminate. Amplifiers integral to an actively crossed-over enclosure are the notable exception. Else, even with just one given speaker load characteristic out of thousands, the load may vary from this, to that of the cable alone. And the open-ended cable may be of any length and construction. Whatever is connected, a high performance amplifier is expected to remain rock stable.

Delivering high frequencies is a problem to amplifiers with high NFB, especially when it is global. The amount of feedback (loop-gain) is reducing in most amplifiers well before 20kHz, and often, in high NFB amplifiers, from 10Hz. The output impedance begins to rise at typically +6dB/octave at frequencies above this point. Thus the reducing NFB makes the amplifier's output (i) look inductive, and also (ii) develops an output impedance that increasingly interacts with any load capacitance (to ground or return side) to cause a lagging phase shift. If the extra phase shift is very much, and occurs at a low enough (usually radio or ultrasonic) frequency, RF oscillation will result. Sometimes this happens with just the capacitance of a scope probe when the amplifier design is being conceived, and the solution is oddly enough, to connect a larger capacitor across the output, or else a large capacitor in series with a low value resistor, i.e. a *Zobel network* (Figure 5.32 i), named after Dr.Zobel of Bell Labs. This makes more sense when all the components involved are viewed as their equivalent circuits at high audio and ultrasonic frequencies.

Figure 5.32

Output Networks.

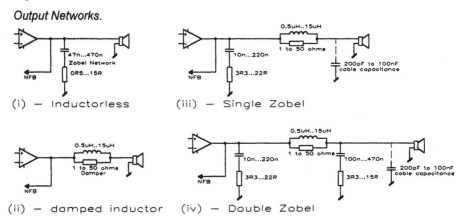

(i) — Inductorless   (iii) — Single Zobel

(ii) — damped inductor   (iv) — Double Zobel

For practical use, with speakers (with or without passive crossovers) and/or speaker cables connected to the output, more output 'compensation' may be needed. Nearly all speaker cable looks capacitative but in a complex way. A series inductor (Figure 5.32, ii) is intuitively appropriate as a way of progressively standing the amplifier off from the load capacitance above 20kHz, but also ironic as the solution to another series inductance, the virtual one made by global NFB. Many amplifier designers have found to their surprise that such an inductor must be air cored; unless the amplifier's maximum peak current delivery is very low and the core very large, no ferrite will be fully satisfactory. It is no excuse to say that the same core is used in a passive crossover that the amplifier might be driving, since within such a crossover, the peak current conditions in the coil could be quite different. The inductor may also be damped by a shunt resistor. The value used in practice may have to be different to the computed optimum for at least two reasons, resulting in poor or 'less good' damping above the audio range [21] which may have to be accepted as one of those compromises that measure bad but sounds fine.

Some power amplifiers, often those with reduced or nil global NFB manage with a rudimentary OPN, often comprising just the Zobel. These are so called 'output inductorless types'. They first appeared in the early 1980s after some of the UK's high-end, low budget domestic power amplifier makers identified a consistent and valuable improvement in sonic quality when the output inductor was excised. As the inductor was almost certainly not of the special calibre that is required, this is hardly surprising. The motto '*no component unless essential*' is admirable, but inductorless power amplifiers are renowned for becoming RF unstable and even blowing-up, when low inductance speaker cables are connected. This problem is definitely in the amplifier supplier's court, unless the maker states specific cables that must or must not be used.

The minimum network for RF stability over a realistic range of load conditions happens to require an inductor (Figure 5.32, iii). After years of mis-explanation by others, Ivor Brown appears to be the first [22] to show systematically how this basic RC+RL network helps the load impedance appear more resistive (or at least, less wildly reactive) to the amplifier at HF, when the speaker cable and speaker are modelled.

A second Zobel (Figure 5.32,iv) after the choke helps to curtail the phase and impedance variations further. This secondary Zobel often proves essential via the purely empirical route too. Usually, the resistor is lower and the capacitor larger than in the first Zobel. To be sure that there are no unexpected snags, OPNs require testing over a wide range of conditions. Sweepable computer simulation greatly aids exploring and optimising the stability and the network power dissipation rating needs under the very varied range of load conditions.

To do any good, the OPN *must* be placed outside of the feedback network. Therefore the series resistance (really ESR) of the inductor must be lower than the amplifier's output impedance, assuming the latter is considered of value. This requires a large component that seems out of step with a modern 'high tech' amplifier, particularly compared to the size of say a 1kW switching PSU transformer. The associated resistors and capacitors will almost certainly be bulky if safely rated. However, if Philip Newell's recommendation were taken up, wherein any passive crossovers are located next to (or inside) their amplifier, the large size of parts required in a properly rated OPN would be seen in context.

On the basis of being outside the loop, Neville Thiele [21] makes the point that distortion and particularly squarewave and the impulse measurements should be made prior to this network. But this is never considered in conventional tests, where an amplifier's output terminals are all there is – see section 7.9.

5.6.3 RF protection

Monitoring for RF oscillation at amplifier outputs, with indication and shutdown if it is encountered, is rare, but ideal on at least three counts.

i)   The amplifier, the OPN and the speaker are protected from potential RF burnout.

ii)   Attention is drawn to a problem that is otherwise invisible to most users, and one for which they do not otherwise have the tools to prove it exists.

ii) Assuming the indication leads to shutdown, people and animals are protected from RF radiation from speaker cables.

Output networks may be exploited in a number of ways as detectors and thus active presenters of RF oscillation. This is ironic, as resistors in OPNs are the most common part to burn-out in transistor amplifiers – if the output transistors survive. The author has produced sniffers that detect millivolts of RF up to 500MHz on top of large (100v) audio signals. In the most advanced designs this extends to auto load-compensation, where the amplifier stabilises itself dynamically, using as little compensation as is required.

5.7 DC offset, at output

DC offset is the name for one kind of unwanted signal at an amplifier's output. It is worth considering in depth, as the book makes the case for direct coupling, and DC offset is the Imp of this realm. Whether DC offset (voltage and the resulting current) is a small quirk, a slight nuisance or the sign of a major breakdown depends on the size of the DC voltage. In conventional amps with high NFB and internal direct ('DC') coupling, DC output offset voltage is assumed to be reduced to a low level by the very high feedback factor *at* DC ('0' Hz), often 110dB or more. This is true to an extent. But transistor mismatches, and other production variables, as well as many kinds of fault conditions that disrupt NFB, can cause offset voltages that are non-trivial. With inadequately protected BJT output stages with high global NFB, small DC offsets can cause transistor fatalities (see section 5.5).

When to dial 999

There are no standards for DC offset (error) voltage at the output of an audio amplifier.

A large offset, above say 1 volt (for >50w/8Ω), constitutes a DC fault. **This is a serious condition** – see the next section. *Never plug an unprotected speaker into an amp suspected of having a DC fault.* The telltale sign is a loud thump usually followed by a buzzing, and no music.

Minor effects

Offsets below 1V at the output, are considered medium to small and are almost harmless to the drive-unit *per se*. Dependent on program symmetry, a small offset may slightly reduce or increase headroom in one polarity. Amplifier clipping on any symmetrical large signals will be slightly asymmetric, changing the distortion spectra just before hard clip. If the amplifier's output is direct coupled, the speaker cone or diaphragm will also be slightly off-centered, with similar effects.

Irritating Effects

DC offsets small enough to be considered harmless may still cause disturbing clicks and moderate *plops* when speakers are connected-up. If the amp is wholly direct coupled or has a slow acting servo, such clicks may well originate from DC offsets in preceding sources.

Acceptable levels

The approximate sound level of DC 'clicks' and 'blahhts' can be ascertained from rough calculation. For example, if an amplifier drives sensitive large monitor speakers giving about 104dB(c-wtd) rms SPL for 1 volt rms at the input, then an offset of 0.01v (1/100th) applied by the preceding equipment will cause an SPL that is about 40dB lower, or $64dB_{SPL}$. This is as loud as room conversation, hence potentially disturbing, but not too shocking. But an offset of +0.63v giving $100dB_{SPL}$ would be more ugly.

Occurrence

A DC offset across the output terminals does not normally occur in amplifiers with output transformers – which are usually those with valve output stages. But it might occur if shorted transformer windings or bad connections are at large.

In amplifiers with a series output capacitor (usually single-rail transistor types, or OTL valve amps), offset can occur but is limited by electrolytic capacitor leakage and any bleed resistor.

Servo-fix?

Some amplifiers use servo circuits to continually 'auto-null' any output offset voltage. Servo circuits have a finite range and rate of adjustment. They are no substitute for DC protection. But if there is no overt DC fault, they can ensure that offset is kept below say +/– 5 mV over time and independent of temperature. Because the integration (ie. averaging) time of a servo's feedback loop needs to be longer than the longest repetition period of music signals, namely low bass, any such system will take time to settle after switching on.

Some practical sonic snags of servo circuits are detailed in section 3.4.1.

5.7.1 DC (Fault) protection (DCP, DCFP)

DC fault protection circuits have long been a feature of professional power amplifiers. The intention is to protect (i) the speaker(s) from damaging DC voltages and (ii) preferably disable the amplifier channel. For users of domestic amplifiers which have no DCP and which are connected to speakers costing 'a pretty penny', the fact that all transistors will one day fail and that most will fail short-circuit, putting 50v or more of DC on the speaker, should be sobering.

The reason why DC above a few volts is damaging to drive units is because some part of the voice coil is pushed out of the gap. A high swing power amplifier can push the *entire* voice coil out. Whatever is out of the gap looses the cooling effect of the usual proximity (typically within 0.2mm) to a substantial heatsink, the magnet, as well as cooling by air pumping. This part of the coil's power handling is therefore reduced, and if driven normally, the insulation and glue in this section cooks. With a large offset over say 20v, the coil is quickly burnt out, and either rubs or breaks, whereas with smaller offsets, say 1 to 10v (according to the power rating) the voice-coil's insulation will be slowly degraded, over weeks or months.

The requirement for protection is simple enough: If more than between +/– 1v to +/–5v DC appears (standards vary) on the output terminals, the output should be disconnected. This shouldn't happen on any legitimate bass signals, however low and loud. Yet it must happen with DC, within a few seconds.

In theory this can be accomplished simply enough if the amplifier already employs an output muting relay. A one or two pole low pass filter is used to detect DC, not signal. The snag is that the relay won't usually open under ideal zero-current conditions; there will be an arc as the contact opens. Thus the contacts are degraded as soon as the relay performs in anger. Any gold-flashing or other precious metal will then have been lost. For this reason, a relay in the output path has long been anathema in high-end domestic power amplifiers.

Without an output relay, the next simplest way to curtail DC at the output would seem to be to mute the input. But this presumes it comes from the source, and assumes a healthy amplifier. It won't help if one or more OPS transistors have blown 'short', or are very leaky, two likely causes of a DC fault.

The next simplest way is to use a 'crowbar' (thyristor clamp) to short the output. This will save the speaker (hence speaker makers fit them to their wares !) but it can increase the amount of destruction in a partly zapped or sick amplifier – if it ever operates.

The third approach is to switch off the supply. It is easiest and safest to do this at a point where just one supply line need be switched. This is usually the AC mains supply. The DC protection in this case must then be either passive, or else powered separately. The apparent snag with this approach is that the DC supply in most power amplifiers contains enough energy for it to take a few seconds for the supply to discharge. Fortunately, the average bass driver (the driver most likely to be subjected to DC) has no trouble with a power dissipation and voltage that is tapering off rapidly. If the supply capacity is very large, a high surge-capacity dumper resistor may be switched in.

One important practical feature is often absent from those amplifiers that do have DC protection: A test facility. In some of my own designs, users can test that the DC protection circuit works at close to the lowest detection voltage, by moving a header. In future computer controlled systems, DC protection testing should be included in power-up or daily test routines.

5.8 The output interface

In 1978-80 Otala and colleagues [26] reminded the world that "... under certain conditions, the loudspeaker reaction to the drive signal can propagate in the feedback loop of (the) power amplifier and intermodulate with the drive signal itself." They named this 'Interface Intermodulation Distortion' (I.I.D.). Testing was carried out using a setup similar to the 'two-amplifier' output impedance test in section 7.4.2.

The obvious solution is to have a low output impedance, so the reactive energy is attenuated and absorbed. But if this is attained by global NFB alone, there is a risk that if the feedback ever saturates (due to an interference impulse say), there will be a high level of 'IID' – unless the open-loop (feedback-free) output impedance is also low. Even then the amplifier becomes a less pure voltage source.

This was seen as a problem for high NFB. Yet it follows that an amplifier without global feedback will exhibit increased IID, unless its output stage's output impedance has been lowered by the 10 to 1000 fold that NFB would have attained. This means using a *lot* of output devices, unless the transconductance is high. Thus the Sziklai OPS suits amplifiers with no global NFB, whereas a very large number of L-MOSFETs would be needed. The exception is when the amplifier is partnered with a drive unit that has low reaction, alias benign loading, hence employs conjugate networks (see chapter 2) and/or pole pieces to cancel Back-EMF.

5.8.1 Muting systems

Output muting, comprising a series relay, is often held to be desirable in professional amplifiers, particularly in large installations, where dozens or even hundreds of speakers cannot be connected manually after every switch-on. Then if there is an unpleasant sound at switch on/off, for whatever reason, there will not be a problem if the muting is fast enough at power-down, and the un-mute delayed enough at switch-on. The relay used for output muting may find multiple use, for DC protection, for redundancy between bridged channels, and in computer controlled systems, to enable the speaker impedance to be separately tested.

Fault conditions aside, it is possible to switch an output relay without zapping the contacts. For example, manually muting or not connecting program before powering up and down, will keep current off the contacts. But this is purely voluntary, and without appropriate sequencing of signal muting at power up/down, relay contacts can become badly pitted, or they may even weld together, while switching on a particularly high speaker current peak. It follows that without cast-iron protection against non-zero-current operation, output relays in high quality amplifiers should be pluggable items, easy to replace or remove for cleaning and inspection. Even without current, if an amplifier is left off for some days or months or even a few minutes in some atmospheres, the relay contacts may be degraded by atmospheric pollution. On this basis, some computer-controlled amplifier designs may regularly exercise and clean the relays, by repeatedly powering up and down.

Solid-state switching with MOSFETs would overcome problems of contact degradation, but economically and sonically satisfactory designs have yet to be demonstrated.

Domestic amplifiers eschew series output muting. Shunt muting at the output is sonically benign but tempting fate. With some topologies, the turn-on pulse is intrinsically small. Experiencing only a slight 'plut' at switch on is also more likely when amplifiers operate from dual supplies; or are bridged; or from supplies which ramp-up slowly at switch on. Linsley Hood once advocated matching the reservoir capacitor values in dual-rail supplies. The turn-on behaviour of audio has probably never been studied; it belongs in the realms of chaos and anti-catastrophe mathematics. Weird sounds after switch-off can only be empirically engineered away.

Whether or not there is an output relay, incoming signal should still be muted in one of the normal ways (which may be sonically transparent, i.e. in shunt, and need not be a 'hard' mute), so substantial program is not directed at the output stage until it has settled, and so program is muted before the supply rails have fallen very far, at switch off.

5.9 Output stage, cooling requirements

The maximum *junction* temperature of silicon power transistor is about 200°c, while for a plastic small signal transistor, a 150°c maximum is more typical. *But these limits should be infrequently approached, since the rate of wearout is much higher than at a lower temperature.* For as in chemical reactions, after Arrhenius, transistor wear and stress halves and reliability doubles for each 10°c reduction in junction temperature. It is measure of the very high intrinsic reliability of MOSFETs, that they remain more reliable than BJTs while sustaining much higher junction temperatures, up to 250°c. This is possible because lateral MOSFETs in particular do not suffer so readily from leakage current-induced thermal runaway.

The capacity of the heat conductance (or rather the lack of it) is commonly expressed in *°c per watt*. If the power to be dissipated is at all high, this *thermal resistance* in the path must be very low. If not, any attempt to dissipate the power will cause stressful or destructive junction temperatures. Also, the highest thermal resistance in the path limits the significance of lowering the others past a point. If a transistor has a thermal conductance from its junction to the heatsink of say 5°c per watt (so 50 watts implied a rise of 250°c), then *no amount of highly conductive heatsinking will greatly improve the poor conductance within, no matter how visually impressive.*

Without knowing the thermal conductance of the transistor innards, the interface and the heatsink, the examiner or prospective purchaser cannot determine the significance of a particular heatsink temperature. In general however, with 'typical' thermal resistances, if a heatsink rises much over 75°c, the reliability of associated BJT output transistors is likely to be noticeably foreshortened. With MOSFETs, higher temperatures can be withstood commensurate with the higher maximum junction temperature. But they are not recommended unless the heat is directly ducted away, as few other, surrounding electronic components are rated for temperatures above 70°c, and most must be operated below this temperature to allow for their own dissipation and reliability considerations.

This is a good juncture to mention that what is inside a standard transistor case can vary widely. One TO3 metal can may contain parts that can handle 250 watts; while the *die* inside others handles only 35w, say.

After the power device, the device mounting interface is next most important. What is being used here is harder to disguise.

Insulation is required for the Lin topology, and most other types where the mounting bases are either the supply rails or the output, depending on the OPS sub-topology and device. For example, the mounting base of audio power BJTs is *always* collector; for L-MOSFETs it is *always* source; and for D-MOS it is *always* drain.

The traditional insulation has been mica + grease. Some kind of spreadable, heat conducting paste is essential for all metal-metal interfaces, or to anything else which isn't pliable, because surfaces that appear smooth to human eyes are like mountain ranges in reality, and without grease, only the small 'peaks' make direct contact. Silicone grease or oil should preferably not be used (despite its recommendation by Toshiba) as it creeps, attacks electronic components and degrades contacts. And complete removal may prove impossible. A zinc-oxide-filled silicone-grease-free paste is best.

Makers using CE type topologies (see chapter 4) can readily ground their heatsinks – to 0v potential at least. Other makers have arranged blown heatsinks that are suitably insulated so they can operate 'live', either at rail or output potential. Then only grease or a dry-mounting, non-insulative substance is required, e.g. graphite foil. This arrangement can save typically 10°c, i.e. with otherwise identical conditions, the junction of a bare-mounted device can run 10°c cooler. This is significant in terms of doubling potential reliability; output transistor device failures that might have occurred after three years will be set back until six years have elapsed – at least in theory. Without adequate protection it may be all very hypothetical.

Where the topology does not allow bare mounting, the maker can regain the advantage at a cost by installing more devices, to spread the dissipation. Flexible, silicone rubber-based insulators ('silpads') have developed since the early '80s, to the point where the most advanced, loaded types, can offer thermal resistances, hence temperature gradients, that are lower than greased bare-mounts. These high performance materials (e.g. Keratherm) are generally smooth and light coloured. Today, the only advantages of bare mounted devices are the absence of a problem in the event of a punctured insulator, and the assembly time saving.

Aluminium oxide happens to be one of the best electrical insulators and yet best conductors of heat, a rare combination. If the mounting plane is aluminum, and smooth and flat enough (this involves machining costs), and the anodising is also smooth and flat and consistently thick enough, then a low thermal resistance with high voltage insulation is possible with nothing further to add except thermal grease.

Correct and consistent torquing of paralleled output devices is a commonly hidden keynote of quality manufacture that is likely to prove reliable in the long run. The optimum torque should be reset after burn-in. Early and medium term failures in redundant designs with lots of paralleled parts may be caused by one device loosening off. Domestic amplifiers can 'loosen off' if you move house enough. And any amplifier will if it gets regularly vibrated by low bass. Competently engineered power amplifiers employ suitably-rated compression washers to maintain the torque, while the fastenings are solidly locked. Where power devices are clamped by bars, the clamping screws should be lock-nutted and also adequately fine threaded; the very coarse screws used by one US maker prevents proper torquing. Another method long-ago developed but slow to catch on, is to use springy steel clips to 'mount' power devices. This offers the benefit of a steady pressure (in place of torque) that won't loosen with vibration – assuming stable fitment.

Figure 5.33

Power device attachment with clips is increasingly employed by large scale makers. © QSC.

5.9.1 Heat exchange

In the first transistor amplifiers, *heatsinks* were the chassis or side plate, or else springy copper clips. Sometimes, the radiating surface was optimistically painted black. Optimistic, because a matt black finish only *aids* radiation from a heatsink when the surface is above 70°c, while of course, much above that, a germanium transistor amplifier would probably have expired.

By the early '60s extruded aluminium heatsinks with various fins had begun to appear. The T03 metal can case had also developed, and Figure 5.32 shows the heat path, with the bottleneck made apparent by exploding the tiny chip hidden under the lid.

Figure 5.34

The heat path begins in a bottleneck. The size of the heatsink may have only a limited effect on the junction temperature.
© Crown Inc.

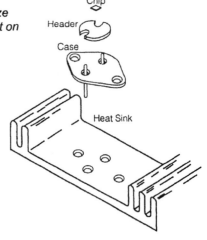

Power delivery increased but eventually, by the early to mid '70s, the heat produced by even class B designs exceeded the natural dissipation capabilities of the size of fins you could fit onto the rear of a 19" rack mount case, let alone a smaller sized domestic amplifier. In those days, 6" (150mm) equipment cooling fans were called 'miniature' and were also relatively expensive. But from this point onwards, professional power amplifiers began to employ fans to stay cool. Amongst the first tunnel cooled amplifiers were the historic Turbosound *Fanamp*s, built by Tim Isaac from the cost-no-object power supply of a old mainframe computer. Each *Fanamp* utilised a short section of tunnel. The later Hafler *P500* was similarly cooled, while other makers postured with longer tunnels. The significance of the thermal gradient on the transistor's reliability and work-load sharing took a long time to dawn.

Meanwhile, in 1983, Rauch Precision launched the *DVT 250s*. It employed a blown lateral 'pinfin' (Figure 5.35) with stubby cast fingers. With this, 2kW of MOSFET dissipation could be kept to below 50°c above ambient in a depth of just over two inches (≅60mm).

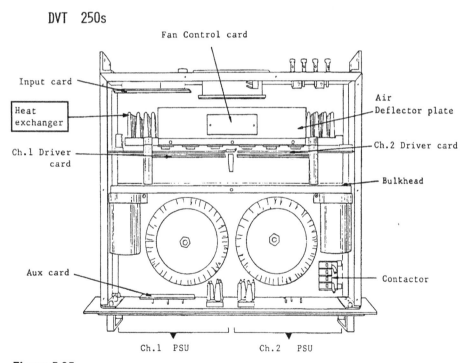

Figure 5.35

The pinfin heat exchanger, originally designed and widely sold by Pantechnic, who introduced L-MOSFETs to thousands of sound-system DIYers, was a low cost, very high performance heat exchanger. Here, in the Rauch DVT 250s, air was blown onto the pins by the fan, top, centre. The airflow is after standard C (see section 9.6.1), but it could be converted to standard A by a change of front panel.
© Owner MS&L.

In Crown's *Micro-* and *Macrotech* series of industrial/audio amplifiers that followed in 1985, the removal of waste heat from the output transistors was further facilitated by the metal cased output BJTs being able to be bare-mounted to a copper bar. This has higher heat conductivity and also higher specific heat, than aluminium. Its capacity may therefore be a double-edged sword, since it will be harder for it to recover from thermal saturation if the cooling is at all inadequate. In the Macrotech amplifiers, the bar was then welded to a short lateral heatsink formed of convoluted copper wafer, to open the thermal path out rapidly into a large surface area. In recent years, ARX Systems have spearheaded the use of an all-aluminium welded honeycomb to make an even lighter touring amplifier. ARX's series is an example, like the Chameleon (see section 4.6.3), of totally conventional amplifier technology 'cocking a snoot' at amplifiers employing classes G, D and switching power supplies, in pursuit of light weight. It must be left to listeners to decide whether competitions over levity may often defeat their purpose by becoming unsatisfactory in bass weight. Other makers, e.g. again Rauch Precision in the P-series, and also BSS Audio and MC^2 Audio in the UK, have all achieved unusually high power dissipation handling with surprisingly small, totally different and original heatsinks, all of differing, intuitive design. HH's ill-fated S-500D with its 'force-cooled dissipator' module (1977) may just have been the inspiration.

Figure 5.36

In 1977, HH's S-500D was the precursor of modern, compact high power amplifiers. It used car-radiator fins for compact heat-exchange. But the design failed for other reasons, including BJT cross-conduction.
© MAJ Electronics.

Today, some aspects of heat exchange can be readily modeled on an analogue circuit simulator, with R-C networks, with T_{amb} as a current source, etc. Each obstacle in the way between the waste air and the transistor junction is a resistor.

5.9.2 Thermal protection

'Overheat' or 'thermal' protection and warning systems are essential in most high performance amplifiers. Class A should need no protection against overdrive. The need for protection is obvious enough. Parties get held. Fans *do* jam or wear out, slowing imperceptibly. Carelessly placed clothing or curtains can cover vents, chimneys or louvres. All but a large, high ceilinged or draughty room will eventually get very hot when a few kW of amplifiers are working hard. In a smaller enclosed space, like a rack cupboard, unless forcefully ventilated, the air temperature will eventually reduce the ability of the amplifier to deliver power, as the thermal headroom is used up, and the SOA shrinks. A well protected amplifier should then either mute itself, or better, turn its signal drive down until the temperature starts to fall back.

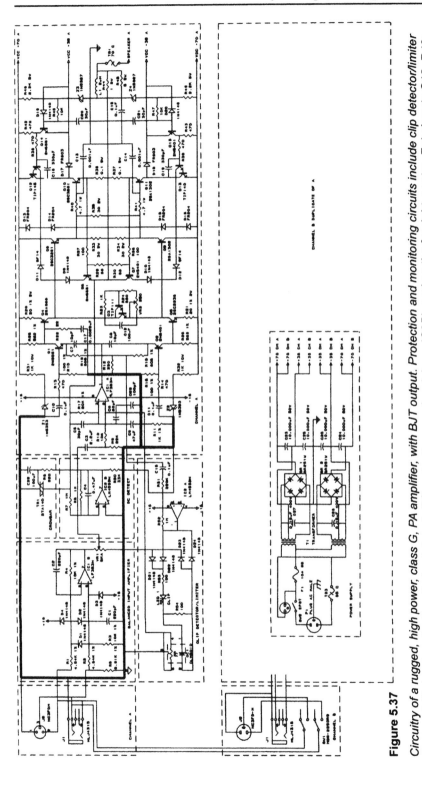

Figure 5.37

Circuitry of a rugged, high power, class G, PA amplifier, with BJT output. Protection and monitoring circuits include clip detector/limiter (bottom left); input protection diodes (above); DC protection crowbar (above IC2B); and on the far right, centre, Zobel parts C13, R40; damped choke L1, R44 and thermal protection switch. Note also symmetrical circuitry based around an IC op-amp driver - see section 4.4.3.
© Crest Inc.

Thermal protection is also needed in all but class A and D amplifiers, because the thermal capacity required for some kinds of music programme and speaker loads can greatly increase dissipation, beyond expectations. Since given heat exchanger design costs according to its capacity, capacity is rarely in excess of needs. Where is the line to be drawn? Makers are bound to use as little as they believe is safe, usually for the often limited range of circumstances they can envisage.

In best equipped professional designs, thermal protection is not taken lightly, and there is a failsafe system if the temperature passes a level above the normal protection threshold.

5.10 Logical systems

The protection systems of an audio power amplifier have to combine in a logical way, with the designer thinking through all the combinations of events. Apparently, this step is often missed. For example, the detection of overheating should not switch off the power if the cooling fan then stops running. And the coincidence of having equal but opposing polarity DC faults on two channels should not result in the DC alert being cancelled in a shared sensing circuit.

The 'CPU' (Central Processing Unit) which combines, exchanges and prioritises the protection, sensing and other monitoring signals may range from crude but effective discrete BJT logic, to a cluster of CMOS, through to a microprocessor.

The design of the the protection circuitry and CPU should be intrinsically safe, i.e. not able to pass a 'safe' signal if a supply rail is absent, say.

5.11 Output transformers

Transformers are necessary in nearly all amplifiers that employ valve output stages. The exceptions are the few OTL types that are made, e.g. Harvey Rosenberg's (see Ref. [2], Chapter 1). In all other cases, the transformer is used to match the high source impedance (typ. 2kΩ to 50kΩ) of the valve circuitry's output node(s), to a conventional, nominally say 15Ω to 5Ω speaker or drive unit. In many cases, the transformer is also used to combine the conjugate halves of the output stage – see section 4.2.2. It may also be used to set operating conditions, e.g. by using taps creatively, as in the 'Ultra-linear' connection.

Transformers are also used in large, fixed installations, to step-up the voltage swing of *any* given amplifier (usually solid-state though) to 70v or 100v rms. Speakers then tap off different power levels along the line. For lowest cost, LF response is commonly compromised. While the response may be adequate for the intelligibility of the human voice, such 'hundred volt line' transformers are beyond this book's remit. If transformers must be used to raise the voltage swings of direct-coupled solid-state amps, they should be protected against being driven at frequencies below the operable and thence into magnetic saturation – which would adversely load the OPS – by suitable high pass filtering, ie. LF cut. 'Below the operable' includes anything above $1/4$ volt DC. If DC offset voltage is not guaranteed to remain below say +/–100mV ($1/10$th volt), a DC blocking capacitor will need to be placed in series with the primary winding.

References

1 Getreu, Ian.A, *Modelling the BJT*, Elsevier, 1978.

2 Bailey, Arthur R, *Output transistor protection in AF amplifiers*, Wireless World, June 1968.

3 Jarl, R.B. and R. Kumbatovic, *A test set for non-destructive safe-area measurements under high voltage, high current conditions, AN 6145*, RCA, 1979 (or 1st ed. 1977).

4 Linsley-Hood, *Expert witness*, EW + WW, Aug 1995.

5 (Nameless), *Application note X3504*, March 1991, Toshiba (UK).

6 Pass, Nelson, *The Pass Zen amplifier*, The Audio Amateur, Q2, 1994.

7 Jung, Walt.G, M.Stephens and C.Todd, *Slewing induced distortion and its effects on audio amplifier performance*, 57th AES, May 1977.

8 Jung, Walt.G, *Slewing induced distortion, parts 1-4*, The Audio Amateur (USA), 1977.

9 Jung, Walt.G, *An overview of SID and TIM*, Audio, Jun-Aug 1979.

10 Self, Douglas, *High speed audio power*, EW+WW, Sept 1994.

11 Garde, Peter, *Amplifier first-stage crieria for avoiding slew-rate limiting*, J.AES, May 1986.

12 Jung, Walt, *Audio IC Op-Amp Applications*, p.43 *et seq.*, Howard Sams, 3rd edition 1986.

13 McCombe, John, and F.R.Haigh, *Overhead Line Practice*, MacDonald, London, 3rd edition 1966.

14 Duncan, Ben, *Simulated attack on slew rates*, EW+WW, April 1995.

15 Self, D.R.G, *Distortion in power amplifiers, part 4* , distortion 6, Electronics World, Jan 1994.

16 Baxandall, P.J, Chapter 2, p.113 *et seq.*, *Audio Engineer's Reference Book*, Ed. Michael Talbot-Smith, Focal Press.

17 Baxandall, P.J, *Testing the amplifier/loudspeaker interface*, Speaker Builder (USA), Q5, 1988.

18 (Nameless), *Understanding power amplifier specifications*, AN-123, Burr-Brown, 1983

19 deCelles, John, *Audio amplifiers utilizing: SPiKe protection*, AN 898, NSC, Oct 1993.

20 Duncan, Ben, *Amplifier clip detector*, EW and WW, Feb 1993.

21 Thiele, A.Neville, *Transient response in feedback audio amplifiers*, AES preprint 4029-B5, April 1995.

22 Brown, Ivor, *Between amplifier and speaker*, EW+WW, Feb 1995.

23 Severns, Rudy, *MOSPOWER Applications Handbook*, Siliconix, 1984.

24 Giles, Martin (ed), *Audio/radio handbook*, p.4.54 to 58, National Semiconductor, 1980.

25 Hardcastle, Ian and Basil Lane, *High power amplifier*, Wireless World, Oct 1970.

26 Otala, Matti and Jorma Lammasniemi, *Intermodulation at the amplifier-loudspeaker interface*, Wireless World, parts 1 and 2, Nov and Dec 1980.

Further reading

27 Becker, R.B.H, *High power audio amplifier design*, Wireless World, Feb 1972.

28 Marshall, T, *Damping factor*, Wireless World, Oct 1974.

29 Metzger, Daniel L, *Electronic Components, Instruments and Troubleshooting*, Prentice-Hall, 1981. ISBN 0-13-250266-6.

30 Perkins, Cal, *Power amplifiers: design*, S&VC, Mar 1985.

31 Piper, J.R.l, *Output transistor protection in class B Amplifiers*, Wireless World, Feb 1972.

32 Roddam, Thomas, *Transistor Amplifiers for Audio Frequencies - Principles of Design*, Iliffe, London, 1964.

33 Stochino, Giovanni, *Non-slewing audio power*, EW+WW, Mar 1996.

34 Stuart, J.R, *Audio amplifier load specification*, Wireless World, Dec 1975.

35 Tharma, Poothathamby, *Transistor Audio Amplifiers*, Iliffe/Butterworth, 1971.

36 Walker, Peter J, *Audio amplifier load specification*, Wireless World, Dec 1975.

The power supply

"Rhythm ... is fundamental to the spirit"

(lyrics from tribal house, 'U-dig')

6.1 Mains frequency (50/60Hz) supplies

All power amplifiers require a suitable power source. As *all* earthly amplifiers are less than 100% efficient, the amount of power (in kW) required to be 'on tap' is more than will be delivered to the speakers.

The historic perspective

The first, primitive audio amplifiers were powered from batteries (lead-acid accumulators). Around 1928, indirectly heated valves, the heaters of which could be operated directly from the by then quite widespread public AC mains supply, were pioneered. At this time, the conventional AC to DC (HT) supply topology was established, comprising:

(i) An isolating transformer. This changes the voltage with simplicity and high efficiency, to the required level, usually higher for valves and lower for most transistors. The transformer is also essential for safety. The incoming AC is galvanically isolated (unlike the connections of an auto-transformer or Variac), so *whichever way around* the live and neutral connections are wired, the amplifier electronics will *not* become live to the mains AC potential.

(ii) A rectifier. After the voltage transformation, the AC can be converted to unidirectional current flow, the first step to getting a clean DC supply. Although before the silicon diode revolution (*ca.* 1960), rectifiers were expensive, rectification of both negative and positive cycles, called *Full Wave Rectification* (FWR), is essential to avoid saturating the core of the transformer with DC. By using a centre tapped transformer, just two rectifiers (rather than four) are able to provide FWR in 'bi-phase mode'. To this day, rectifier valves are 'double' for this reason.

(iii) Smoothing. Valve power amplifiers draw relatively little current; valve rectifiers were limited in the surge current they could handle; and HT reservoir capacitors much above 1μF have been historically expensive. In turn, the reservoir capacitance (C) in conventional valve amplifiers with valve rectifiers is small, from 2μF in the 1930s, up to 50μF in more modern cousins. Alas, most classic valve amplifier circuitry is far more sensitive to hum on the supply than modern solid-state designs. This is called 'poor power supply (noise) rejection' (poor PSR). To overcome this a series choke (L) was followed by a second, larger smoothing capacitor (C). Sometimes the first reservoir capacitor was preceeded by another choke (L), i.e. 'L-C-L-C' smoothing. As each choke limits the current surge drawn, a capacitance so fed could be made larger than without a choke. Working in synergy with the choke, the capacitance could also greatly aid the removal or 'smoothing out' of the 100/120Hz *ripple – including the spray of harmonics across the audio band,* that is left after rectification.

Solid-state developments

In the transistor electronics that followed after 1960, DC rail voltages were lower and currents proportionately higher – often twenty times – for a given power capability. This made chokes both far more bulky, and costly. Also, their manufacture was relatively labour intensive, and more prone to inflation. Simultaneously, far higher value electrolytics (e.g. 10,000μF) were needed to store the same energy as in a typical valve amplifier's much smaller reservoir capacitor. The reason is that

$$\text{Energy stored} = \{ (C \times V^2) / 2 \} \text{ Joules}$$

For example, 100μF charged with 575v in a valve amplifier contains 16.5 Joules. Exactly the same energy is contained in 10,000μF (100 times more capacity) charged with 57.5v (a tenth of the voltage) in a transistor amplifier. The same was true of a great deal of electronics. And so large valued, low voltage electrolytics had to be developed, and in time, the price per μF tumbled.

Silicon rectifier diodes also arrived in the early '60s. Their performance and ratings developed quickly, and prices came down. Compared to any valve rectifier, they could handle far higher momentary surge currents, often twenty times their average, steady rating. This meant a choke wasn't needed to protect the rectifier from the repeatedly high surge current flowing into a large capacitance. The upshot was a move from the relatively gentle LC, 'C-L-C' or L-C, L-C smoothing of the valve era, to purely 'capacitative smoothing' (0-C), a brute force method of creating clean and reasonably stable DC supply rails. This method is the one used in the vast majority of solid state amplifiers, and much other electronics, although EMC provisions seem bound to change this situation in the coming years.

Figure 6.1

Dual (+/–) DC rails

50/60Hz dual-rail supply.

232

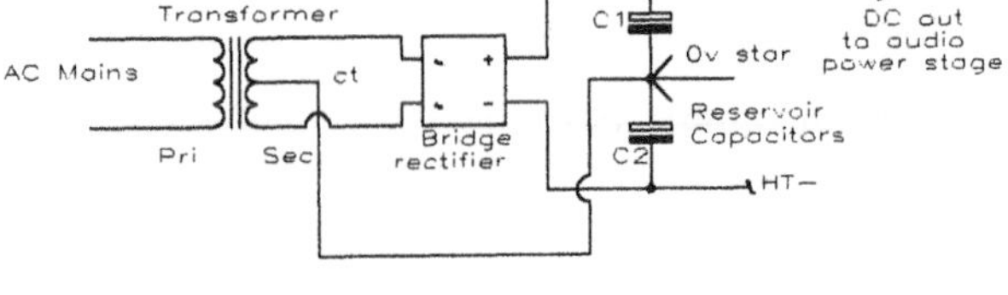

Modern elements

Figure 6.1 reiterates the key parts in the most common of modern 'passive' power supply topologies used by solid-state power amplifiers, alias the *dual* or 'split'-rail, *capacitively smoothed, unregulated* psu. This, as noted in section 4.3.5, helps grounding, as return supply currents can more readily be kept wholly separate from the signal ground.

Transformer

The transformer changes the mains voltage to the usually lower voltage needed for the DC supply rails. This part is usually the heaviest item in a high-performance power amplifier if the supply operates *at* (not just with) 50/60Hz.

Size and weight increase in rough proportion to the VA rating. *But size for a given VA rating can vary greatly with the technology and design finesse, and so the solidity of a power supply cannot always be assessed by gazing at the size of the transformer.*

There is no 'right' size of transformer for most audio amplifiers. The professional wants a transformer that won't ever burn-out or have to thermal-out, when worked hard. But too heavy a transformer is out for touring. A variety of VA rating values can be calculated. The mean power dissipation that determines heating will depend greatly on signal PMR (30 down to 9dB maybe), and the output stages' class. Only in class A can the worse case power demand be reasonably tightly forecast. In the other classes, makers find that the transformer can be downrated to about 33% of the amplifier's maximum continuous average power rating. In some budget domestic amplifiers, the transformer-reservoir capacitor combination has been tuned by listening, to resonate in the bass with a particular type or model of speaker box. This 'one note' approach is not really high performance.

Figure 6.2

This remarkable picture shows how just the power supply part of the QUAD current-dumping circuit - shown later - was re-engineered by Pro-Monitor Company, in London, for high audio quality with tri-amped monitors. On the lower left is the soft-start PCB with relays and surge-limiting resistor. The main toroidal transformer is obvious enough, in the centre. On the right are auxiliary supplies, and smaller toroidal transformer. Above, are the reservoir capacitors for the three, dual supplies. These exit to the amplifier enclosure, on heavy duty, military connectors. On the left is an RF interference filter on the incoming AC, and on the right, the auxiliary logic and control electronics, to ensure for example that the HT supplies are not fired-up until the DC power plugs are all engaged, to avoid arcing damage to the contacts. A second case contained the similarly seriously re-engineered audio stages.
© Pro Mon Co Ltd.

In high-end amplifiers, the transformer is often oversized by as much as tenfold, to improve sonic quality. This helps stiffen the supply, i.e. lower the source impedance. An oversized core is less sensitive to partly-saturating when there are small amounts or bursts of 'DC' (even harmonic voltages) on the AC supply. Otherwise, 'DC' on the AC supply can cause quite loud acoustic hum, let alone affect sonic quality.

Bridge

The universal bridge rectifier comprises four silicon diodes used to establish a uni-directional current flow. It must be rated to withstand a high surge or inrush current, not just every time the amplifier is switched on, but also every time the mains supply sags even slightly between cycles.

Although the four-pack rectifier device is called a 'bridge' and wired as such, in all amplifiers following the Figure 6.1 circuit (which is the majority, by far), it is actually operating as a dual bi-phase rectifier. This occurs because of the transformer centre tap (ct) that becomes 0v, the centre rail; or output in the CE topologies (see chapter 4). Each rail is fed alternatively by just two of the rectifiers, as in valve supplies, except that for solid state amplifiers, we have a mirror-image, negative rail.

In some amplifiers, individual bridge rectifiers *are* employed in their named role, for the positive and negative supplies, which are derived from isolated single secondary transformer windings, without centre taps. The centre tap is derived from the commoning point of the two DC outputs. This arrangement may be used to reduce the rectifier's utilisation, so it can handle higher current without excess temperature rise.

Reservoir capacitors

The reservoir capacitors (C1,2) in Figure 6.1 smooth out the DC voltage in a brute force fashion. The minimum reservoir capacitance needed for sonic accuracy at bass frequencies is relatively large, around 10,000µF for a relatively low power delivery of 100w$\Rightarrow$8Ω. Higher power rated professional and also class A and no-compromise amps, may employ as much as 250,000µF per rail in the quest for bass purity during dense, high energy passages. Large capacitors, above about 10,000µF with modern forms, offer diminishing returns. High performance power amplifier supplies may use multiple, smaller capacitors in parallel, to attain high capacitance while overcoming inductance and ESR (dynamic resistance) limitations, or else special types that perform closer to an ideal capacitor. Examples include amplifiers made by Arcam, BSS Audio, QSC, as well as esoteric types [5].

When lightly loaded, the ripple on the supply is typically a few mV. But during and after a bass transient, the 100/120Hz ripple voltage increases, to as much as 1v. Some of this leaks through and contaminates the signal, and the increased noise may be noticed as a loss of sonic detail.

Batteries

Batteries have not figured in the world of high performance audio power amplifiers for many years. But since 1989, they have seen increasing used by high-end DIYers to power their signal sources, after tiring of the poor quality of their local AC power. Also since the late 1980s, 'higher end' auto amplifiers have been developed, and while most cars are unsuited to any attempt at high quality sound reproduction, 12v and 24v 'auto-grade' amplifiers have seen increasing use, both due to the remote or mobile nature of a listener's residence; and by those harnessing alternative energy. Sonic quality from lead-acid batteries is reported to be of a uniformly high order. This may be due to the total absence of ripple harmonics, RF and other noise; or their low supply impedance; or both.

6.1.1 50/60Hz EMI considerations

The conventional but refined 50/60Hz power supplies inside high-end and professional audio power amplifiers are models of low EMI, at least on the basis of steady-state measurements at the output of the power amplifier that the supply feeds. In reality, the conventional power supply adds its share of pollution to its own power source. Moreover, the AC mains has been getting more polluted as time goes by. This has had two effects.

First, since the mid 70s, the sonic improvements wrought by reducing local pollution of the mains and attenuating the other EMI residuals, have been increasingly noted or accidentally stumbled across by domestic amplifier DIYers, and esoteric makers.

Second, European EMC legislation. Other than cleaning up some common emissions, it includes a clean-up of mains harmonics, that will affect the design of high power amplifiers in ways that are yet to be seen, but that will also, in time, clean up the worst of AC mains pollution.

Transformer

The transformers in high-power rated audio amplifiers can easily radiate particularly strong AC magnetic fields, at 50/60Hz and at the odd harmonic frequencies, e.g. 150Hz and 250Hz in UK. These near, strong fields are expensive and awkward to attenuate more than a few dB by shielding. In high quality power amplifiers magnetic field leakage is kept to a minimum by:

(i)   knowledgeable and precise design and winding.

(ii) by avoiding E+I laminated transformers, using *Toroidal* or *C-core* forms instead. For historical reasons, near ideal toroidal transformers have been highly developed for audio in the UK, whereas US and Japanese makers have focused on C-core transformers for premium applications.

(iii) by placing copper screens around and over core gaps.

Figure 6.3

Inside QSC's *Powerlight.*
Parts of the EMI common-
mode filter inductors are visible
on the left, in a shielded sub-
enclosure. Right centre is the
HF transformer. The copper foil
seals the magnetic gap. Note
also broad PCB tracks for
efficient, low inductance
current flow at 100kHz. The
class G power stage is behind
with power devices mounted by
spring-clips.
© QSC Inc.

In addition, in a well designed amplifier, the audio circuitry is rendered insensitive to magnetic fields by suitable wiring and layout. The keynotes:

(i) Avoid conductor *and other low resistance* (<500Ω) loops.

(ii) If loops are unavoidable, increase the resistance if possible (e.g. putting 100Ω resistor in a low current-handling 0v wire), and/or keep the area of the loop as small as possible. Decoupling capacitor loops *have* to have low impedance – so compactness is paramount.

(iii) Keep sensitive circuitry, nodes and low resistance loops *as far as possible* from the field.

(iv) Use an 'H' (magnetic) or inductive probe to find out how the field lies. Do not guess or theorise.

Rectifier(s)

Ordinary silicon rectifiers are quite sluggish. Like early power transistors, their inability to shut off the current flow limits their effectiveness at frequencies much above 5 to 20kHz. Ironically, rectification produces harmonics which contain just such frequencies, albeit in small amounts. Sluggish switching involves the production of small bursts of RF energy twice every cycle, i.e. with a prf of 100 or 120Hz [1]. This RF is commonly able to 'jump across' the transformer, to appear on the mains cable leading into the amplifier, from which conduction or slight radiation to other equipment is possible [2]. These long-existent RF emissions may be reduced, both to meet EMC regulations, and in high-end equipment, to enhance sonic quality, either:

(i) by using faster, cleaner rectifier diodes. These may be fast types with *soft recovery*; or even better, the Schottky type. Or,

(ii) by suppressing the RF. The bursts of oscillation may be damped by a small (e.g. 10nF) capacitor or R·C network. ESR must be very low as the noise source has a low source impedance, and the suppressors must be placed close to the bridge terminals. This is awkward with the 50/60 Hz bridges in most large amplifiers, which are 'industrial' types with push-on blade connections. Capacitors in such

236

positions with long (above 1/2" or 12mm) leads will not be very effective at high RF.

Capacitors

Large, electrolytic reservoir capacitors in choke-free power supply circuits contribute to the pre-existing harmonic distortion on the mains supply. The effect is worst when :

(i) the capacitance (μF) is high, and

(ii) the power supply series impedance is low relative to the mains (source) impedance. See sections 9.3.2 and 9.3.3.

(iii) the capacitors are hot.

These precise conditions are *also* often conducive to sonic quality, although not necessarily for identical reasons.

PFC

The harmonic generation, bad power factor and consequent poor power conversion efficiency of the conventional, passive 50/60Hz power supply cannot be fixed by the active PFC methods available to 'active', switching power supplies. Instead, 1930s power supply technology, the choke, possibly even a swinging choke, provides a synergistic answer [3]. Such a choke may be connected in line with the primary and embedded in the transformer winding, hence invisible as a separate part.

6.1.2 Surge handling

Even if there is no large signal (in a class B, A-B or related amplifier), the capacitor-smoothed power supply draws a high current 'blip' every time the mains supply voltage increases after a sag. In Europe, for manufacture today, this is now covered by EMC legislation. PFC does away with the effect.

Switch-on surge is worst when an amplifier's reservoir capacitors are fully discharged, but still warm from hard use, and when the amplifier has a high-current-capable, 'oversized' transformer ('power' rating regardless), especially when connected to a low resistance supply. When switched on, this combination can cause the ordinary incandescent house lights to dim, and circuit breakers to trip. In conventional, passive power supplies, there are commonly three approaches to curtail surges.

i) Switched soft-start. Using either semiconductor switches (Thyristors, Triac, MOSFET) or a relay, to short-out a series surge-limiting resistor, or ntc thermistor. If switched-out after power-up, a thermistor is able to fully cool, so it provides consistent surge limiting when the power is next cycled. While appearing solid and reliable, relays controlling soft-start circuits have been known to suffer welded contacts, or otherwise stick-on, or else fail to energise, due to supply and load variables. If the amplifier is driven with the soft-start element lodged in line, inadequately-rated and poorly sited surge limiting resistors commonly overheat and burn the PCB and/or surrounding parts. Their own solder joints may go dry and the parts loosen.

Solid-state switches are more reliable in this regard, but without any negative feedback, these simple, apparently elegant and even trivial circuits can prove unpredictable across populations, with catastrophic results.

ii) Ntc thermistor. This is simpler and largely self-protecting. It has the disadvantage (for the more efficient or MTBF-conscious designs) that the thermistor runs hot, adding dissipation inside the amplifier case. The ntc also imposes a lower limit on the mains source resistance, depending on how warm it is. The ntc's working resistance reduces directly as more current is drawn, and indirectly as the amplifier's innards warm up, but with a lag – else it would not be much use as a surge limiter! Thus for highest sonic quality with a class B or A-B or related amplifier, ntc thermistors (excepting those in active, switching supplies – see later) should ideally be switched out after powering-up. Go back to (i).

iii) Sequencing. Professional amplifiers (e.g. Harrison, Carver) have been fitted with daisy-chain power sequencers. Groups of amplifiers are linked by a control cable. When power is first applied, only the first amplifier in the chain initially switches on. After a pre-defined settling time, say 1/2 second, it passes the 'On' command on, to the next one in line, and so on. This supplementary scheme works so long as the surge of each individual amplifier is already low enough not to cause a problem.

6.1.3 Actively-adaptive 50/60Hz PSU

The investment of space and weight and funds in a given quantity of power supply parts (transformer, capacitors) can be most utilised or available to meet the prevailing load conditions by using the transformer like a gearbox. The idea of adapting a solid-state power amplifier power supply transformer's current and voltage ratios to best suit the demand, is not new, but was forgotten in the 'silicon power-rush' (1975-1985). Its first mention in recent years is by one Tomlinson Holman, then making APT hi-fi amplifiers, in 1980 [4].

Today, Crown's goliath-power-rated VZ series, again based on industrial amplifier experience, are the best known exponent. Figure 6.4 shows the basis of their elegant scheme, which relies on the dedicated bridge (or bridged-bridge) with its floating DC supplies, for its simplicity. The two batteries represent one floating psu for one self-bridged channel. When switch SW.1 is open ('LV'), the diodes D1,2 cause the supplies to appear in parallel, so halving the voltage but doubling the available current and halving the source impedance of the supply. This is ideal for low impedance and difficult speaker loads. When closed, the diodes are back-biased and the batteries re-appear in series.

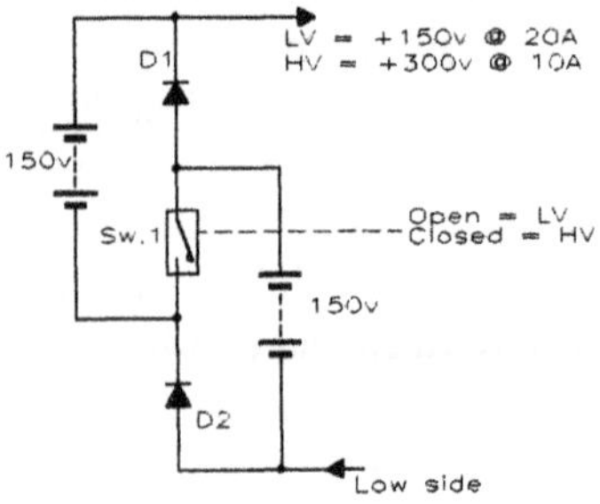

Figure 6.4

Adaptive PSU – Crown VZ System

```
                          LV = +150v @ 20A
                    D1    HV = +300v @ 10A
         150v
                          Open = LV
         Sw.1             Closed = HV

                    150v

                    D2
                          Low side
```

In Crown's VZ amplifiers, the switch is electronic. The default condition is high current, low voltage. This reduces dissipation for low to medium level programme to about 25% of what it would otherwise be. The switch closes to double the voltage ('HV') whenever the signal reaches 80% of the low level clip threshold. This scheme is Crown's own answer to class G, and doubtless, other maker's patents. A significant sonic limitation when used as in the VZ series at AC mains frequency, is that if the transient requiring the HV condition, occurs at the wrong moment, then nothing will happen for up to 8.3mS, or 10mS in 50Hz territories during which the signal would be clipped. The HV condition stays locked for not less than 200mS.

6.1.4 Regulated 50/60Hz 'passive' supplies

In class B, A-B and related amplifiers (i.e. most class G and H units as well), the lowish but finite impedance of the passive power supply interacts with the signal-dependent current to create an error signal rich in harmonics, on the supply rails [5] [6]. Even with a moderately stiff supply, the error voltage can be sizable. If this error signal, with its comparatively strong associated magnetic field couples with any of the amplifier circuit's sensitive nodes [7], it will add a gross distortion, a kind of *modulation noise*, to the signal. Cherry's solution was to reduce loops and otherwise actually *increase* HF supply impedance, with series chokes. Typically 10 ohm resistors may be used to raise low-level ground track impedances.

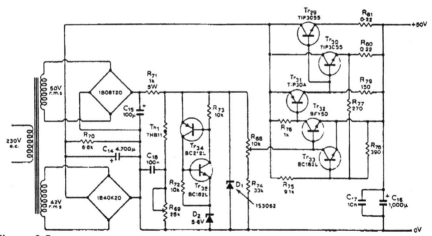

Figure 6.5

A regulated power supply from the early 70s, for a medium powered amplifier. The regulator could help reduce hum and maintain power supply specifications against supply sag. It would take some years before amplifiers had reached the stage where the quite hard-to-measure but sometimes easily heard reduction in noise modulation and error signal on the rails was considered important. Note dual bridge rectifier arrangement on left. The circuitry to the immediate right comprises regulator control circuitry including short-circuit sensing and thermal trip. The regulator's output stage is on the right; note 'Quadruple' of BJTs. Today's best BJTs are far faster, tougher and cheaper.
© EW, from Becker Article, Feb 72.

Another approach is to use electronic (i.e. active) means to reduce the supply impedance. Sonically, this is the real benefit of *regulated* supplies. A regulated power supply is like an ideal audio power amplifier, a near perfect voltage source, with near zero output impedance i.e. its voltage does not hardly at all fluctuate with loading. Regulated supplies became cheap and affordable enough for audio power amplifiers in the early silicon boom 1965-72. They were incorporated for show (e.g. by QUAD, in their 303 of 1966), then their use faded, as from the early 70s, designers increasingly found how the son-of-Lin-based circuits, with their LTP, dual supplies, follower outputs, could have high ripple and hum rejecting properties. Thus raw unregulated supplies (Figure 6.1) would do, and became universally accepted.

In most cases, regulation against mains voltage fluctuations, and to a lesser extent, the filtration of 100/120Hz ripple are relatively minor benefits of regulation, in comparison to the impedance reduction afforded. As with an amplifier with high global NFB, active regulation can create very low source impedances, as low as tens of micro-ohms at mid to low audio frequencies. This is equivalent to having a notional, perfect, cupboard-sized 200 Farad capacitor with an ESR of $25\mu\Omega$. A suitable output capacitor array attaining an ordinary size (say $15,000\mu F$) can expand the low source impedance regime into low RF [8]. This may aggravate some magnetic effects, yet greatly reduces the formation of harmonic error signal voltages on the rails, and also changes the harmonic structure [9]. The significance of this is that CE and other OPS topologies (see chapter 4), also those circuits that use low or zero global feedback, do not benefit from the very high intrinsic PSR afforded by followers. Thus supply error voltages may audibly pollute the audio signal, and are worth reducing if the power amplifier's PSR cannot be improved upon.

6.2 Supply amongst channels

For amplifiers with more than one channel, there are broadly three power supply options. This can apply almost irrespective of the kind of power supply technology.

(i) Dual Mono construction. Each amplifier has its own supply, fully isolated. This is pursued because in stereo, the two channels may be handling completely different signals at every instant, and it has been found that isolation of shared components can greatly enhance image depth and stability. To pursue this properly, they should ideally not even share the same mains cable, *or even be connected to the same electrical grid system.* Thus *truly* isolated supplies are practically limited to those employing batteries, or else synchronised switching.

Down from pure isolation there are sequences of arrangements with increasing sharing of the parts, viz:

(ii) shared power grid, mains cable and mains hardware. Most so called '100% isolated supplies' are in this state. This is a robust enough condition. If one output stage develops a hard short across its supplies, or to ground, the other channel(s) will usually be wholly unaffected.

(iii) Shared transformer primary. Sometimes called a 'split supply' but then easily confused with 'split' as in dual (+/–) supply. By implication, there are separate, 100% 'isolated' secondary windings. But the separation is only galvanic or 'ohmic'; the two secondaries are intimately linked through their common core. To some extent, both will experience and transmit each other's perturbations. AC secondary fuses (in pairs – fuses 1c, 2c in Figure 6.14) with good discrimination are required to protect the transformer, without the other channel loosing its power also, by blowing of the primary fuse (A or B in Figure 6.14).

(iv) Shared transformer. Two bridges can be hung across one secondary. The diodes act as 'OR' logic elements. If real rectifier diodes were perfect, the amplifiers would not 'see' each other. This scheme cannot be used with CE type OPS topologies, where the transformer secondary is floating and 'hot'.

(v) X-OR'd connected. A variation of (iv). Here, two isolated 'stereo' supplies are connected to each other by rectifier OR'ing diodes. These cause a small loss of efficiency and reliability, in return for a great improvement in redundancy, as if either of the individual power supplies fails or sags too much, the other will step in as required. The sonic effects are most likely chequered, as isolation is lost on transients. Again, this scheme cannot be used with CE type OPS topologies, where the transformer secondary is floating and 'hot'. See (vi) ground caveats.

(vi) Common supply. Simple and cheap. The entire resources of one power supply, probably twice as big as that which could be provided per channel, are available to both channels, at all times. For good sonics, great care is then required with grounding. Multiple nested ground star points are a must. Not acceptable for some professional users, since if one power stage develops a hard short (rail to rail), then unless there are DC supply fuses (but see section 5.5.4), one channel can bring down both. On the other hand, there is less to go wrong, and the economy of consolidation.

Figure 6.6

The QUAD-405 contained the elegant 'current-dumping' design (see chapter 4). But it performed hugely below its sonic potential, due to the mental limitations of the designers, see Figure 6.2. This picture displays a common power supply, poor wiring practice (e.g. looming, bad noding, sub-standard, unpolarised connectors) and sub-standard components, such as TV set resistors and spring-loaded speaker 'connections'.
© Hi-Fi News.

6.2.1 Bridge benefits

Bridging (see sections 4.5 and 4.5.1) amounts to balanced drive. It alters the current pathways in the amplifier. In conventional CC (follower) topologies (see section 4.2.6), it keeps hostile currents out of the ground system. It can also improve PSR with some supply configurations.

Bridged amplifiers are also able to make better use of the power supply's capacity, by abandoning the conventional dual power supply and the centre tap. In a conventional class B,A-B and most class G and H solid-state amplifiers working from dual rails, *one half of the power supply is idle half the time*, i.e. for every other half cycle. With a dedicated bridged output stage, employing a floating supply (Figure 6.7), utilisation is every cycle, since all the current passing through the speaker is returned to the opposite rail. It has no where else to go. This allows the transformer and reservoir capacitors to be half the accustomed size, for any given rating factor. AC power consumption is also slightly reduced.

Figure 6.7

Bridge PSU.

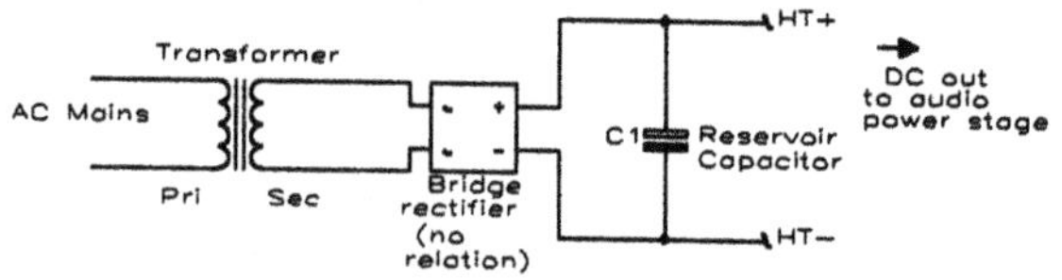

The small disadvantage is that rectification is also performed in (its own) true bridge configuration, where two rectifier diodes are conducting in series at all times. This doubles the small rectifier losses, compared to the dual supply (Figure 6.1) with dual, bi-phase rectification, where the current for each polarity passes through only one rectifier diode at a time.

6.2.2 Operation with 3 Phase AC

Operation from 3-phase AC mains has seen some limited use, in high performance touring amplifiers, created by Oceania, in New Zealand. Amplifiers for fixed installations utilising a communal 3 phase to DC supply have also been proposed by a US-UK professional team in recent years. In all cases the higher voltage (e.g. up to 485v between phases in 240v territories) increases the danger of electrocution in the worst case of mishandling, but there are useful benefits. The polyphase supply frequency is 150Hz/180Hz, so the smoothing effect of any given capacitor array is improved, and recharge is applied three times more often. If one phase dips, the other two soon average out the effect. If all three phases are always connected, the transformer size may be reduced, or its ability to transfer power or energy is increased, in proportion to the increase in frequency. Sonically, bass is more solid. The logical extension of this is to pass up to frequencies above audio.

6.3 Pulse-width power (PWM PSU)

Section 4.7 recounts Bob Carver's application of Class G. But his 1979 review of amplifier efficiency was relatively farsighted, in embracing the PSU as well.

Origins

The power supply technology in Carver's amplifiers up to the present day is a development and refinement of the 'phase-controlled back-slope' power supply that has been long established (since the late '60s) in industrial electronics and TV sets [10].

Triac control

The essence of Carver's scheme (from the *'Cube'* to the present Carver Professional range) was to control the mains transformer's duty cycle, by switching with a triac (Figure 6.8). The triacs' switching rate is constant, but the period over which the triac remains energised is controlled in part, by the incoming audio signal. The conducting pulse width is small under standby conditions, but increases progressively in line with the immediate power requirements. There's a slight lag in duty cycle adjustment (Carver's patent cites 200µS), which is bridged by the *small* reservoir capacitors. Large capacitors aren't necessary because the energy stored in the transformer's magnetic flux (and in a subsidiary resonant inductor, Carver's 'Magnetic Field Coil') is available between pulses, to bridge the gap.

Figure 6.8

Carver's PWM PSU.

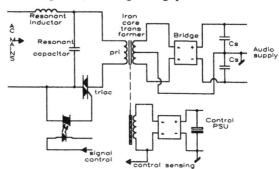

Magnetic cycle

This is where Carver's phrase 'Magnetic Field Amplifier' comes close to being truly descriptive: In effect, full use is being made of magnetic energy ($L\,di/dt$) stored in the transformer, that would normally be dissipated. The outcome is that a 50VA transformer (as in the original Carver M400A 'Cube' for example) can provide over 1kVA for short periods, enough to generate up to 750 watts of audio burst power! This phase-controlled back-slope PSU also exhibits regulation, so the DC output voltage can remain steady in spite of sag or surge on the AC power line.

EMC problems

The main snag with this approach is EMI. As with the conventional PSU in Figure 6.1, Carver's PSU in Figure 6.8 draws high peak currents. But the mechanism is subtly different and the effect potentially worse, because each time the transformer is pulsed on, it has to be charged with flux from scratch. Recall, if you can, the way the lights can dim when a traditional high power amplifier with no soft-start is first energised? Now imagine this kind of inrush current recurring twice every mains cycle. In practice, the peak current draw is limited by and dependent upon the mains supply impedance. Still, when one channel of Carver's classic PM 1.5 for example (a model now superseded) is driven into clip with a pink noise signal into 4 ohms, *the peak current rises from 10A* (just below clip) *to 60A*. For currents of this magnitude, if the supply wiring resistance isn't low enough, the mains voltage waveform may have 'chunks' taken out of it [11]. Owing to the regulation inherent in the Carver amplifiers, they should be largely unaffected by the resulting sag in the rms mains voltage. *The same can't be said necessarily for ordinary power amplifiers (with unregulated supplies) and other audio equipment sharing the same supply, as in most systems.*

Critique

Carver's approach has seen circumstantial criticism, because the potential for weight, size and cost reduction in the PSU has been pushed. So like the class G output stage's heatsinking, the PSU is rated solely to cope with the power requirements of full range and substantially uncompressed music programme. And whilst small and light, Carver's PSU can be less efficient than a 50/60Hz PSU, at 56% and rising only slightly when driven hard.

The benefits of Carver's PWM scheme is that it uses fewer and simpler parts than HF switching supplies. The triac used for switching is also inherently more 'blowout proof' than any transistor, which bodes well for field reliability.

6.3.1 HF power supplies (SMPS, HF switchers)

As with 3-phase power, the *raison d'etre* of switching supplies is that if only the AC supply's frequency was higher, then size and cost of the passive energy conversion elements (transformers, chokes and capacitors) could shrink in more than linear proportion, and efficiency increase. Aircraft use 400Hz supplies for this reason.

Figure 6.9

Fifteen years after Carver's first foray, this recent, lightweight American design employs a PWM (so called 'digital') power supply together with a class G output stage. The combination of high-efficiencies allows a 300 to 600w/ch amplifier to reside in 2u of rack space, run cool, and weigh little more than a solidly-built preamp or processor of the same size. The irony is that there are amplifiers made in the UK, employing conventional power supplies and class B or A-B output stages in elegant ways, that are simpler, have the same or even better power density (watts available per 'rack unit') and for touring, add little more weight, once the rack and cable weight is included. © Soundtech Inc.

HF changes

As frequency rises, the volume of copper and core material needed for the mains transformer shrinks, as does the potential copper loss. As they are being recharged much more often, the reservoir capacitor sizes can be far less (μF) for a given result. For example, 20μF recharged at 80kHz (80,000 times a second) is notionally equivalent to about 16,000μF in a passive 50/60Hz power supply. At HF, capacitors have more stringent ESR specifications. Electrolytics have to be special 'low ESR' or 'HF' types. Inductors used to double-up the hf noise filtering will likewise be smaller in value and compact.

Early push-pull type SMPS operated below 16kHz. As high voltage switching BJTs improved, and then MOSFETs (and even IGBTs) arrived, switching frequencies in industrial units and more recently, in audio power amplifiers supplies, have been pushed to 800kHz and beyond. Figure 6.10 shows the classic push-pull arrangement. It has long been used to provide DC power for hungry computers. With variations, it has (for example) featured in the late '70s in Sony's Class D Hi-Fi amplifier (an appropriate place), then in the '80s, in Yamaha's PD-2500, in Carver's M2.0, and in Peavey's DECA 1200, another appropriate place.

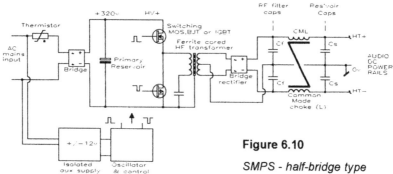

Figure 6.10

SMPS - half-bridge type

Operation

The AC mains is rectified and the peak voltage stored in reservoir capacitors. Theis high voltage DC is then 'chopped' by suitable switching transistors, so it appears on the transformer's primary as an HF square wave. The switching frequency is 100 to 5000 times higher than 50Hz. The waveform emerging from the transformer's secondary is again rectified and smoothed, or rather, filtered.

Design hurdles

These are at least threefold. First, the power factor may be no better than the 50/60Hz PSU, because the current flow into the reservoir capacitor(s) is no less constrained to the peak period of the AC voltage waveform. In fact, without a transformer's winding resistance in the way, the peak current is potentially higher unless the maker has fitted an ntc surge-limiting thermistor.

Second, while 'chopping' or switching DC to create AC is superficially attractive, it's electromagnetically messy. The whole power supply is bathed in and liable to radiate harmonics and RF spuriae ranging up to 10MHz and above. At the time of writing, the long-term effects of European EMC legislation on the few audio power amplifiers using conventional SMPS remains to be seen.

Third, as one set of transformer losses disappear, others come increasingly into play above 40kHz, notably *skin effect* and other effects caused by eddy currents. Presently above 200kHz, and assuming capacity centered on 1kW, diminishing returns set in steeply. Even MOSFET switching losses rise. These can be ameliorated with heatsinks and fans but that partly defeats the weight and space gains.

Flyback

Lab Gruppen of Sweden have developed their own scheme, employing a development of flyback technology, and making this work at up to 3kW, as opposed to the usual practical limitation of about 300 watts. In this scheme, there are primary and secondary switches. The primary switch is off while the power is delivered through the secondary side of the transformer. The rail voltages are stabilised in the process.

As in Carver's scheme, and others, use is made of energy storage in the transformer core. On the downside, only one half of the core's magnetic capabilities are used, i.e. single ended magnetisation. Also the switching frequency is practically restricted to below some 50kHz. Efficiency though, is claimed to be about 80%.

SMPS outcomes

Practical efficiencies range from 85% (Peavey's DECA series) to about 55% (Carver's PM-2.0). Much depends on the switching frequency, and whether size or efficiency have been given ultimate precedence at the design stage.

Despite transformer shrinkage, all HF switching supplies, unless power factor corrected, may require almost the same physical volume of reservoir capacitance that has been 'saved' on the secondary, displaced to the primary. Here, electrolytics rated at 400v DC (to be safe at the peak of 240v AC rms +15%) AC, aren't so small. But they do store plenty of energy, e.g. 1000μF @ 320v = 51.2 Joules.

Makers

The following makers have used (in the past decade) or continue to use, or have recently launched either SMPS or related HF switching supplies:

BGW	US	Professional
Carver	US	Professional
Chord	UK	Domestic
Lab Gruppen	Sweden	Professional
Peavey	US	Semi-professional
QSC	US	Professional.
Sound Tech	US	Professional.
Yamaha	US	Professional.

6.3.2 Resonant power

Tesla's dream

Nikolai Tesla was exploiting HF resonance for power conversion in the USA a century ago. Then in 1985, in aftermath of financial deregulation, the city of London had an interference and current-draw problem with suddenly 8,000 PCs in one building. The problem was solved by a top British power electronics consultant, working from an ex-government establishment. Part of the solution was to modify the PCs' SMPS to make them more efficient and cleaner with the least effort. This involved resonance. Rauch's founder, Jerry Mead, heard about the benefits of the solution as it was installed in 1986, and the first audio power amp to employ an HF switching resonant power supply was Rauch Precision's model *DVT-300s*, launched early in 1987. It was a technological and sonic success.

Figure 6.11

Inside the Rauch DVT300s, which in 1987 was the world's first audio power amplifier to employ resonance in an HF switching supply. The amplifier, rated at 600w⇒4 ohms, is split symmetrically down the middle, except for the aux supply's small 50Hz transformer on the right side, immediately behind the bulkhead. The power supplies are at the front. Looking at the right channel supply, the two primary (left) and four, huge (10,000μF) secondary reservoir capacitors (right) are clearly visible. The ferrite-cored transformer, operating at 80kHz and handling 1.2kVA, can be seen on the left of the latter. The two large axial capacitors below it are the resonant elements. The 'black box' on the left, centre, is the switching MOSFET's heatsinking, which receives cool air drawn in from the front, by the fan in the centre. Behind the bulkhead is another short-tunnel heat exchanger, for the class A-B L-MOS output stage. This amplifier, later re-vamped as P600, has been extensively used in the UK and Europe for touring and club-installatons, and has given operators almost a decade of sonic and technological superiority over anything offered by US makers.
© MS&L/Rauch.

Configuration

Switching supplies using resonance have been developed as a cleaner, more efficient way to convert power at high frequencies. It involves the production of a quasi-sinusoidal current waveform, which is switched at the point of zero current (or voltage, depending on topology). Whereas PWM and push-pull SMPS PSUs employ a fixed period with a variable 'on' time, the resonant PSU's 'on' time is fixed, while the period varies. Figure 6.12 shows a typical resonant power supply topology; one of several configurations.

247

Figure 6.12

Resonant Switching Supply.

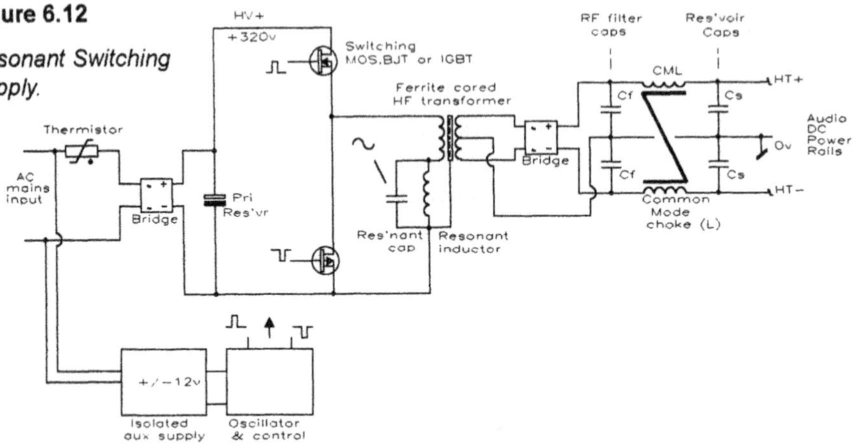

Note the main circuit elements are the same as the push-pull SMPS in Figure 6.10, except for the middle, where the primary DC voltage is switched in connection with a resonant tank circuit, like an HF version of Carver's 'magnetic field coil'. This section of the circuit is also similar in some ways to a class C amplifier (used at RF), which is efficient because it operates without bias and doesn't try to be linear, instead relying on L×C resonance to tidy up the waveshape. Series resonance is also efficient because zero current- (or in some cases zero-voltage-) switching ideally keeps dissipation in the power switches down, to around 1% of the output power.

Practice

The table below summarises typical losses, yielding a nett efficiency of around 92%. The knock-on benefit of such high efficiency is synergistic. The cooler any part of an amplifier runs, the cooler it all runs. And cooler you keep the device, the longer it lives.

Losses in a Resonant PSU

Part(s)	Loss (watts)
Primary bridge	16 w
Resonant LC tank	8 w
MOSFET switches	40 w
Transformer	13 w
Secondary bridge	32 w
Total loss:	109 w
Total output power:	1300 w
% efficiency:	92.2%

Model = BSS Audio, EPC-780 (see upper front cover picture).

248

Lowered EMI

Resonance is cleaner. Since the AC in the transformer is not too far from being gently sinusoidal, and thanks to zero-current switching, fewer components and less work is required to keep EMI emissions within EMC, VDE and FCC limits. The graph below shows the HF noise fundamentals at the output of a typical amplifier.

Noise Level in –dBvr vs. Freq. in Hertz

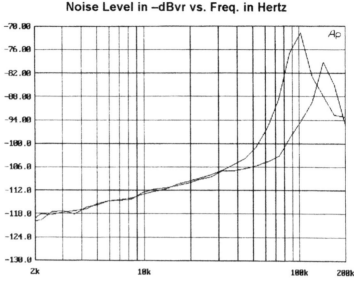

Figure 6.13

Typical residue on the outputs of dual power supplies with nominal 100kHz resonant switching supply. Note slight difference in the fundamental frequencies, and rapid reduction of noise in audio band, to highly acceptable levels.

As with push-pull SMPS, a resonant PSU's operating frequency can be as high as 200kHz before losses caused by reverse conduction, ringing and parasitics rise disproportionately. At this juncture, efficiency and the avoidance of heat directly conflicts with the reduction of transformer size and weight. As it stands, resonant PSUs offer a potential quadrupling of power density, typically 25w/in³, compared to 6w/in³ for traditional SMPS type.

Makers

At the time of writing, audio power amplifiers employing resonant supplies are predominantly professional and all known to the author are designed and manufactured in the UK.

Maker	Historic Models	YFI
Rauch (MS&L)	DVT 300s*, P120, P600.	1987
BSS Audio	EPC-760 & 780.	1988
Chevin	A1000, 2k, 3k + others.	1991

YFI = Year First Introduced.   * = no longer made.

6.3.3 The higher adaptive PSUs

Bob Carver's most recent creation, the 'Lightstar' amplifier, appears to have taken the next evolutionary step, past the VZ (Figure 6.13). Here, the 'digital power supply system' is said to provide 'high energy into any loudspeaker load, regardless of impedance.' Clearly, the power supply's V-I ratios are subject to control by the amplifier. Once with a switching supply, the idea surfaces that is that it is easier to cleanly 'wiggle' a unipolar voltage from +20 to +100v than to wiggle a music signal the same 80v from +40 to −40v, through zero volts. BGW are doing something similar – their new SMPS is working at 840kHz with 'signal tracking'.

6.3.4 HF switching summary

Tradeoffs

All of the 'smart electronic' techniques described above depend on fast, high repetition rate switching to provide improved efficiency and/or reductions in the weight and size of the transformer. And each in turn generates VHF noise, to varying degrees. At best, the results have been visible as a thickening on the scope trace at the amplifier's output, and an occasional spike, typically a few μS wide. It looks bad, it inevitably contributes to a worsening of HF intermodulation figures, and it might be audible as HF distortion and 'veiling'. But it often has no audible ill effect. Then again, if you're unlucky (or untidy), the noise radiates from, or is conducted by, the mains cable, and gets inside all the adjacent equipment you care to probe.

EMC

The escape of such RF noise is now tightly legislated against in European EMC legislation. The long-term effect of this, on the RF cleanliness of amplifiers with switching supplies has yet to be seen. Noise has previously been suppressed by design to within reasonable limits (e.g. FCC type) by all the reputable makers of amplifiers with switching PSUs. *But for all the point of sale legislation, it only takes one component to later drift or go open, or an earth connection to loosen, for any individual amplifier to turn into a powerful RF noise generator. There is no annual EMC check.*

Buyer evaluation

For amplifiers employing anything other than a 'stone age' 50/60Hz PSU, it's doubly valuable to make a 'goods inward' inspection of RF levels present at the outputs and around the mains lead before installation.

The higher parts count in 'smart', active power supplies potentially reduces MTBF (reliability) figures. Then there's the speed with which overstressed small components can expire. In real life, most audio output stage transistors continue to blow with a short between their main terminals. Often, a shorted speaker connection is the culprit. It follows that to be properly protected, a 'smart' PSU needs to be able to cope with a direct short across its outputs.

In comparison, overloaded 50/60Hz PSUs do at least give off a warning buzz and some interesting smells, and more often than not, a fuse blows before any real damage is done. They also exhibit a soft, gradual current limit, whereas 'smart' supplies have definite 'hard-knee' limits which can never be exceeded, however momentarily. On the other hand, 'smart' PSUs open the way to sophisticated protection. With suitable circuitry, any HF switching PSU can be powered down within a fraction of a millisecond, should its life be threatened.

HF sonics

Cleanly executed switching supplies improve sonic quality. The immediate audible benefit is in the bass end: music with low frequency components that are unluckily related to the mains frequency aren't syncopated any longer by the supply recharge rate, which is now *above* audio. Also, the supply has far more than the the the usual amount of energy storage, for the same *physical* volume of capacitance.

6.4 Power supply (PSU) efficiency round-up

In common with the output stages looked at in chapter 4, practical power supplies are less than 100% efficient. *An amplifier's overall efficiency is the product of the two:*

$$\% \text{ overall} = \%\text{OPS} \times \%\text{PSU}$$

Because neither are more than 100%, each contributes to reducing the nett figure. For example, if both were 50% efficient, nett efficiency drops to 25% ($0.5^2 = 0.25$).

Power factor

To discover either the overall efficiency of a power amplifier, or just that of its PSU, we first need to measure the current drawn off the AC supply when driving a known number of watts into a defined load. Excepting those amplifiers fitted with 'unity' PFC, both 50/60Hz passive and switching power supplies present a highly non-linear as well as reactive load to the AC power line; the periodic AC current generally leads the AC voltage. PSUs using certain 'smart' electronic techniques (effectively now outlawed for sale in Europe) may present even more complex non linear loads.

For these reasons alone, power input requires careful definition, to screen out the effects of *non-unity Power Factor* (a perfectly resistive load has a power factor of 1.0; for many amplifiers, PF is around 0.8 to 0.6) and associated high peak currents. The latter aggravate losses in cables, leading to voltage droop, hence power and efficiency losses. For the most part, these effects are essentially external to the amplifier. They're also liable to be significant only if the incoming line power cabling has too high a resistance; or the socketry in line is dirty or loose. But, with high peak currents, the cable gauge needed may be many times that suggested on the basis of safe current rating alone. The efficiency comparisons that follow assume a competent, low resistance installation. An audit of the accuracy of power input measurements would need to take account of auxiliary circuitry, subtracting the power drawn by fans, lamps, LEDs and relays.

First analysis

Beginning with the passive type shown in Figure 6.1, some waste heat is dissipated by every major part: In the transformer, rectifier bridge and the reservoir capacitors (C1,2). In the textbooks, the bi-phase bridge circuit has an efficiency of 81%. In practice, a typical high power amplifier containing a generously rated ('stiff') supply of the kind illustrated, yields 70 to 78%, depending on loading. For a variety of reasons efficiency rises if the amplifier is driven hard into a low impedance load.

The capacitor-smoothed (0–C) circuit typically pulls a peak current that's at least 3x greater than the rms current consumption. Charge transfer is bundled up into the small period when the incoming voltage at the bridge exceeds the voltage on the reservoir capacitors. If the windings are rated to handle this high peak current, then rather like a Class A-B or particularly class A amplifier OPS, handling music, the transformer ends up bigger, while its capacity to supply power is under utilised for the remaining (say) 80% of the time. If the transformer is made small to save on size and weight, I^2R heating losses rise, and efficiency suffers.

Beset with a circuit that's 30 to 70 years old (in various guises), amplifier designers are left to score small gains in efficiency by adopting refined 'C' core or toroidal transformers, by computer optimisation of these; and by taking steps to keep down the ESR (internal DC resistance) of the capacitors. Dissipation can be reduced by suitable uprating. In turn, PSU component temperatures are reduced, particularly when the amplifier is driven hard, so enhancing reliability. However, beyond a point, this is defeating in other ways, as weight, size and cost are also uprated, generally ahead of any reasonable budget.

At least the passive power supply is simple and reliable. It needs no elaborate cooling aids, while interference (EMI) radiation is restricted to low frequency magnetic fields, reservoir capacitor surge, and rectifier RF noise.

6.4.1 Amplifier efficiency summary

Note: Amplifier output stage efficiency is covered in chapter 4

Classification	OPS	PSU	Overall Efficiency
Type 1:	Standard	Standard	Average
Type 2:	High	Standard	Above avg.
Type 3:	Standard	High	Above avg.
Type 4:	High	High	High

Efficiency summary

There are many different *kinds* of 'efficiency', and many ways of measuring each. Each class of amplifier has an broad optimum power/volume ratio (w/ins³, or w/mm³). Although (for touring PA) fuel and truck rental can be saved by using the smallest and lightest amplifier available, all is not what it seems: Some kinds of efficiency suffer when size and weight reduction are pushed too far. Then there may be tradeoffs to the kind of programme that can be handled, as well as higher operating temperatures, and environmental side-effects on the AC supply, say.

Strictly, high efficiency in audio power amplifiers means a high ratio between output power and wasted power. It's not synonymous with small size and levity, although it may contribute to both. On this basis, two power amplifiers placing slightly different emphasis on these factors may be quite different.

Fortunately there *are* combinations of power, size, weight, reliability and sonic quality to suit most purposes. It remains for amplifier users need to define exactly what they need, and to thoroughly evaluate shortlisted models under real operating conditions before purchase.

6.5 Power supply fusing

Fuses are 'sacrificial links' – wires that will melt, break or vaporize harmlessly at and above a defined current. They are strongly associated with power supplies. Three types of fuses are commonly employed in audio power amps.

Type F (Q, QB)

The ordinary type. Responds without delay to currents above about *double* the nominal rating. A current equal to the nominal rating might blow over minutes or hours, or not, as tolerances are quite wide. Sometimes called 'quick blow' or 'QB'. The fusewire is straight and plain.

Type T

Time lag, delay or 'anti-surge'. This type *does not* prevent current surges. Instead, it blows quite quickly but after a delay. Thus it resists blowing on excess currents that are short lived. In other words, it prevents nuisance blowing when a regular but short lived surge current is the norm – precisely what occurs with most power amplifiers when first turned on. If visible, the fuse element may be thicker than a plain wire, and coiled; or it may be part springlike; or there may be a bob of metal on the element. But these appearances are not always present.

Semi-delay' types are specialty types, rarely encountered, at least in audio power amps. They provide a half-way house.

The above types are rarely fast enough to reliably protect transistors or even rectifier bridges, from chain destruction, as one failure causes others. Indeed, cynics have described transistors as devices designed to protect fuses !

FF

Rapid or ultra-fast type. Also known as semiconductor fuses. These respond up to 100 times faster than the T and F types. Only these kinds stand much chance of reliably saving otherwise unprotected output transistors from over current. And even their ability to protect semiconductors is not assured. To begin with, semiconductor makers do not make the fuses, and the fuse makers do not make fuses to protect particular transistors, but to conform to neat round Euro-numbers!

Casing

All the above fuses are normally encountered as miniature cartridge links, in all but the highest power ratings. Cartridge fuses may be made of glass, but the glass can shatter if the fusing energy is high enough. This is rarely dangerous, but can cause some inconvenience, as glass fragments can choke-up a panel mounted holder. As a rule of thumb, if the psu transformer is rated above 100VA, or the amp rated at above 200 watts, all mains and any other fuses in the high current path to the output, should be ceramic bodied types, known as HRC (High Rupture Capacity). These can interrupt fault currents of hundreds of amperes before bursting.

Wide tolerances

When replacing or specifying fuses, it is important to be aware that the ratings are quite nominal. There are surprisingly wide tolerances in the blowing current, and the speed of action. A 5A rated 'F' fuse could take an hour to blow when passing 9 amperes. Another identically rated fuse, but of a different make, type, or batch, could blow in under 1 second at 7 amperes. One thing is certain: ordinary fuses do not blow at all quickly until fault currents are 1.25x to 2x (125% to 200%) over their rating.

Placements

In audio power amplifiers, there are several common locations for fuses. Looking at the simplified universal schematic in Figure 6.14, primary position A relates to power amplifiers used in the UK and some other countries where the mains plug is fused with a rating that should suit the flexible cord's gauge. In the UK, such fuses are 1" (25mm) HRC, type 'F', available with ratings of 1, 2, 3, 5, 7, 10 and 13 amperes. In other countries, the spur or outlet may be fused instead.

Fuse B is in the live power line. It should be rated to prevent the transformer's insulation from overheating, in the event that its output is shorted. Use of a 'T' fuse allows a lower rating to be used than would otherwise be the case. An 'F' type of fuse would need to be grossly over-rated, to avoid nuisance blowing at switch on, when current in the first half cycle might sometimes exceed 100 amperes. As shown, this fuse might alternatively be a circuit breaker, commonly combined with the power switch.

Breaker benefits

Fuses A and/or B are almost certainly mandatory for safety in most country's safety codes. Yet many audiophiles have taken the risky step of bypassing these positions, which are often the sole fire-protecting elements in high-end domestic amplifiers.

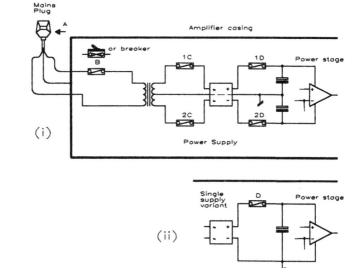

Figure 6.14

Fuse & Breaker Positions in Power Amplifiers.

The outcome has been improved sonics, possibly as a result of avoiding the fuses' thermal modulation. In this and other positions, fire-protection safety may be restored without sonic compromise by replacing the fuse and holder with a suitable circuit breaker having low and stable reset resistance.

Secondary AC

Fuses C1,2 and D1,2 and D are in the transformers' secondary current paths. These positions can be used to protect the output stage and bridge rectifier with more discrimination. That's because some of the current surge at switch-on is drawn by the transformer, as it magnetizes. So the surge at these points C and D is pro-rata lower. The transformer is also mostly protected – though not from a short within itself.

However much fuses are rated to protect an amplifier from abuse or damage the protection will only last so long as the exact correct replacement fuses are *always* fitted. This particularly applies to fuses that are readily accessible from panel fuseholders, and fuses in the 'user' positions

One thing fuses *should* definitely achieve is prevent any kind of fire taking hold in a major part, most notably in the mains transformer. Fuses rated in this way, to only blow when there is a major problem that requires bench repair, are less likely to draw attention to themselves. In well designed equipment such fuses are secure inside the enclosure.

References

1   Duncan, Ben, *AMP-02, part 2*, Hi Fi News, Nov 1989.

2   Dibble, Ken, & Allen Mornington-West, *EMC conductive emission performance of professional power amplifiers*, Proc.IOA, Vol.16, Nov 1994.

3   Angus, James, S, *EMC and the choke fed supply*, Proc.IOA, Vol.13, Nov 1991.

4   Holman, Tomlinson, *The amplifier/loudspeaker interface*, Hi-Fi News, Dec 1980.

5   Ball, Greg, *Distorting power supplies*, EW+WW, Dec 1990.

6   Pass, Nelson, *Linearity, slew rates, damping, Stasis and ...* , HFN/RR, Sept 1983.

7   Cherry, Edward, *A new distortion mechanism in class B amplifiers*, JAES, May 1981.

8   Duncan, Ben, *PSU regulation boosts audio performance*, EW+WW, Oct 1992.

9   Duncan, Ben, *Harmonic convergence*, Stereophile, October 1992.

10 Geddes, Keith, & Gordon Bussey, *The Setmakers*, BREMA, 1991. ISBN 0-9517042-0-6.

11 Perkins, Cal, *Power amplifiers: design*, S&VC, Mar 1985.

Further reading

12 Brown, Marty, *Practical Switching Power Supply Design*, Academic Press Inc, 1990. ISBN 0-12-137030-5.

13 Keith Billings, *Switchmode Power Supply Handbook*, McGraw Hill, 1989. ISBN 0-07-005330-8.

14 Marshman, Chris, *The Guide to the EMC directive 89/336/EEC*, EPA Press (UK), IEEE press (USA), 1992.

15 Marty Brown, *Power Supply Cookbook*, Butterworth-Heinemann, 1994. ISBN 0-7506-9442-4.

16 Miller, R, *Measured RFI differences between rectifier diodes*, TAA, Q1, 1994.

17 Ott, Henry, *Noise Reduction Techniques in Electronic Systems*, Wiley-Interscience, 1976. ISBN 0-471-65726-3.

Specifications and testing

7.1 Why specifications?

Specifications are an orderly and compact way of telling users through figures, about a power amplifier's:

(i) audio performance.

(ii) operating capabilities.

(iii) effect on the environment, and environmental ratings.

(iv) protection, and the 'survival envelope'.

The specifications may confirm that technical standards are met; suggest a certain level of engineering competence, or confirm cognizance of particular needs. In part, they define what is being bought and sold.

Although users do not have to read specs, makers do have to issue them.

7.1.1 Types of spec.

Audio power amplifiers' technical specifications break down into the following categories of information:

1. Audio performance

These aim to describe with how cleanly and linearly the equipment passes or processes audio signals. An ideal device might add no noise or distortion – in the broadest sense of 'undesired change'. Real equipment is imperfect, and measurements, expressed in figures, are used to describe and highlight the gear's relative perfection or otherwise, giving some indication as to how it will sound.

It is important to be aware that no amount of presently available audio performance figures can watertightly describe or predict a power amplifier's sonic qualities.

Even so, traditional measurements that do not (or no longer) predict sound quality are still useful, to confirm that some fundamental and readily measurable aspects of signal handling meet a standard, and any degradation remains ideally below most people's audibility threshold – commensurate with the quality of the measurement.

Theorists who would 'hear with their eyes', are cautioned that no matter how finely and artistically the information is presented, the specifications of audio equipment, including power amplifiers, are regularly truncated to a bare and vague minimum, or can contain guessed figures, concocted by under-staffed makers and pressurised copywriters. As a rule, the more detail of test conditions a particular specification gives, the more believable it is that the maker has actually performed it. It is still possible for terse figures to belie an in-depth test process. Some practice at spec-reading is needed in order to read between the lines.

2. Operating capabilities

(i) Ratings of impedances, levels and loading etc, in the 'spec' tell you whether the power amplifier will work in your system, in particular whether it will work well with your/the particular signal source(s) and speakers. If you have already bought the equipment, this part of the spec may highlight reasons *why* you are having problems.

(ii) Functional ranges that you can get out of any knobs or functions should be defined. In many power amplifiers, there are no knobs. In others, there are just 'set and leave' gain controls. But in amplifiers specialised for touring PA and installed multichannel sound systems, there can be integral limiting and VCA (remote) gain-control functions.

3. Environmental factors

(i) Effects. Until now, very few makers have documented the peak current that is drawn by power amplifiers. This can be dependent to a considerable extent on the qualities of the mains supply. Acoustic noise is not specified either, but many power amplifiers produce loud buzzes when the mains supply is impure, and most amplifiers with cooling fans produce SPLs that range up to the annoying-to-be-with-in-the-same-room.

Other environmental effects have also been largely ignored, namely conducted EMI introduced onto the AC mains supply, and magnetic field emanating from amplifiers and their cabling. The measurement of, and restrictions on such emissions, have been quite comprehensively covered in Europe by 'EMC' legislation – see section 8.9.

(ii) Ratings. Maximum safe, operable levels for temperature and humidity, as well as mains voltage tolerance come under this heading. Mains operated equipment commonly has a range of voltage either side of each voltage setting (115v, 240v etc) over which it will work properly, or over which other specifications are guaranteed.

4. Protection

There should be a description of the different kinds of accident and abuse protection; when, and how fast it operates, how it resets, and why it has no effect on sonic quality. See chapter 8.

7.1.2 Standards for audio power amps

1. 'Consumer' Hi-Fi standards

There is no minimum 'consumer' (let alone professional) specification for audio power amplifiers in the UK, nor in most countries. The historic part-exceptions are Germany and the USA. The most notorious outpourings from these countries in the transistor age are:

DIN 45-500 (1963-66, partly revised 1972), Germany. Set an unrealistic minimum power output (6w/ch) along with crude basic performance specs, viz. frequency response, SNR, distortion over power bandwidth and IMD.

IHF-A-201 (1966), USA. The first update of the Institute of High Fidelity's original 1950's standard.

IHF-A-202 (1978). USA [1]. The 2nd update, entitled '*Method of measurement for Audio Amplifiers*', but also covers preamps, etc. Requires five key specs to be measured by makers including power but sets no minimum power output. The other four - that *must* be cited - are: Any dynamic headroom; frequency response; sensitivity; and SNR. Up to another 15 'secondary disclosures' may be optionally cited, e.g. A-wtd crosstalk.

2. Professional standards

There are even fewer of these.

THX, Lucasfilm, USA. Beginning in 1985, Tomlinson Holman and team set minimum standards of audio equipment performance for use in Lucasfilm's *THX* cinemas worldwide. Note that 'THX' stood for 'Tomlinson Holman eXperiment'. The experiment continues, as a newer, related specification covers approved home cinema equipment. The standards are proprietary, confidential, and are subject to continuous improvement. Amplifier requirements are covered in depth and with some realism [2].

Realistic Amplifier testing: Procedures for professional touring music, Ben Duncan Research, UK. Key audio performance measurements and some realistic survival tests, for touring PA (as well as recording studio) amplifiers [3].

7.2 Why test things ?

Testing is a sieve against the risk of poorly performing, sub-standard or unsafe amplifiers being shipped, or put into use. It involves validation of an imperfectly-edged physical creation, against model results that are some combination of the theoretically predicted, practically anticipated, and statistically-defined.

Testing is mostly performed after birth (manufacturing test). It may also be performed after major repair or as part of a annual check-up (servicing test).

7.2.1 Test tools and orientation

The most apposite test equipment for audio power amplifier testing will depend on circumstances. The golden rule is:

'Caveat Testum Finitum'

which broadly translates as 'Beware testing is finite' – in the sense that it will never tell you everything.

Testing is described in a methodical order developed by the author. There is no point in testing any kind of distortion until you know the AUT is basically healthy. So knowing bandwidth, and that it has reasonably low hum and noise, must come first.

Figure 7.1. shows a brochure specification written by the author in 1987 which lists all the apposite measurements – not necessarily in the order of test – and shows how test conditions can be concisely listed. The level of detail and accuracy of definition is still a model for most professional makers. Yet a truly comprehensive specification today could readily cover several pages with new tests, permutations and the associated test condition information.

Figure 7.1

A model audio power amplifier specification.

7.2.2 Realtime test signals

Sinewaves

Sine or *sinusoidal* waves (Figure 7.2 shows various visual manifestations) are the *de facto* universal test signal. They are repetitive, continuous (essential for convenience in many conventional tests) and simple to analyse. Although music is made from bits of sine-waves, individual sine waves are not at all like music signals, except for the odd occasions when some kinds of music contain steady and quite pure tones for a few moments.

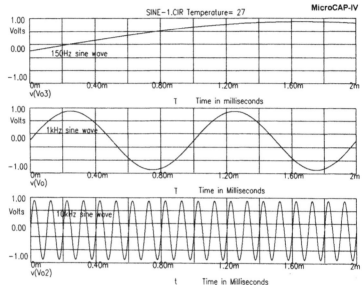

Figure 7.2

Sinewaves at low, mid and high frequencies, showing progress over 2 milliseconds. For the visual monitoring of frequencies swept over audio band, three 'scopes are required, with different timebase settings, to keep more than one, but not too many cycles, visible.

Squarewaves

Square waves (Figure 7.3, overleaf) are also continuous waves and even less like music signals than sine waves. They are used to examine amplifier's transient behaviour, both small and large signal.

Static Vs. dynamic

(Continuous) sine and squarewave are commonly described as static test signals – in the sense that the pattern per each successive cycle is the same. Even swept sine waves are commonly static, as the frequency or amplitude sweeping is driven by the conclusion of each measurement during which frequency is stable.

The test signal types that follow are considered dynamic – as the amplitude and frequency of the signal at each successive cycle (or inter-zero-crossing period) is different, even if it should repeat after some number of cycles.

Figure 7.3

Perfect squarewaves are not found in music but are still useful as they yield considerable transient information, and for real world use testing, approximately simulate highly clipped input signals. The lower graph magnifies the 1ms corner (marked ><) X100, and shows how even a nominally perfect square wave has a slight 'tip', but it is well and quickly damped.

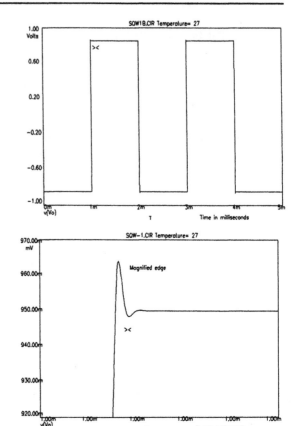

2 and 3 tone sinewaves

Two, three or more continuous sinewaves that differ in size, frequency and phase, may be mixed in many combinations, to get a signal that more closely simulates music (Figure 7.4). Such *multitone* signals are particularly used for testing for *intermodulation* products ie. IM distortion.

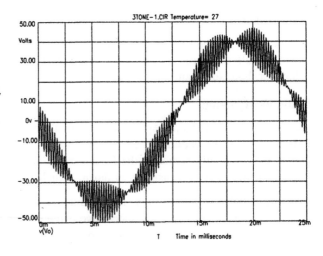

Figure 7.4

Here, three tones at 40Hz, 4.6kHz and 4.8kHz are mixed in a 38v:5v:5v ratio. This test, devised by Ivor Brown, is described in section 7.7.4 .

Tone-bursts, burst waves

A tone burst is a sinewave that stops and starts (Figure 7.5), so it has no steady state. It may be seen as a combination of sinewave plus squarewave. In fact, burst waves are produced by 'gating', where a 'logic signal' *that is a squarewave* turns a sinewave on and off. The point immediately after the sinewave resumes and the point after it ceases has provided a wealth of data about audio equipment response in the time domain. Energy storage/discharge as well as dynamically-conscious protection (e.g. compression, V-I limiting) may be investigated by varying the periodicity and MSR of the squarewave.

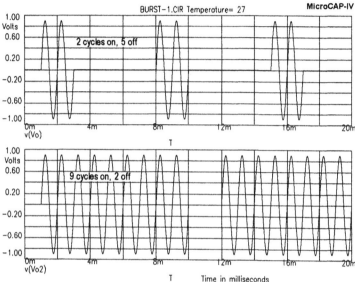

Figure 7.5

Burst tones with different 'on/off' alias 'mark/space' ratios.

Pseudo-random signals

By mixing three or more sinewaves and giving them levels, phase relationships and frequencies that vary in time, a signal can be produced that looks very much like music, as it appears on a 'scope screen (see Figure 2.17, in section 2.4.4). The PMR can be adjusted to make the 'music' have worst case qualities in either direction, and be used for real or virtual testing. If the waveforms are controlled by a computer, then they are known while appearing random, allowing error analysis that is comprehensive. See section 7.7.7.

Noise

Noise in the context of testing is usually hiss – a collection of incoherent and generally random signals that seem toneless.

White noise

The noise from 'hissy' sources, particularly a mistuned VHF ('FM') radio are impure examples of noise that contains all frequencies equally, analogous to white light.

White noise is also the sound of steam escaping from a coal-fired railway locomotive. Energy in each successive octave band doubles (+3dB/octave up-slope). This seems 'top heavy' to human ears. A different kind of noise, closer to music, is usually needed or is more useful for test purposes.

Pink noise

Pink noise or its close derivatives is the test signal most apposite to power amp warm-up, survival testing, thermal testing, burning-in, and special tests in conjunction with speakers, e.g. measuring drive unit power ratings.

Analogous to pink or reddish light, the energy *per octave* is equal in *pink noise*. When pure and averaged over time, this noise has a 'flat frequency response' on a log frequency scale. In practice, the ear readily perceives patterns in the *pseudo* type of pink noise, which is made by *pseudo-random* digital clocking. Otherwise pink noise is naturally occurring, e.g. waterfalls, and may be collected (not generated) by amplifying natural sources (called $1/f$) in chemically 'dirty' semiconductors. White noise is also naturally occuring. After collection and amplification it can be converted to pink noise by passing it through a −3 dB /octave low pass filter.

Real pink noise sources differ in their sonic qualities. Measurable qualities such as PMR and symmetry also differ. With natural sources in particular there can be 'lumpiness', caused by the very low, random periodicity (rate) of sub-sonic peaks. *Unpredictability and variability are both the strength and the weakness of noise as a test signal.* Pink noise approximates the behaviour of dense music – as if all the music in the world were mixed into one signal! Like live music, it has occasional high peaks that will be clipped.

7.2.3 The test equipment revolution

Traditionally, comprehensive testing would require individual generators for most of the test signals just described, variable-bandwidth AC voltmeters, distortion analysers, and many other items of equipment, connected in different combinations for each test. Since 1985, multipurpose instruments connected to, and controlled by a PC (computer), have greatly simplified testing and data gathering, and has expanded the repertoire of tests. The *de facto* world standard has been *Audio Precision*, with *System One*. Here, the tedious taking of measurements at many different spot frequencies and levels, writing them down, and typing them up, or cumbersome mechanical plotting, has been superseded by fast sweeping, where data is taken at many, appropriately spaced points – but only after adequate time for settling, and if successive samples converge closely enough. Numeric and graphical hardcopy is provided (almost!) at the push of a button. Curves may be overlaid, so the effects of test permutations and also unit-to-unit variations can be collated and seen in context. When used with a standard VGA monitor, the AP *System One* plots up to 4 curves x 2 channels (or data fields) or 8 curves x 1 channel, in eight colours. Graphs may also be expanded, references changed, and the axis scales (log/lin) and aspect ratio may be changed, opening up a very wide choice of presentation options.

7.3 Physical environment

The IEC 268-1 standard suggests the environment in which the amplifier under test is placed should be within the following:

Ambient temperature	15 to 35°c
Ambient humidity	45 to 75%
Ambient air pressure	860 to 1060 mb

Changing the ambient air pressure if it is outside these recommendations seems unlikely, but it is worthwhile noting that loudspeaker performance (including impedance dips) can be considerably changed by both air pressure and humidity.

With critical tests and comparisons, the ambient temperature around the *AUT* (Amplifier Under Test) may need justifying, and should certainly be recorded and maintained throughout testing.

7.3.1 Mains measurement and conditioning

To measure the rms voltage of the mains, a meter with true rms (trms) sensing must be used. A meter that reads average (even if it is calibrated in rms) will misread because of the mains' waveform distortion. This is endemic. However, *the rms reading is only useful for gauging average power consumption* and the PMR.

For calibrating supply voltages and making close, accurate and *transferable* power measurements (on the majority of amplifiers which have unregulated output stage power rails), the mains' *peak* voltage is a better point of reference, as nearly all DC supplies charges to a voltage that is closely proportional to this level. Otherwise, the mains supply is conventionally set (with a Variac) to centre on the official rms voltage when testing an amplifier. If the tests are broken off, then continued an hour, day or week later, then in the interim, both the supply voltage and waveform distortion will have randomly changed. So the same true rms voltage (say 240v) no longer provides the same peak voltage. The DC power rails could thus be 5v different, and so then the amplifier's spec, particularly its power rating and clip point, will have changed – making a nonsense of test data that cites output power or even voltage to more than 1 or 2 significant figures. Figure 7.6.shows a full-wave rectified, peak reading, mains monitoring circuit for use with a DVM. The output may be switched to pass straight AC for comparison, with the DVM also switched to AC trms.

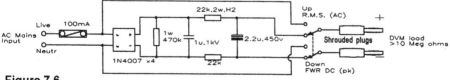

Figure 7.6

Peak mains (DC) reading sensor. Switching allows the rms and peak (full-wave rectified DC) voltage values to be compared.

265

The AC mains used for testing must be reasonably noise free, i.e. 'clean'. It is worth underlining that the value of the mains supply impedance can increasingly affect the measured performance of audio power amplifiers above about 100w/ch, most particularly conventional, non power-factor corrected supplies, with over-sized transformers and reservoir capacitors. Chapter 9 gives information on this, mains power installation, and conditioning. If these instructions are followed, voltage drops caused by the AUT's own loading may be made insignificant or at least greatly reduced.

But there are still bad days, when the incoming AC supply voltage changes continuously and erratically, seemingly the moment you start testing. Voltage variations are dealt with by either manually monitoring the supply, and operating the Variac by hand when making critical measurements (such as distortion *Vs.* level tests); or by employing a servo-controlled variac, reactor or other AC power regulator, the faster acting the better. But not adding noise to the line is even more important.

7.3.2 Power amplifier preconditioning

Power amplifiers of all kinds should be warmed up or 'run in' before testing. Some amps run hot and warm up without signal; others need driving, else testing may be unrealistic.

Transistors *Vs.* valves

Solid-state electronics may not need to warm-up to work, but warming-up is nonetheless desirable before testing, firstly because raw semiconductor behaviour is highly temperature sensitive.

For amplifiers incorporating valves, the heaters will take 2 to 3 minutes to heat up before which there is no electrical output. Beyond this the behaviour of valves *per se* is less affected by temperature, but other parts may be strongly affected and may take some hours to reach equilibrium.

Temperature effects

The values of most passive components (excepting best ceramic capacitors) are also temperature dependent. Although the performance of a good circuit design *should* be independent of temperature over a range, due particularly to negative feedback, temperature may still affect performance in practice, for example, because of thermal gradients. In real use, power amplifiers run warm after a while and the performance that matters most is the performance under this condition.

Secondly, in all amplifiers, the heat exchangers (if solid state) or glass bulbs (if valve) take some tens of minutes to reach thermal equilibrium, within say 1°c. In many instances, the small signal circuitry (which in solid state equipment usually reaches equilibrium with its own temperature production in minutes) will be warmed up to the greater equilibrium only in time – ultimately reaching the heat exchanger temperature in many (not-so-cleverly thought-out) professional designs, where forced hot air passes over the small signal circuitry.

Valve amp warm-up

In valve amplifiers, pre-conditioning is necessary to reach thermal equilibrium across the chassis, which may take longer than in solid-state equipment, as the 'heat-exchangers' are made of glass, and the small-signal parts are frequently located the other side of the metal chassis, itself an initial heat shield but also ultimately a re-radiating surface, after it gets hot. This may take at least an hour.

Transistor warm-up

In order to heat up much at all, both solid state and valve amplifiers working in Class B and all higher derivatives (A-B, G, H, etc – see chapter 4) must be driven with a continuous test signal into a load towards the minimum ohmic rating. With conventional class A-B (or B) amplifiers having a fixed supply voltage, which may sag a little under load, the highest normal, legitimate dissipation in the output stage occurs when an amplifier is driven (assuming a continuous sine wave) into its rated load at –3.5dB to –4dB below clip (alias –4dBr or –4dBvr). This 'third of rated power' condition also causes the most rapid warm-up. Although an unrealistic condition if music is played below clipping, a solidly designed amplifier should certainly survive, and operate for a while, before protecting itself.

Typical conventions

At *Stereophile* magazine, high-end domestic amplifiers are pre-conditioned with sine wave drive at 1/3rd of full power (i.e. $\cong$ –3.5 dBvr) for 30 minutes.

The domestic *THX* specification (for home theatre) requires the AUT to handle 1/10th of full power (i.e. driven to –10.5dBr, where 0dBr is full voltage swing into rated load) for 3 hours 'without substantial change in measured performance'.

The test standard for professional touring PA amplifiers [3] is more gruelling. Amplifier are first switched on for 30 minutes with no signal. Key performance figures are taken. The AUT is then driven for one hour with pink noise, band-limited as in the AES/EIA speaker test standard [4] and with a 10dB +/–1dB PMR, and set just below visible clipping. Key performance figures are then re-measured. With a resistive load of 5 ohms, and both channels driven, this is a stringent but fair test.

Class exceptions

Classes A, D, G and H are the exceptions.

For Class A, no signal is needed. The amp is simply switched on and left until the temperature has stabilised. This no drive condition results in highest dissipation.

For Class D (PWM type), % efficiency is almost invariant with drive. A sinewave at –4dBr may be used for convenient standardisation but 0dBr would give slightly more and the highest, dissipation.

For Classes G and H (at least in two-rail circuits), maximum dissipation in the output stage occurs at higher levels than class A-B, at typically 56% full power, alias –2.6 dBr.

7.3.3 The test load

To be meaningful, most tests require the AUT to be loaded. The *test load* is usually a passive device or circuit that has to remain suitably stable in value and consistent over a 100,000:1 frequency range (from 1Hz to 100kHz) and the 300 million fold range between up to 300 volts and 3μV, while having load characteristics that the tester can justify. Excellent low resistance, very stable ohmic connections are also needed. Slight dirt or corrosion in the connections can limit the resolution and repeatability of distortion and other linearity tests.

The resistive load

Near-pure resistors are the staple reference load for most tests. The resistance used will inevitably be wirewound to handle the power. It must be sufficiently pure for the impedance to remain within tolerance at 20kHz, and to some lesser tolerance above. A reasonable tolerance over the audio band is +/– 1%, i.e. within +/– 80mΩ for 8 ohms. For an exact resistive load to be not more than 8.0Ω+1% at 20kHz, ie. 8.08Ω, the load's inductance including connections must then be less than about 600nH. Even if the load's net inductance is low enough, the loop-inductance of a metre of connection cable from the AUT can add this much. Fortunately, many real speaker cables and voice coils have this order of inductance, so this particular error gets us closer rather to, than further from, reality.

If a test load is made from regular wirewound resistors, their reasonably small inductance can be reduced by paralleling. If bare, plain wirewound resistors wound without inductance cancelling techniques are used, such as nichrome heating elements, the inductance can be largely cancelled by connecting one unit as two halves, by using a centre tap, and wiring as seen in Figure 7.7. This scheme creates opposing magnetic fields in the same axis [5]. If the original resistor's value was R, then the effective resistor is R/4, as the two halves are in parallel. So two 1kW, 57.6Ω elements each connected thus, and then in parallel, would give 7.2 ohms, and assuming adequate cooling, would handle up to 2kW. Resistance variation from 20°C can be kept well below 5% when operated below red heat.

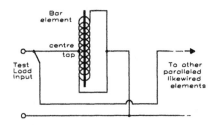

Figure 7.7

Low cost test resistor. Uses an everyday nichrome heating element. The centre tap connection cancels much of the parasitic inductance. It also quarters the original element resistance. Using combinations of readily available (750w, 1kW, 1.5kW, 2kW, 3kW) elements, two or more parallel-connected units may be required to produce standard load impedances e.g. 15, 8, 4, 2, 1, 0.5Ω.

Load specifications

The old DIN 45-500 standard assumed that the loudspeaker's minimum impedance is 80% (0.8x) of the nominal rating. It said nothing about phase angle.

In 1975, Peter Walker [6] suggested that a high performance audio power amp should be able to drive with its full output *voltage* capability, the range of reactive loads implied by

$$R \pm jX$$

without exceeding it rated distortion (%THD+N). In this equation:

R = amplifier's rated minimum load

X = 0 to +∞, ie. a reactive load that can be leading or lagging, hence either a coil or a capacitor, ranging as low as zero ohms up to infinity, in series with the resistive portion, R.

This expression is the simplest model for the kind of reactive loading developed, and applies to a single kind of drive unit, so it is most appropriate to a bi- or tri-amped system. As a measure of the counter-intuitively stressful nature of the series reactance, the equation requires (assuming constant output voltage) an amplifier to deliver *half its maximum rated current at zero output voltage* – at which juncture the output transistors are seeing the full rail voltage. This is the most difficult phase angle condition.

Stuart [7] reckoned on the following enlargement of Walker's formula, to define the minimum impedance dip:

$$jXR/(R+jX)$$

This is based on combining one condition,

$$R/2$$

i.e. a minimum resistive load of half the rated

with another:

$$R//JX$$

where :

// = in parallel with

X = ranges 0Ω to ∞Ω as above; and

R = the amp's rated load capability.

The combined equation again ranges over the worst phase angle condition, and demands that an amplifier delivers its *full* rated current at zero output voltage. This tough requirement is however based on the behaviour of real studio monitor speakers at the time. Fortunately, the worst case phase angle would not normally be encountered for very long with program. If an amplifier can drive such a load for at most a few seconds at a time, with a 10% duty cycle, all will be well. With all-BJT amplifiers, the usually necessary SOA sensing protection circuitry will, if well designed, employ some delay to cope with such transient excursions.

If a relaxed specification is deemed good enough, because an amplifier is supplied with, or known to be designed to drive, particular speakers that are known to offer a defined, benign loading, the definition is simply changed so X (the 'imaginary' reactive part) ranges from R upwards, and not down to zero. [8].

Between 1983 and 1987, Otala and colleagues established that the impedances of high performance speakers using ordinary moving coil drivers could typically dip to a sixth of their nominal impedance [9]. In other words, an amplifier intended to drive *any* 8 ohm loudspeaker, should be rated to drive as low as 1.2 ohms, without hiccuping – at least in bursts for a few seconds at a time.

Today's THX specification for domestic equipment sets a given voltage (say 20v rms) at which the amplifier must handle loads from (say) 2 ohms to infinity, and also drive a worst case phase angle of (say) +90° at the rated 8 ohms. What the amplifier is *not* required to do is drive the worst case phase angle simultaneous with the lowest impedance.

Other transducers

The impedance behaviour of speakers using transducers other than the moving coil kind, is quite different again. Ribbon speakers have an unusually low impedance across the band of around half to one ohm, that is of course principally resistive, then inductive – due to skin effect as well as length.

Electrostatic speakers are renowned for difficult loading, being principally modeled by a damped capacitor in series with series inductance (Figure 7.8, i). This kind of loading was omitted from the IHF-A-202 standard. In the UK, an 8Ω +2.2μF capacitative load test (Figure 7.8, ii) has been long established [37] as a licence to drive ESLs. Still, it is rare for the majority of audio performance tests to be performed into these loads. For a start, the 8Ω+2.2μF load destroys many domestic and even some professional amplifiers when they are swept much above 20kHz in the 'top' 0 to –15dBvr of their range. Instead, these tests are mostly reserved for checking RF stability with HF square waves, and for abuse survival testing.

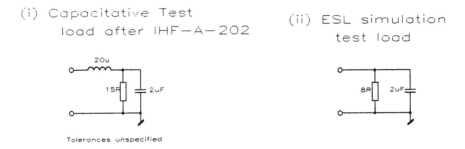

Figure 7.8

Capacitative test loads for ESL simulation and abuse testing.

Why no 'real' load

With few exceptions, loudspeakers are not made to be driven with continuous signals, at least at their highest rated power – which is usually based on audio *program*, with much higher PMR than a sinewave. Typically, a drive-unit's or speaker's continuous rating (in *watts* $_{avg.}$) is a fifth to a fifteenth of the program – or highest – rating.

Despite of the fact that continuous signals are wholly unrealistic as a measure for audio power, they are repeatedly necessary for performance testing, at least for a few tens of seconds at a time, in both conventional visual inspections of scope traces, and in computer-driven swept tests. Using loudspeakers as test loads with test signals also creates unwanted, high SPL, anti-musical sounds. Muffling almost any speaker even slightly, changes the impedance characteristics. And the drive unit or voice-coil must be readily field-exchangeable, to replace inevitable burnouts.

Even if there is a spare room to put a test speaker into, even slight movement of the speaker or other objects could substantially change the speaker's impedance, which is dependent (amongst other factors) on reflections in the space. This suggests a loudspeaker driven with test signals needs to work into a defined acoustic. And different speakers' impedance patterns vary endlessly. Which should we choose and why ? Seasoned designers and reviewers have their 'rogues' gallery' of speakers which have proven a hard load, or even fatal, to past amplifiers.

The upshot is that performing audio power amp measurements with real speakers as loads requires a high investment in hardware and acoustically treated space (which may even defeat the test's realism) , and is of limited practicality and best avoided altogether at high levels, say anything above 5 to 10 watts with most domestic drivers; and above 75 to 150w with most pro-drivers.

The SLS load

An SLS (Simulated LoudSpeaker) load is a network of passive, reactive and damping components (capacitors, inductors and resistors) that more-or-less behaves like a given loudspeaker, at least under the kind of test conditions it was designed for.

Like a resistive test load, an SLS can be designed to withstand any duration and power level of continuous test signal. But with even modestly high powers, above 50 to 100 watts, specially made, physically bulky inductors are needed, as even 'high power' conventional ferrite cored types begin on the slope to saturation, causing the load to be unacceptably level dependent. This could mean a lower powered amplifier seeing an easier load than a higher powered one. For fair measurements, the power handling *Vs.* impedance curve of an SLS should be measured before first use, so the maximum voltage and power levels for consistent loading are identified, and then marked up on the load.

A simple SLS to simulate a single drive-unit (Figure 7.9) still needs a conventional, high power-rated resistive load. This defines the DC resistance, and remains the sole part dissipating significant power (as heat). The capacitor may be electrolytic

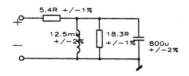

Figure 7.9

IHF-A-202, General Test Load.
This test load is the basis of a simulated electro-dynamic speaker, for improved test realism without sound. Pure, highly rated and calibrated parts are required for consistent testing.

but must have a high ripple current rating, and to achieve this and also be adequately rated for handling high swing amplifiers, also a high voltage rating, up to 550v DC. The inductor must practically be an air- or steel-cored type for consistent, saturation-free inductance over the range of current (0 to 75A, sometimes more) and frequency (at least 1000:1)

Advanced SLS

More advanced SLS may be constructed [10] using diodic-resistive meshes to simulate the driver's non-linearity with level, and harmonic current draw. These are most appropriate not to power testing, but to evaluating %THD, %IMD, etc, especially dynamically. Real speakers may not survive this and if the noise is adequately contained, their electrical loading and thermal rating are likely affected.

Visual monitoring

An essential part of the test load is to have 'scopes connected at all times, to monitor both the output voltage, and if possible current. The voltage reading connection is the normal one, showing (voltage) clipping, any RF instability, and anything else untoward. The current reading connection is optional. It may be used to show if current-clipping occurs, during any testing into reactive loads. To capture RF oscillation, the scope(s) should have at least 35MHz, and ideally 100MHz bandwidth. The test events may also be taped – a domestic stereo cassette tape machine can record V-I data. If the AUT should expire during a test, the conditions leading up to this (at least below 30kHz) can be be reviewed.

7.4 Frequency response (Bandwidth, BW)

Frequency response is a measure of amplitude, or relative gain, against frequency (Figure 7.10), alias bandwidth. For audio, the approximate centre of the audio band, say 1kHz, is taken as the reference point (0dB). For gathering useful information on audio amplifiers, the bandwidth measurement may (and *must*) be made into a variety of rated loads, including no load. The '*power bandwidth*' measurement of conventional electronics is over-simplistic and is wholly eclipsed in this and later sections.

In common with all high performance audio equipment, audio power amplifiers should have an 'essentially flat' response under all operating conditions. The response may *roll-off* monotonically (continuedly) beyond the audio band. At high

Figure 7.10

This model bandwidth curve is offset below the 0dBr line for clarity. At low frequencies there is no sign of droop at the Audio Precision's 10Hz limit, at least on this scale. At 20kHz, the response is down less than -0.25dB, falling off with gradual acceleration to -6dB at 200kHz.

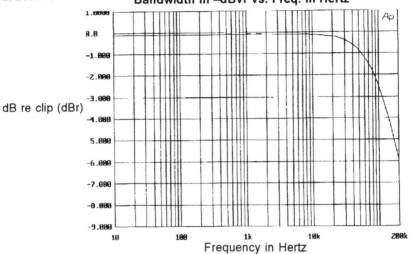

frequencies, this is unavoidable. But there should be no rise (+dB), and no peaks (+dB) nor dips (–dB) in the response, even where rolling-off. As in Figure 7.10.

Assuming an audio power amplifier is intended for full-range use, it should respond up to 20kHz, ideally with much less than –0.25dB of rolloff. This seems stringent, but rolloff is cumulative, so with just two units in the audio chain, and two cables and a speaker *each* meeting this specification, this short audio path alone, will impart a total 20kHz rolloff of –1.25dB. In percentage terms, +/– 1.25dB is above +/– 15%, and while barely audible as a change in loudness in the midrange at moderate levels, a droop or rise of this size is quite noticeable when it affects a portion of the frequency range, and also when it affects low and very high frequencies. Ideally the response should not then drop like a stone, but continue, either flat or with no more than –10dB rolloff, up to 100 or 200 kHz.

For a good 'transient' response, which describes how quickly and neatly a signal settles down when the level changes suddenly, a slow or modest rate of rolloff is necessary. Looking at Figure 7.10, the initial rate of reduction is ≅ 0dB per octave, gradually accelerating to reach a rate of –6dB per octave, past 200kHz. *Note: the rate of –6 dB/ octave means signal voltage (and current) level halves for each doubling of frequency, in other words reduces at approximately –20dB per decade.* At higher, radio frequencies, a large reduction or attenuation is desirable, so these do not get far if they seep into the signal path. Fortunately, once these higher frequencies are reached, the effect of a rapid reduction or rolloff has little effect on the transient response for audio. The rolloff rate may then increase (to –12, –18dB/ octave, or more) to furnish useful filtration against *all* higher frequencies.

At low frequencies, considerations of excess phase shift and signal delay require the rolloff to be −0.5dB at considerably lower than the 20Hz that is often advocated parrot-fashion – see sections 1.12 and 3.4. It is partly this, and partly due to technological limitations, that many audio test systems do not record below 5, 10 or 20Hz. The HP 3561A analyser (as used by high-end domestic amplifier reviewer Martin Colloms) is one exception: using pseudo-dc digital stimulus and a digital filter, it measures down to 0.005Hz. Otherwise, with most test sets, a healthily low LF rolloff frequency is beyond measurement capability and has to be cited as 'by design', although it may be reliably cross-checked by computer simulation.

Frequency response may be cited as a pair of frequencies, at which the response is −3, −1, −0.5dB down, e.g.

> **'Frequency Response, 10Hz to 50kHz, −0.1dB'.**

Still, the reader is left to speculate as to whether there were any peaks or dips, or other irregularities. A frequency response graph can provide this information at a glance, subject to scaling, measurement and printing resolution.

Frequency response of modern power amplifiers with high NFB is conventionally measured at typically −0.5 to −1dB below visible clip (judged on a 'scope), alias − 1dBr. This needs qualification. Below overload, the frequency response of a power amplifier is assumed to be invariant with level. This is broadly true when the amplifier employs some NFB and omits transformer coupling. *If not, response should be replotted at say −10 and −30dBr.* And when some amplifiers are working close to (or in some cases, as with soft-clip, some way below) their limits, protective functions, e.g. limiting, anticlip, etc can cause frequency response variations that are dynamic, i.e. 'level conscious'. If uncontrolled, such frequency response aberrations may have unpleasant, unsettling or just unnecessarily attention-grabbing effects. *For amplifiers equipped with such protection, it is instructive to plot the frequency response at different drive levels,* above *and below* the indicated onset of the potentially offending protective circuitry.

Some well protected amplifiers mute the drive as the test signal sweeps out to the frequency extremes. The sweep has to begin somewhere and may trigger protection before it can begin. Short of disabling the protection system, the most data can be gathered by setting the sweep to run as fast a possible (eg. with widened settling tolerances), then running from 1kHz upwards (if there is RF muting protection), and having a separate sweep down from 1kHz, if there is subsonic muting protection.

7.4.1 Gain and balance

Gain by derivation

Audio power amplifier gain is normally measured and cited at 1kHz, which is the notional middle of the audio band, and also a tidy number. Gain is often not measured explicitly, as it can be derived from the frequency response measurement, provided the input level is recorded (in dBu). The gain (in dB, a purely relative, dimensionless unit) is simply the difference between this and the 1kHz level (if in dBu), read off from the frequency response graph.

The gain figure has a tolerance, as depending on how much global NFB is used, gain will vary across the range of output loading. The difference is small, say 0.5dB for amplifiers with high, global NFB. Without both, one, or other of these, more significant gain variations of 3dB, or more, may apply across the rated load range. This affects SPL predictions.

Balance

Balance is the *difference in gain* between channels. These are usually stereo, but may be *any* amplifier channels that together must create a coherent acoustic output, eg. bass and mid/high for a bi-amped system.

More than +/-0.1dB of variation between stereo channels can upset imaging in high performance domestic and studio monitoring systems. This is fixed in theory (but never for long) by manually balancing faders or other individual gain controls. In PA systems, larger gain errors can be lived with, but then system alignment is no easier and some parts of a system may be driven beyond clip (or subjected to limiting), while others are under-utilised.

Bandwidth *precedes* balance

If the frequency responses are also not identical, the 1kHz that gain is expressed at, is seen to be arbitrary. The equal loudness contours (see chapter 1) increase the complexity of the meaning of the disparity. *It follows that matching amplifier gains is fruitless* (e.g. for A-B testing) *until the frequency responses are matched to a greater accuracy*, ideally at least three times better (in fractional dB). For example, if frequency response between two AUTs differs +/-0.1dB (over some part of 20Hz to 20kHz say), then gain matching to better than +/-0.3dB is an effort better spent on improving the frequency response.

Transient balance

Balance testing is ordinarily performed in the steady state. Colloms [11] has described a method for dynamic testing. A signal (that may be pulsed) with a similar amplitude and identical phase as the signal emerging at the output, is driven up the

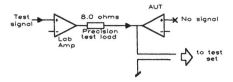

Figure 7.11

Here, output impedance (or 'damping factor') is measured using a second, high quality amplifier for testing ('Lab Amp'). Its output is driven up the AUT's quiescent output, and the signal is viewed here to read output impedance, Z_o and interface errors.

275

output by a secondary source, in series with 4 ohms. Differences in the sum of the signals at each output (that are greater than the tolerance of a perfect summation) show any dynamic imbalance. The arrangement follows that used for damping tests in Figure 7.11.

7.4.2 Output impedance (Z_o)

The significance of output impedance is 'goodness of voltage source'. An amplifier with a very low output impedance will be able to maintain its output voltage without flinching, irrespective of load impedance dips. The output impedance is derived from the bandwidth plots, provided the amplifier's output is measured *at* (or within a *very* small fraction of a milliohm from) its own output terminals, not at any point down the load cable.

Granted this, then the impedance is visible from, and may also be accurately computed from, the (ideally) small difference between plots into (say) 8Ω and 4Ω. For modern, direct-coupled amplifiers with low output impedances, below $100m\Omega$ say, the measuring equipment must have resolution and repeatability to below 0.05dB (about 0.6%). The load impedance must also be stable. Load resistance changes with temperature (*Tempco*) must be accounted for, or designed out of the test.

How is impedance calculated? Say the 8Ω plot has a level of +35dBu, while the 4Ω plot is at +34.9dBu, both at 1kHz.

The source impedance needed to account for this 0.1dB drop or difference is

$$(0.1 / 20 \text{ x } 1 / \log_{10}) = 1.01158 \cong 1.16\%$$

This is 1.16% of $(8\Omega - 4\Omega) = 0.016$ x 4, $\cong 46m\Omega$. Rechecking:

$$\{ (0.045\Omega + 4\Omega) / 4\Omega \} = 1.0115$$

then if you cannot see the approx. 1.16%, multiply result by

$$\{ \log_{10} \text{ x } 20 \} = 0.0993dB$$

which is close to the 0.1dB.

Knowing that (in this case) the first 0.1dB division represents $46m\Omega$, the approximate variation in Z_0 with frequency may be roughly estimated by eye, from the graphs of the two frequency responses (Figure 7.12).

Another way to measure output impedance is to use a second amplifier to drive up the AUT via a test load resistor (Figure 7.11). The AUT is powered-up but not driven. This connection tests the AUT's external current sinking ability, and approximates the conditions with loudspeaker back-EMFs. The second, 'lab' amplifier should ideally have a wider bandwidth, faster risetime and a higher slew limit than the AUT; or else the test bounds should be adjusted inwards accordingly. It need not have an exceptionally low Z_0, but the value of Z_0 should (rather circularly) be known. The resistor (typically 8Ω) limits and helps define the current, and approximates voice coil series resistance. Its resistance must also be accurately known.

Figure 7.12

A typical magnified comparison of bandwidth into 4Ω and 8Ω, that gives information on the Amp Under Test's output impedance Vs. frequency. The AUT is assumed not to be clipping. Other than using the difference to compute numeric values, the amplifier under test is clearly showing an higher source impedance above 2kHz with the 4Ω loading condition. Notice that with magnification, the response is not really flat over much of the band, and that some of the small (0.02dB) wiggles below 300Hz are the test set's own, being typical of even the finest transformer coupling.

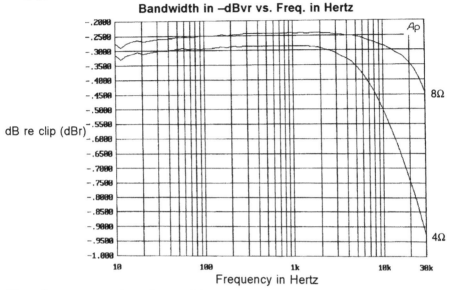

Bandwidth in –dBvr vs. Freq. in Hertz

The unknown output impedance of the AUT, call it Z_x, is calculated by measuring the rms voltage on either side of the resistor, then computing Z_0 from the voltage division. For example, if the voltage at the test amplifier's output is 10.0v at say 1kHz, and the test amplifier's Z_0 is 25mΩ at this frequency, and the voltage sustained at the output of the AUT is 9.90v, then we have a 100mV drop to account for. Current is approximately

$$10v/8\Omega + Z_x$$

Call it $\cong$ 1.2A rms. So Zo must be close to:

$$100mV/1.2A = 83m\Omega$$

If 83mΩ, I is really 10/8.083, and so Zo is actually close to 80.1mΩ.

As many power amplifiers have much lower Z_0, accurate testing (for what it is worth) demands increasingly expensive test tools. For example, 1mV rms in 10v (1 part in 10,000) must be resolved, to read a by-no-means exceptional Z_0 of 8mΩ. *However, the principal value of this relatively little-used yet elegant technique is not one dimensional number gathering, but its ability to show waveform integrity at the AUT's output. The analysis is instructive both as a visual check on a 'scope, and one or more of the usual distortion tests, particularly if an SLS is used as the series load.*

7.4.3 Damping factor

'Damping factor' is an arguable figure-of-merit, that is derived from Z_O. It is approximately an impedance ratio, attempting to indicate how much grip a power amplifier has on the speaker(s) it might be connected to.

Typical damping factors range from 0.1 to 10 for primitive output stages with nil voltage feedback; 10 to 50 for transformer coupled output stages with low to medium global feedback; and 50 to 2000 for direct coupled output stages with medium to high global NFB.

Strictly, *Damping Factor* is the voltage division caused by the nominal speaker impedance working against the source resistance (Z_S), i.e.

$$(Z_0 + Z_S) / Z_S$$

For values above 50 or so this simplifies with increasing accuracy, to just (Z_O/Z_S). For example, an 80mΩ Z_0 yields a notional damping factor of:

$$(8\Omega / 0.08\Omega) = 100$$

when referred to a nominally 8 ohm speaker.

However, Z_S comprises Z_0 *and* the series impedance of the crossover (or any protective series capacitor) *and* the connecting wires. In many cases, the latter two impedances will dominate. In any event, in nearly all cases, they are *wholly* undefined: the power amplifier maker has *no* knowledge of what will be connected. Moreover, the series impedance of the speaker's voice coil, any passive crossover or protection parts, and the amplifier's Z_0, *all* vary in their own way with frequency. The result is that the 'real' damping factor is the difference between the speaker's impedance, and sum of the others [see section 2.3.2]. Even with static signals, there will be complex nested variations with drive, temperature, and aging. It follows that a single figure for damping factor, as used hitherto, is hugely inadequate; reality is closer to a damping 'surface' or 'waterfall' of three, four or more dimensions, viz. impedance ratio *vs.* frequency, time, level and temperature.

Damping factors beyond about 20 are considered by some [12] to be worthless, on the grounds that the speaker's voice coil resistance, always several ohms, is always in series with the ultimate load. On this basis, a high damping factor is still worthwhile, as it is tantamount to good regulation, i.e. low Z_0. But it is anyway a gross simplification to say that the voice coil resistance is in series with the load; in reality, each turn of the voice coil couples individually and proportionately, so the resistance is distributed. Thus, with transient signals (i.e. music), damping counts.

7.4.4 Phase response

Phase, a measure of both absolute and relative difference in signal timing, and also a record of polarity, is nearly always discussed and rated in degrees (°). But remember, phase may also be measured in radians, where 3.142 rads = 180°.

At + or −180°, a negative feedback signal has become purely positive feedback, and polarity has flipped (or 'phase is reversed'), so the signal appears to have moved half-a-cycle to the left (or right). At 360°, the signal is effectively at 0° again ('gone full circle'), but it is also delayed (or advanced) by one cycle.

Phase may be roughly assessed from calibrated dual 'scope traces. Calibration involves testing with a dual signal source having a known, major phase shift, e.g. 180° between sources. A more accurate and immediate measure of specific phase angles near to 0° (or '360°') and 180° is made with an X-Y connection, giving well known Lissajous ('Lissa' jew') patterns. Else a digital phase meter is most likely to be used. Here the zero crossing time of two signals is compared. Such instruments should be used with knowledge of their limits, as they may feign an accurate result, even if the frequency or test signal amplitude is outside limits. A resolution of 0.1° or less and worst case accuracy of 1° is desirable for testing high performance amplifiers. The input attenuation may need to handle up to 250v AC rms.

No widespread standards exist for testing the 'phase response' of audio power amplifiers, except that the input signal may be taken to be the 0° reference. But phase shift at the output relative to input is rarely so explicitly measured. Instead, relative phase at the output, usually assuming 0° datum is at 1kHz at the output, is cited at the frequency extremes. The ideal spec might read:

'Phase response, after warm up, into 8 ohm resistive load:

≤+5° at 10Hz, ≤−5° at 20kHz'

or something similar. If the input is used as the reference, and the amplifier inverts, so 1kHz is at 180°, this may be subtracted, for the convenience of 0° as the reference. Otherwise or additionally, phase may be specified as a graph of linear degrees (°) against the usual log frequency scale (Figure 7.13, overleaf). In either case, the warm up procedure is worth noting, as phase values can change considerably with the temperature of the passive frequency response determining parts.

Alas, all phase curves are meaningless, or at least hiding the reality, until expressed on a linear frequency scale [13, 14]. Once this is done, the natural signal delay that is independent of frequency, hence harmless, becomes a simple, straight line component. The harmful stuff, called *excess phase shift*, has a *non*-linear ° per Hz relationship, obvious when a ruler is placed against the curve. It is the sole cause of curvature when phase response is shown against a linear frequency scale (Figure 7.14, overleaf). The excess phase shift causes waveform distortion ('smearing') and should be kept to a low level in the audio band. A maximum of 10° (of both kinds) of phase shift in the audio band was regarded as the allowable limit in 1973 [15]. Direct coupling (at LF), extended bandwidth and low dB/octave slopes help to avoid harmful phase shift at both LF and HF.

Phase matching (i.e. similarity in phase response) between amplifier channels, and between separate amplifiers, used to drive different parts of a stereo (or higher) loudspeaker system, is rarely measured, yet most important to stereo performance.

Figure 7.13

The upper graph shows two power amplifier frequency responses, wide and narrow. The lower plot shows the corresponding phase response, with the same log frequency scale The two are different maps of the same territory. Within each, the way the information is presented may hide facts that ought to be salient.

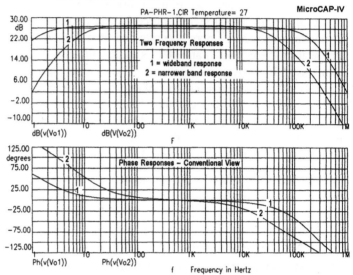

Figure 7.14

Phase linearity exposed. The upper panel remaps the low frequency phase response on a linear scale. The lower panel does the same for the high frequency response. In both cases, wideband amplifier (1) has clearly the most linear phase, as well as having the smallest rate of change in ° per Hz. By comparison, the narrow band amplifier's LF phase deviates significantly up to at least 80Hz. Linear phase shift (an unbending line) on this map implies there is no frequency dependent signal delay, which may otherwise steal clarity, vitality and timing from music.

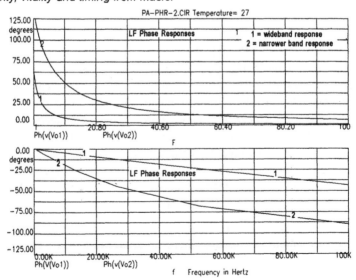

7.5 Introducing noise

For audio power amplifiers, as for all audio, 'noise' (in the most general sense of 'unwanted signals') has a special meaning. Although usually measured with the input(s) shorted, noise comprises any (added, unwanted) signals that are present, whether indeed the, or any, wanted signal is present or absent – *but not any (added, unwanted) signals caused by the wanted signal.* These latter noise forms come under the banners of intermodulation, or harmonic, distortion, noise modulation, breakthrough, and other labels. On the other hand, noise measurement doesn't distinguish much between hiss, hums, buzzes, crackles, whistles and popping sounds; often, they are all lumped together as 'noise'.

Potentially comprising any or many significant frequencies ('hum' for example is predominantly made from signals at up to 10 frequencies), noise is commonly specified as an average across the audio range, in – (minus) dB.

For power amplifiers, the average noise level is nearly always referred to *0dBr*, the reference full output level. In older specifications, it may be referred to 1 watt or other wholly arbitrary levels. Another possibility is a level in dBu. This gives an uncompetitive-looking low minus number, and if used, the amplifiers' voltage gain (in dB) must be added by the reader, to 'see' dBr figures. If the AUT has a gain control, the setting should be cited, as this will influence noise.

To be at all meaningful, the noise specification must cite the bandwidth of the measurement. The norm is 20 (or 22)Hz to 20kHz. To show the effect of excluding any hum (or that there isn't much), a second measurement may be made with a 400Hz high pass filter.

If the average noise is tested without any filtering in the audio band, it's said to be *unweighted,* often abbreviated *unwtd. 'A'* weighted noise measurements, identically filtered to those used for acoustic SPL measurements, are far narrower and often cited by low-quality or figure conscious makers because they give numerically higher (better looking) noise figures. There is also some measure of psychoacoustic justification. But any added precision in predicting actual noise annoyance that such 'A' wtg or other *psophometric* filters (such as CCIF, CCIR etc) offer, takes no account of the vast and scarcely defined range of operating and ambient SPLs, and room and speaker frequency responses, and ambient noise spectra, which altogether have a rather dominating effect on the practical audibility of noise.

In amplifiers with gain controls, particularly integrated pre-/power-amplifiers, the panel control setting will alter the ratio of the noise contribution of the stage(s) preceding the gain control. Generally, an honest specification will employ a worst case condition, in theory the maximum level setting. However, hum and EMI are often highest around about the midway gain setting, particularly if the pot wiring runs by the mains transformer.

7.5.1 Noise spectra

Modern, mostly PC-based test-sets, can produce *spectral plots* of noise. With plain analogue techniques, the bandwidth of each of the constituent measurements is normally 1/3rd octave (like a 27 band – or so – graphic equaliser).

With DSP, the bandwidth can be narrowed to at least 3Hz, which is still $1/_{10}$th octave at 30Hz. This means large noise variations between adjacent frequencies may be seen with far greater acuity.

In all cases, peaks are typically found at one or more of the following frequencies where there is *coherent* noise:

50 or 60Hz	mains line fundamental
100 or 120Hz	mains 2nd harmonic
150 or 180Hz	mains 3rd harmonic
200 or 240Hz	mains 4th harmonic
250 or 300Hz	mains 5th harmonic
500Hz to 10kHz (typ)	Electric motors (variable)
16kHz	Video monitors
32kHz	Video monitors
25 to 200kHz	SMPS fundamental
50kHz to >200kHz	SMPS harmonics

Other than these peaks, 'real' (stochastic, random) noise is distributed across the audio band. Usually, it rises in proportion to frequency at between +3 to +6dB per octave. Figure 7.15 shows some of these effects.

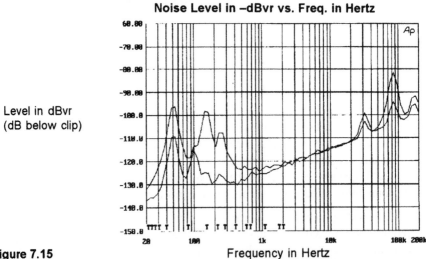

Noise Level in –dBvr vs. Freq. in Hertz

Level in dBvr
(dB below clip)

Frequency in Hertz

Figure 7.15

Noise spectra of two power amplifier channels, plotted with swept 1/3rd octave bandpass filtering. 'T' (Timeout) shows a noisy value. Truncated peaks are due to a limited number of points being used to speed-up plotting. Using a ruler, note the nominally +3dB/octave rise in the background behind the peaks, indicating this part will be perceived as halfway between pink and white noise. Note also interchannel differences, due to layout.

7.5.2 Breakthrough and crosstalk (channel separation)

Signal(s) may leak between channels. When wiring and layout has asymmetries, then high level signals in the path may leak into other parts of the signal path or to other channels. *Crosstalk* (or '*X-talk*') is reserved to describe leakage between two, usually stereo channels while *Breakthrough* is more *ad-hoc* leakage. In power amplifiers, which word is used will depend more on usage than form. In multi-channel amps, breakthrough is worst between adjacent channels, and at maximum levels. It's also rarely symmetrical; Ch.2 may pollute Ch.1 more than Ch.1 affects Ch.2 or *vice-versa* (Figure 7.16).

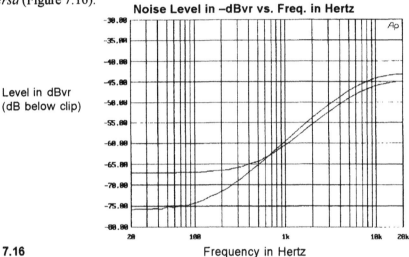

Noise Level in –dBvr vs. Freq. in Hertz

Level in dBvr
(dB below clip)

Figure 7.16   Frequency in Hertz

Crosstalk between two channels is very often asymmetric, as here, with one having 8dB more leakage below 100Hz. Note the +6dB/octave slope that lies behind the limits at high and low frequencies. This slope indicates capacitive, alias electrostatic, crosstalk.

To test, one channel (either) is driven within a dB or so of maximum level, with nothing connected to its outputs. The signal appearing or 'breaking through onto' the other channel(s) is then measured (in turn). The inputs of these undriven channels must be terminated (e.g. with 47Ω at the input socket). For realism the driven channel should be (silently) loaded, so any breakthrough induced from high level output current is included. But unless a specific real usage is being evaluated, this shouldn't include any noise induced from outside the AUT. So the driven load's wiring must be carefully dressed away from the wiring that is sensing the undriven channel output(s).

Alternatively, all channels but one are driven, and the leakage measured on the undriven one. This is obviously more taxing. To be meaningful, the levels used must be cited. As ever, a good spec defines the procedure used and settings.

Like noise, breakthrough and crosstalk are usually specified as an average, and the pros and cons of peak *Vs.* rms measurements, spectral plots and even weighting, apply again. Usually, crosstalk and breakthrough figures decay (get worse) with increasing frequency. The THX specification for home cinema considers less than −50dB of crosstalk acceptable at 20Hz and 20kHz. *While −50dBr is in itself just about passable for music, the rate of reduction (in dB/octave) that is implied, for the crosstalk to be ideally below −100dB at 3kHz, where sensitivity is far, far greater, is an unlikely −18dB/octave.* As with CMR, more is always playing safer. An achievable unweighted figure for a well laid-out stereo power amplifier of any rating, with separate channel powering, is −120dBr at most (if not quite all) audio frequencies. At this level, there is no disguising the fact that what is being measured is probably noise, and that the actual breakthrough may be far lower and inaudible, being lower than the noise. For unless special DSP analysis is used, tests can't read below the noise, so breakthrough figures are never normally better than the noise figure taken under the same conditions. So all breakthrough and crosstalk figures are, in effect, '+noise'. A model spec might read:

'Crosstalk+N, unwtd rms level at Ch.2 output, when Ch.1 driven at 0dBr into 8 ohms, Ch.2 input terminated in 47R, includes residual noise, −105dBr @ 3kHz; <−90dB, 22Hz–22kHz (see also spectra).'

7.5.3 Understanding CMR measurements

With well designed power amplifier input stages, CMR is typically between −60 and −100dB at low frequencies, usually reducing with a constant slope of +6dB/octave, coming in at HF or above the audio band if the CMR is weak, and below 1kHz if it is strong (Figure 7.17).

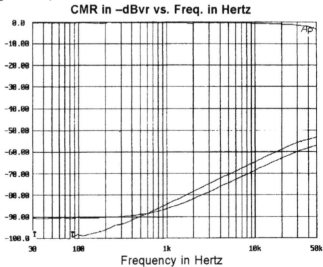

Figure 7.17

Lower plots shows how CMR reduces monotonically at high frequencies, and may or may not be constant below 1kHz. The upper plot (0dBr) is the concurrent differential signal at the output. The effective CMR is the difference between the two. Without this consideration, the effect of the amplifier's limited bandwidth (due to RF filtering roll-off in the signal path) is to make the CMR appear to get better above some point.

The reduction is commonly disguised and lessened by the equipment's own hf rolloff, and also by any inappropriately-set filtering in the test set. A single figure (the bland '–60dB' cited by some makers) disguises the variation with frequency.

CMR (or CMRR) figures expressed in (–) dB are somewhat approximate, at least unless absolute levels and measurement bandwidth are stated, because practical CMR measurements include the power amplifier's noise. Even if the CM signal-to-noise ratio *is* favoured by probing inside the equipment to directly read the output of the actual 'debalancer' stage, the residue of equipment with a good, high CMR can be seen to be mostly noise across at least part of the spectrum. In effect 'CMR+N' is being measured, where '+N' is noise. The ratio between the two inevitably varies with frequency. CMR figures also ideally need putting into context by using decibels referred to clip (dBr, dBvr).

7.5.4 Measuring CMR

Originally, CMR was a simple, spot measurement, something like 'CMR = –60dB'. Today, with *Audio Precision System One* and similar modern test instruments, it has become routine to plot CMR *Vs.* frequency. The AP *System One* measures CMR via a $^1/_3$rd octave bandpass filter, which can be swept up to 200kHz. Any test set's CMR measuring capability is practically limited by its generator. In the case of the AP *System One* test set, it's typically –110dB, and still –100dB above 20kHz.

To avoid unnecessary +N contribution, CMR should ideally be measured at the output of the input stage. But this point isn't so readily accessible. The easy route is always to measure the signal at the amplifier's output sockets. This has the disadvantage that noise produced inside the equipment is added to the residue, so a good, high CMR figure is decreased unless the internally produced noise is very low. CMV residue is favoured over noise at all frequencies by using a high enough drive level when testing.

It is helpful to note that CMR testing ('CMTST') input signals may be made much larger than normal operating levels, ultimately in direct proportion to CMR. This tests the common-mode voltage handling on the amplifier's input stage; too much input level will cause gross non-linearity (common-mode clipping), and CMR will reduce dramatically. The range of levels over which CMR may be meaningfully tested may be established by individually plotting 10Hz, 1kHz and 200kHz CMR *Vs.* CM drive level.

For equipment that can handle it, +18dBu is one *de facto* standard adopted by testers, being the highest drive level that can be used to sweep up to 200kHz with the *Audio Precision* test set. 0dBr is then (18dB + amplifier's gain) below this, ie. if the output level was 0dBu and gain +22dB, then CMR would be –40dB. This is tantamount to saying the amplifier has a *hypothetical* output of +40dB, even if it is not capable of 77.5 volts rms. This reference level has accordingly been dubbed dBr*H*. For a 200kHz sweep, bandwidth is set at 500kHz. With lower or higher reference levels, and/or with a measurement bandwidth that's narrower (or even wider), and with different measurement techniques, CMR+N can measure several dB different, making a nonsense of fine comparisons.

Beyond a point, CMR becomes hyper-critically dependent on component matching. Component values commonly vary with temperature, so CMR is also temperature dependent. For this reason, when CMR exceeds about −50dB, the ambient temperature and warming-up preconditioning procedure must be specified.

Overall, equipment manufacturers need to take more care to characterise their CMR test conditions. A bald 'CMR = −60dB' is not good enough any longer. An honest spec might read:

'CMR $\geq$ −60dBr @ +18dBu drive over 20 to 20kHz, with 10Hz to 500kHz BW. 0dBrH = +40dBu; CMTST method; T_{amb} = 21°c; warm-up proc. = PRO-PA standard.'

Even this can be misleading at high frequencies, as in practice, the real HF CMR above 1kHz can be limited beyond about −50dB by the imperfect balance in line cabling.

7.6 Input impedance (Z_{in})

Input impedance, either differential or otherwise is taken on trust and rarely checked. Yet a high CMR, better than −40dB is increasingly pointless unless the input impedances are also matched across the frequency range. Measurement is relatively straightforward, using a defined source resistance. Fixing and keeping the matching is a completely different matter.

7.7 Introducing harmonic distortion

Harmonic distortion concerns the spurious 'growth' (or the corresponding change in waveshape) of a music signal caused by its passage through most kinds of audio path or device. The added parts, called *harmonics*, are always exact *integer* multiplications (x2, x3, x11, x18 etc.) of the original, incoming signal's frequencies.

In audio equipment, unasked-for harmonics are commonly produced as a result of signal amplification (ie. gain or loss) being different on the alternate positive or negative signal polarities, ie. gain is changing with alternate half-cycles. The resulting distortion may be cancelled out, by passing signal through some kind of 'equal-but-opposite' circuitry. This is an original reason for the existence of some of the various push-pull, differential, and bridged topologies – see chapter 4. Beware! Such schemes may presume accurate complementarity in the active devices, which like an asymptote, is often approached in print, but never quite delivered. Rectification in an oxidized or a badly plated contact is another potential cause of asymmetry. Distortion caused by such asymmetry is (perhaps surprisingly) mainly composed of *even order* harmonics, ie. 2nd, 4th, 6th, 8th, 10th, etc.

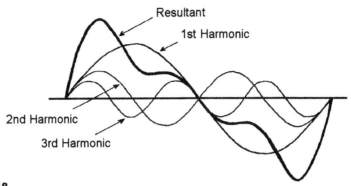

Figure 7.18

The fundamental ('1st harmonic'), 2nd and 3rd harmonics and the resultant waveshape, which depends on the three different amplitude and phase relations.

Another fundamental mechanism is that the gain of stages within any AUT (Amplifier Under Test) is commonly greater at medium levels than at higher or lower levels. As large signals have to pass through small and medium amplitudes to reach a higher amplitude, any changing gain-with-amplitude inevitably bends them (more than smaller signals) and the type of harmonic distortion produced is mainly odd order harmonics, ie. 3rd, 5th, 7th, 9th, *et seq.*

Harmonic distortion is *but one* outcome of this bending. It helps to understand that some kind of *amplitude-related non-linearity* in the AUT is *always* the first cause; that harmonic distortion is but one symptom of the sum of the non-linearities in the AUT's signal path; and that other kinds of distortion (e.g. see IMD, DIM, SID) are no less products of it. In all the above cases, it is exacerbated by large signal excursions. In everyday amplifiers with high global NFB, this and other causes of distortion are forced to be very low, but it still intrinsically rises with level until the NFB 'snaps' at clip.

Crossover distortion, which occurs in all energy saving output stages based on Class A-B or B, arises about zero volts, the origin of all signals in such stages, and where polarity alternation occurs. Here there is the opportunity for both kinds of irregularity, and in turn crossover distortion harmonics are frequently a mix of odds and evens. Negative feedback, if relied upon to overcome gross non-linearities like crossover distortion, has the effect of preferentially reducing lower order harmonics towards invisibility. *But higher harmonics then appear, or are strengthened, in their place.*

Other fundamental causes of non-linearity are thermal modulation (notably in semiconductors, and fuses), and signal contamination.

Harmonic distortion is relatively straightforward to measure, and being rated in %, it seems easy to comprehend. It was the first kind of distortion (or non-linearity) to be widely measured – since about 1935. It has been presented as a leading part of most audio amplifiers' specifications since then, often as the sole indicator of relative absence of non-linearity.

In the pioneering power amplifier circuits, *circa* 1925 [16], harmonic distortion was gross, at above 25%. In the half century between 1935 and 1985, typical percentage total harmonic distortion (%THD) of high quality amplifiers working in the top 30% (0 to 10dB below clip) of their capacity fell by about a thousand fold from around 10% to 1%, to some 0.01% to 0.001%, or down to 1 part in 10,000 to 100,000, at least with transistor amplifiers employing powerful distortion reduction techniques, including liberally applied global NFB; or intrinsically linear stages with lesser feedback. A number of refined, modern audio power amplifiers manifest considerably lower harmonic distortion over most of their operating range, that's also largely below the measurement limit of the most sensitive test equipment. Such figurative excellence, not forgetting that there are no free lunches in engineering, may have no particular or definite relation to audio quality. It is a double edged sword when the more harmonic distortion has been reduced, the more that other, previously masked deficiencies (such as jitter, ugly buzzes, RF intermod) have been exposed. When these are excised, other causes re-surface or appear.

In the past decade, after a break of some 20 years (say 1964-1984), valve amplifiers of varying sophistication with 0.1% distortion at best, and up to 10%, are again manufactured relatively *en masse* and re-employed for domestic and even home recording use. Levels above 0.5% undoubtedly mask the harmonic subtleties of the music, but the same levels and the constitution of the distortion can also nicely mask any and many 'nasties' too. The point here though, is that the implied numeric reversionary increase in distortion level (even allowing logarithmic measure) is not at all what many ears hear. *Size is not all !* Consider also valve instrument amplifiers (*backline amps*), where distortion that is musically valuable has been nurtured. Yet in both cases, harmonic distortion measurements remain useful and are still performed by quality-control conscious makers, as a means of characterisation.

One reason for disparity is that conventional '%THD' harmonic distortion figures crudely totals up the contribution of all the harmonics (and noise). The sonic effects of different harmonic patterns are ignored and lost. In the 1950's [17], tests were performed to establish the % levels of the individual harmonics that could be heard. For example, above 200Hz, more than 0.5% 2nd harmonic was audible. But the acceptable levels for even a few of the admittedly countless different harmonic combinations do not appear to have been tested. However, Olson noted that increased bandwidth (14kHz from 4kHz, say) made the lowest audible levels of both even and odd harmonics lower, i.e. made linearity requirements more stringent.

Today, the maximum acceptable levels amongst sensitive listeners would be at least 10 times lower. In part this is accounted for by the greatly increased resolution and transparency of the music source signal, and the surrounding equipment; and even possibly greater confidence in the listener's ear.

7.7.1 Harmonics: the musical context

"Trumpets ... sounding the 3rd and 5th ... were more vibrant and ... gave a wonderful echoic effect."

Samuel Pepys, 1633–1703

The harmonics being discussed so far – those caused by errors in electronics and transducers – are similar in their essence to the overtones referred to by musicians. But there are some crucial differences.

First, terminology. In music, the fundamental (of a given note) and its overtones are all called partials. They are part of an instrument's sound, causing it to have a unique timbre or tone colour, according to the partials' particular pattern.

Looking at the next table (over), and assuming modern musical convention, the interval that musicians call an octave aligns exactly with the 2nd harmonic in audio terminology. Subsequent octave intervals also align exactly with electronic harmonics, but the series is geared ratiometrically, ie. 2nd, 4th, 8th, 16th, etc. Whether the other, intervening electronic harmonics happen to align with musical intervals is almost a matter of chance. As can be seen from the right hand column, third harmonic distortion is very close to being a twelfth - indistinguishably so for most people. But the fifth is quite off tune from what musicians call a seventeenth. Even if the seventh harmonic was aligned with A flat, the nearest interval, it would sound rather unusual. Being about a third of the way towards the next note though, it is however wholly discordant. As the harmonic series progresses, this condition applies progressively to the majority. It follows that above the 8th harmonic, most even harmonics are as sonically unwelcome as odd ones. But that below this number, the evens are all benign and the odds excepting the seventh may be in some cases.

Harmonic differentiation

The next table compares harmonics in audio electronics with the intervals and natural harmonics recognised by Western musicians. *(after Nelson* [18]*)*

The first column is the musical note. The interval (the numeric ratio of frequency) between any 'n' notes is uniform, so the starting point is arbitrary.

The second column gives the cumulative interval from the first note.

The third column gives the musical name for the interval. It's all relative - so a twelfth is whatever note that has a frequency that is 2.9966x higher than another.

In the fourth column, harmonics made by audio equipment are placed according to the musical interval they are nearest to.

The fifth column gives the frequency.

In the final column, Cents are percent (%) away from the next nearest note, so half way between two notes is expressed as either + or −50 cents adrift. [18]

Note	Interval (ratio)	Musical interval	Audio terminology	Frequency in Hertz	Cents (%) error
C	1.000	First	Fundamental	261.6	0
C#	1.0595			277.2	
D	1.1225	Second	-	293.7	
D#	1.1892			311.1	
E	1.2599	Third	-	329.6	
F	1.3348	Fourth		349.2	
F#	1.4142		-	370.0	
G	1.4983	Fifth		392.0	
G#	1.5874		-	415.3	
A	1.6818	Sixth		440.0	
A#	1.7818		-	466.2	
B	1.8877	Seventh		493.9	
C	2.000	Octave	2nd harmonic	523.3	0.0
C#	2.1189			554.4	
D	2.2449	Ninth	-	587.3	
D#	2.3784			622.3	
E	2.5198	Tenth	-	659.3	
F	2.6697	Eleventh		698.5	
F#	2.8284			740.0	
G	2.9966	Twelfth	3rd harmonic	784.0	−1.9
G#	3.1748			830.6	
A	3.3636	Thirteenth	-	880.0	
A#	3.5636			932.3	
B	3.7755	Fourteenth		987.8	
C	4.000	Double octave	4th harmonic	1046.5	0.0
C#	4.2379			1108.7	
D	4.4898	Sixteenth	-	1174.7	
D#	4.7568			1244.5	
E	5.0397	Seventeenth	5th harmonic	1318.5	+13.2
F	5.3394	Eighteenth		1396.9	
F#	5.6569			1480.0	
G	5.9932	Nineteenth	6th harmonic	1568.0	−1.9
G#	6.3496			1661.2	
A	6.7272	Twentieth		1760.0	
A#	7.1272		≅ 7th harmonic	1864.7	+30.0
B	7.5510	Twenty first		1975.5	
C	8.000	Triple Octave	8th harmonic	2093.0	0.0

Showing how harmonics can occur in practice, in music, the table below charts the measured harmonic content of a very well known instrumental sound – that of a plucked electric bass string.

Harmonic spectrum

- of round-wound, open E string, on Fender Precision Bass Guitar

Harmonic	Note	Frequency	*of* fundamental: %	Level (dB)
1st	E	41.2Hz	100%	0.0
2nd	E	82.4	390%	+11.8
3rd	B	123.6 (123.4171)	51.5%	− 5.8
4th	E	164.8	40.8%	− 7.8
5th	G#	206 (207.652)	21.5%	− 13.3
6th	B	247.2 (246.942)	7.1%	− 23.0
7th	D	288.4 (293.665)	4.6%	− 26.6
8th	E	329.6	43.0%	− 7.3
9th	F#	370.8 (369.994)	16.6%	− 16.6
10th	G#	412 (415.305)	5.7%	− 24.7
11th	A	453.2 (440.0)	0.23%	− 52.6
12th	B	494.4 (493.883)	0.24%	− 52.4
13th	C#	535.6 (554.365)	2.35%	− 32.6
14th	D	576.8 (587.33)	0.1%	− 59.5
15th	D#	618 (622.254)	0.26%	− 51.9
16th	E	659.2	0.12%	− 58.8
17th	F	700.4 (698.456)	0.06%	− 64.5

Testing conducted by Stereophile *technical staff. Player: John Atkinson.*
(X) frequencies in brackets are the exact, ideal frequencies in the even- tempered scale. [see p.83, May 1992 edition]

As implied earlier, distortion figures for high performance audio are much less connected to what you will hear than results for bandwidth, noise, breakthrough and CMR. One problem with %THD+N tests is that the equipment doesn't discriminate like ears. The low harmonics (2nd, 3rd, 4th) aren't very audible at low frequencies, while at mid and high frequencies, they are more audible, but not necessarily objectionable. But THD testing lumps these in with high harmonics (7th and above) which are highly objectionable and very audible at most frequencies, even in very small doses. This is how an amplifier with (say) 0.3% %THD+N can sound better than one with 0.01% THD.

When an amplifier introduces harmonics of its own, the effect they have can vary widely. If, in sparse music, they arise at frequencies or intervals where the music at that instant has nothing to add, they may be very noticeable even when very small – and all the more so if the harmonics are discordant ones. If the harmonics are purely benign (all 2nd say), they may be almost unnoticeable even if relatively large, say 3% at low bass frequencies. But even if an amplifier only produces small amounts of benign harmonics adding to identical partials already in the music, the content is altered. For example, consider a bass note with 0.25% of the third partial. If the amplifier alone contributes 0.05% to this (as 3rd harmonic distortion), the partial is 20% 'larger' than it was. The outcome is a subtle change in the note's sound, yet it is easy to miss that the change in the meaning might be catastrophic. This is particularly likely with bass instruments (lowest piano notes, synths and the 41.2Hz E string of the double acoustic and electric basses) where the natural second harmonic is the dominant part of the natural sound we would hear.

With the large number of frequencies present at once in much music, a high percentage of harmonics added by some audio electronics and transducers might be visualised as a swelling. The music will seem 'louder'. In non-purist performing and recording, compressors are commonly used creatively (as harmonic distortion generators) to get this fattening effect – however unwittingly. A mixture of benign harmonics commonly adds richness.

On their own, as shown clearly in the earlier table, the second, fourth and eighth harmonics are entirely consonant. The second on its own is not heard so much as distortion, even at high levels, but as added 'fatness' or 'body'. The third harmonic on its own is also mostly musically pleasing and interesting - being a fifth interval, an octave up. But it can get nastier in the absence of the 2nd above 1kHz. With the second and third together, bass may seem enhanced, but the distinctive natural timbres of many instruments will be lost, causing for example, a fine grand-piano to sound like a spinet ! [19] The sixth is similar: Both the 3rd and 6th harmonics are in key with 6 out of the 7 'white' notes. By contrast, the fifth harmonic (a musical seventeenth) varies, being out of key with over half the notes. But when the 5th harmonic is combined with the third (as noted by Samuel Pepys 300 years ago), a loud, brassy, almost vibrato quality is added.

The 7th and all higher odd harmonics, and all even harmonics above the 8th, are perceived as increasingly objectionable, and fatiguing. This can arise even when they are not directly audible, being ultrasonic. Even in very small doses (below say 0.001% or 1 part in 10,000), these *high harmonics* may be audible as 'edge', 'grain' or 'hardness' to critical listeners. And these need not be made by bad analogue or digital electronics alone; *the modes* (partials, some of them harmonic) *of a small church bell reach up to at least the 45th harmonic* [20], which is in the mid treble region – where hearing sensitivity is still quite good.

Past these general indications, the audibility of harmonics is not at all readily predictable since it depends on their relative levels, phase relationships, frequencies, the effects of the sound system's other defects, and the levels of everything else going on, ie. the auditory masking threshold [7, 21].

7.7.2 Harmonic distortion (THD, %THD+N)

Total Harmonic Distortion (THD) produced by an amplifier is defined mathematically as 'the ratio of the square root of the sum of the squares, of the rms value of each individual harmonic; to the rms value of the output signal'. In other words, measure and sum the rms voltage of each harmonic, then express the total as a ratio against the *stimulus* signal, the fundamental.

The *de facto* method

The *de facto* universal distortion measurement arrives at the sum of the harmonics produced by inference, by measuring the average signal level across the audio bandwidth, after recovering the output of a single, continuous, highly pure sinusoidal test tone (at say 0dB) through the AUT, and after nulling this, the *fundamental*, to a depth of at least −80dB and preferably to below −120dB. Even assuming infinite rejection of the fundamental, as all the noise in the band is measured, this test is most accurately known as 'THD+N' or '%THD+N' where the '+N' means 'plus whatever noise'. For brevity, '%THD' or just 'THD' are often used and nowadays may be taken to be synonymous with '%THD+N'.

In %THD+N test sets with modern facilities, rejecting the fundamental requires a highly tuned, auto-nulling and fast settling, sweepable rejection filter. It must itself have low %THD, as well as high input CMR.

In the first *distortion factor analysers* (as they were once known), there was no notch, but just a very steep HP filter, −75dB at 400Hz. The test stimulus was always at this one frequency. Harmonics up to the 15th could be accurately resolved, i.e. to 6kHz. Irrespective of the test set, the recovered level comprising harmonics plus any noise, strictly called the *distortion factor*, is described in percentage terms. Occasionally it's expressed in dB. For example, 0.1% THD = 0.1 part in 100 = 1 part in 1000 = −60dB.

Bandwidth limits

As with many other measurements, true noise and often hum can't be excluded, and the window is necessarily set quite wide open to capture and see the whole truth, e.g. since the 4th harmonic of 20kHz and the 8th harmonic of 10kHz are both at 80kHz. Noise and hum introduced by a wider test bandwidth increases the 'distortion' reading most if the real distortion parts are smaller. For example, if the AUT really has 0.0001% THD, but there is 300mV of hum in the 30v output signal, then the %THD+*N* measurement *will* read a steady 1%. On the other hand, if real %THD is high, then the noise level of a high performance power amplifier will not affect the reading – unless it has a *bad* noise sickness.

Test signal

%THD testing requires a continuous sine wave test signal ideally with nil harmonic distortion. The best generators have very low %THD – below 0.0004% is achievable, but don't always have the above x3 (10dB) advantage they need so as to not interact with the %THD of some of the amplifiers being tested.

Frequency variation

Almost regardless of the amount or kind of NFB, %THD nearly always varies with frequency. Typical plots of %THD *Vs.* frequency with high global NFB, look similar to a CMR response rising at HF, from a low plateau at LF (e.g. Figure 7.19). In AUTs employing input or output transformers, or any with inadequately rated power supplies, defective reservoir capacitors, or fuses or semiconductors suffering thermal modulation, and/or low NFB, %THD+N may also rise or fluctuate at the bottom end. Peaks or dips coincident with 50/60Hz power line frequencies or their harmonics, or other EMI frequencies, indicate contamination, with the inflexion depending on whether the test signal and contaminating signal phases added or subtracted.

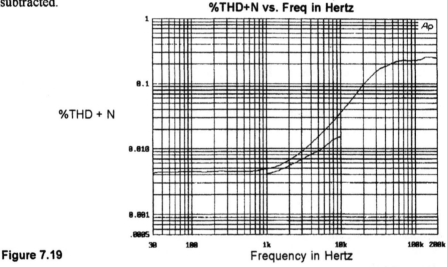

Figure 7.19

Classic %THD+N Vs. Freq. plots. The rise in the upper curve is partly arrested by the standard 80kHz bandwidth filtering and the AUT's own rolloff above 20kHz. The lower curve shows how the figure is not much lower with narrow band 400-22kHz filtering, with the sweep deliberately restricted to the meaningful 1kHz-10kHz area. The small difference shows that the '+N' is not the major component: the plot here is mainly harmonics.

Useful specification

To be remotely meaningful, a single distortion figure must state the range of frequencies over which it's valid or bettered. The drive level, test bandwidth and the output loading of the AUT must also be cited. A wide bandwidth (the AP *System One* allows 22, 30, 80 or 200kHz) lets in more noise, making the figure look worse, but when plotted against frequency it can provide an insight into the cause of the distortion, e.g. whether *dynamic*, like slew limiting. While a test signal is swept (or nulled), the residue on the oscilloscope must be studied, in order to record descriptions of the distortion residue's waveshape, which commonly varies over the frequency range. This too can give insight into the distortion mechanisms at work, and the degree to which % random noise or % 50/60Hz hum is what is really being measured. The residue (particularly if checked wideband) may also show up tiny bursts of RF oscillation as well as clearly displaying any crossover distortion, and the early onset of clipping spikes.

In practical testing, consistent sweeps within −0.5 to −1dB of clip depend on having a stable AC mains supply, where the peak AC voltage does not much change; or on astute operation of a Variac coupled with being on the lookout for the onset of clip − which is easy enough when the %THD residue is displayed on a scope. In turn, a good specification will give this much detail:

'**Harmonic Distortion, %THD+N, taken at −0.5dB below visible onset of clip re. 319v +/-2v peak AC supply, over 10 to 80kHz BW, AUT set in stereo mode, both channels driven, into 8 ohm resistive load, see %THD *vs.* frequency graph. Residue mainly noise below 1kHz; slightly but increasingly spikey above.**'

Dovetailing

If two plots with narrow and wide bandwidths respectively can be overlaid with a carefully chosen overlap (usually 10 to 30kHz), the best of both worlds is attained, with a little visual or manual interpolation. The same technique of dovetailing overlaid plots made with different bandwidths may be used at LF, if hum frequencies cause the insertion of the usual 400Hz HP filter to change the %THD readings.

Which test level?

In conventional power amplifiers with high, global NFB, %THD is normally measured at the highest level, just below clip. This point is chosen as it is worst case in one sense and yet it usually gives the lowest figure.

In power amplifiers employing low or negligible NFB, overall distortion (crossover distortion is one potentially diametric exception) rises in some kind of integer power to the signal level, so the distortion rises at say +6dB/octave, then it may accelerate +12dB/oct, etc. As music signals likely average −10 to −20dB below clip, distortion figures taken at lower levels will likely be lower and more in accord with what is mostly heard. To see how %THD varies at different frequencies against level, three measurements *vs.* frequency at the different levels may be overlaid in steps of 5dB.

With conventional transistor power amplifiers with high global NFB, a higher drive level (provided it's still enough below clip) may paradoxically appear to reduce distortion. But this is only because the distortion is so low that noise dominates the broadband THD measurement at most levels. The 'real' distortion (excluding noise) at normal average operating levels 10dB or more down from clip, will always be lower, meaning the spec is conservative.

Dynamic THY testing

%THD may also be measured 'dynamically'. *This has nothing to do with dynamic types of distortion, nor trtuly dynamic test signals.* Rather, the level (amplitude, rather than frequency) of the pure tone (now at a given fixed frequency), is swept (usually upwards, and in many little steps), typically from –30dBr to clip (≅0dBr) and a bit beyond. Starting at low levels, %THD+N initially falls at –3 to –6dB per octave with increasing drive level, as the '+N' is stripped away. It then reaches either a minima just before clip minimum, or else a low plateau some dB before (Figure 7.20). *Note that the slope of the noise is just an artifact of the linear % scale; if %THD+N is shown on a log (dB) scale, any noise dominating below clip is show as a constant or 'floor', as it really is.* In amplifiers using high global NFB, the onset of the final distortion rise is abrupt, beginning at true clip. In amplifiers using low NFB, or any form of soft clip, limiting, etc, distortion rises gently from levels some dB before the onset of true (hard) clipping.

With the common linear percentage THD scale, AUTs with high noise reach a sharp 'V' just below clip; those with low noise (+N) relative to %THD have a flattened region below clip (as in Figure 7.20) that may be bumpy, due to residue developments with each step, or test equipment autoranging. In this region, AUTs employing

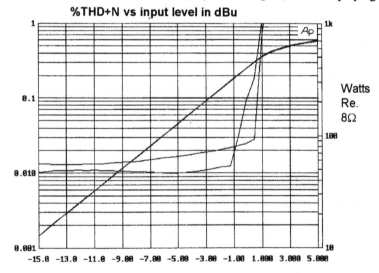

Figure 7.20

Showing typical 'dynamic' %THD ('THD, dYnamic') behaviour of one channel into 8 and 4 ohm resistive loads - in the two almost right-angle-shaped plots. Loading can be seen to cause the clip point (defined as passing X% THD+N) to change. The amplitude scale (abscissa) is the input drive in dBu. The right hand scale is the power delivery in watts re. 8Ω computed from volts, giving the almost 45° straight output power plot. The power scale is logarithmic, and the straight line signifies linear input/output conversion. The bend at the top indicates post clipping. An honest maximum power rating is commonly taken at the point where %THD reaches (say) 0.1%, here about 400w for 8Ω, and slightly less than double this for 4Ω – for which the right hand scale reading must be doubled; ideally there would be a dedicated 4Ω scale.

sliding bias or class G, H or other *amplitude-adaptive* circuit techniques, may exhibit two or more %THD+N minima. To reveal these a dynamic %THD test (THY) may be repeated with overlays at several frequencies and across a range of load conditions. As frequency is fixed, and assuming it is 1kHz, then a narrow bandwidth setting (like 400 Hz to 22kHz) may be used for best resolution.

Class 'D' distortions

Not quite yet, but in the future, the harmonic distortion of digital amplifiers is set to become an issue. Some fresh psychoacoustic 'wildcards' include the counter-intuitive properties of digital noise, the high potential for ear fatigue, and the damage to bass (cf. treble) notes' accuracy caused by *jitter*, and the way in which any sampling process can 'fold down' so harmonics appear at *non* integer frequencies.

Wider context

People who know little about audio often say *"Amplifier distortion is nowadays inaudible since distortion in speakers is way higher"*. Yet even on the very limited basis of comparing one-dimensional %THD+N figures, it is not always true; modern ESLs have distortion at typical listening levels as low as 0.01% 2nd and 0.05% 3rd harmonic respectively, while under the same conditions, even ordinary speakers with moving-coil drivers can manage 0.2% THD, if of modern design.

Acceptable levels

In the past quarter century, below 0.1% THD+N across the audio band has been considered broadly the minimum acceptable in an audio power amplifier. Without considering the harmonic structure, such figures can be misleading, and any comparisons with other amplifier species (let alone with speaker %THD for example, above) are bound to be inept.

Summary

For now, and in spite of its variable disconnection from what is heard, %THD+N remains the test or figure that more than any other, provides a single (or simple) figure of merit to indicate there are no gross problems with the AUT. For example, a low %THD in a wide bandwidth, at 0dBr into a load, demonstrates low noise, low hum and high slew. One difficulty is that the converse is not automatically true.

7.7.3 Individual harmonic analysis (IHA)

"One reason for disparity is that conventional '%THD' harmonic distortion figures crudely totals up the contribution of all the harmonics (and noise). The sonic effects of different harmonic patterns are ignored and lost."

Recalling this keynote from earlier on, surely the individual harmonics can be measured? Yes. Originally the levels (amplitudes) of the individual harmonics were measured with a *wave analyser*. This is a tunable wideband filter, operating like an everyday *superheterodyne* (domestic) radio receiver to stably achieve a far narrower (higher Q) passband(-width) than conventional audio filters. The individual components of a complex waveform (e.g. a distortion residue) can then be individually tuned into while all other frequencies – and noise – are rejected by 70dB or more. This was the way in which the mathematical abstraction, pure THD (without any noise) could be approached: measure each harmonic individually, and sum them vectorally with a pencil on squared paper. Without automation, such a test was too time consuming to be performed under enough conditions to give more than the odd 'spot' picture. Also in the heyday of manually operated wave-analysers, weighting (alias psycho-acoustic EQ) proposed to make %THD+N figures more realistic [22, 23] was not widely implemented, presumably because it was too much trouble to teach everyone a new, highly abstract picture, when it would not have affected the more ordinarily performed %THD+N test results.

In the intervening years, general spectrum analysers have offered realtime results, but few have the relatively extremely narrow bandwidth (ideally 1Hz or less), the low-end frequency range, the deep rejection to ideally >>–100dBr, and high voltage rated balanced inputs, to suit audio power amplifier testing. Since 1990, high performance and versatile audio spectrum analysis has been possible using DSP within the *Dual Domain* version of the Audio Precision *System One* test set.

Individual Harmonic Analysis (IHA) may be presented as a conventional spectrum readout (non-realtime with AP *System One*) (Figure 7.21), or as individual harmonics' % *Vs*. frequency plots (Figure 7.22). This latter facility has long been available with *Bruel and Kjaer* (B&K) test sets, but limited to the higher % harmonic levels more typical of loudspeakers. Both kinds of presentation, particularly the latter compared with the similar one dimensional %THD+N *vs*. freq. plot, can be eye openers, and differences seen are generally more meaningful. It is little documented that harmonic spectrae can change substantially in the top –20dBr, while the %THD+N figure stays consistent.

IHA spectra derived from the AP *System One* or other DSP-aided instruments may be averaged for higher resolution, even below the (true, stochastic) noise floor. Hum and other interfering signals can also be spotted at lower levels than with conventional 1/3rd octave noise testing (see *SNR*, a prior section).

7.7.4 Intermodulation (% IMD, Intermod)

Having tested an audio power amplifier with a single sine wave, and perhaps found the harmonics low enough, it is taken for granted that music, some of it comprising ten or even hundreds of signals comprising even more simultaneous component frequencies, can co-exist without creating added 'noise'. Yet *this should be the case, so long as the signal path is linear*.

Nature of interaction

Inter-modulation is the descriptive title for the event when two or more simultaneous signals do interact, to create others. The signals created are always products of the difference or sum of the two or more interacting signals, and most are *not* harmonics.

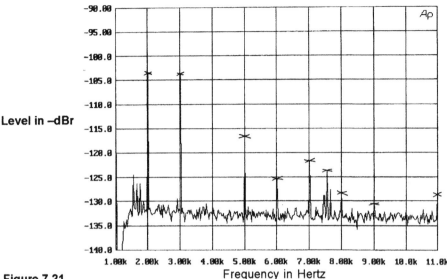

Figure 7.21

Showing the 1kHz spectra of a high performance amplifier with high, global NFB. The X's mark the peak of the spectra which are otherwise indistinguishable from the gridlines in monochrome, as here. The Audio Precision DSP-aided test-set rejects the 1kHz stimulus, and here takes an average of 64 samples of the other harmonics. The AUT has dominant equal 2nd and 3rd at about -104 dB. There is no 4th, and no 10th. Note how the odds (3rd,5th,7th,9th) are decaying monotonically but not the evens. The DSP averaging process also gives a highly averaged view of the noise floor Vs. frequency.

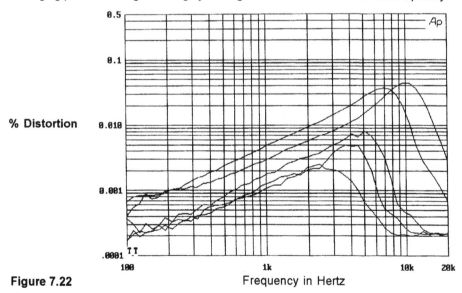

Figure 7.22

Showing the behaviour of individual harmonics 2nd to 6th Vs. frequency. In the AUT tested here, all follow the same pattern, but the order of the magnitudes does not follow the harmonic order - as co-revealed in the spectra.

299

Although as much as 3% intermodulation distortion was permissible in valve (tube) amplifiers, pre-1965, intermodulation distortion caused by solid-state amplifiers is almost wholly discordant, and in high levels is fatiguing and very unpleasant. It follows that intermodulation tests might be more sensitive indicators of sonic performance than %THD+N testing. Note however that the non-linearities that are the first cause of both kinds of distortion, *are one and the same.*

Terminology

Intermod products have even and odd orders; Second order comprises the sum of two original frequencies. Third order is the sum of one original, and one doubled frequency component; or if three signals, then three originals. Fifth order is the sum of one doubled and one tripled frequency, if two tones, or if there were 3 tones, one original and two 'doubleds', i.e. doubled ones.

The integer multiplied frequencies are of course the result of the involvement in the intermodulation, of the harmonics of the individual tones, and even the harmonics of the intermod products, multiplying *ad infinitum*. If the frequencies being subtracted sum positive, they are *even order*; if negative, odd order. So the combination of two tone frequencies (f1+f2) is second and even order, while the products (3f1+2f2) would be called '5th order, odd'. As with harmonics, the symmetrical non-linearities create the odd order products, and *vice versa*.

SMPTE/DIN method

The first standard intermodulation distortion test was developed in the film industry in the late 30's [24]; Sound recordists and producers will be already be familiar with 'Simp'tee' (*Society of Motion Picture and Television Engineers*) as a designation, since the same society has also established the *de facto* timecode. Today, the SMPTE IMD test (as re-defined in 1983) is similar to the DIN 45403 standard.

The test usually comprises a low frequency. 60Hz is the SMPTE norm, presumably chosen by the US originators to simulate power frequency hum. The 60Hz (several other frequencies up to 250Hz may be used in the DIN test) is mixed with a second, HF signal, usually at 7kHz. In the SMPTE standard, this signal is 25% of the LF component, i.e. they are in a 4:1 amplitude ratio. The signal emerging from the AUT is stripped of the LF tone by a steep HPF. An AM demodulator then recovers the true rms level of the sidebands. These sidebands *are* the intermod products, 60Hz (or whatever was the LF) either side of 7kHz, in this instance at 6940 and 7060Hz. Last, the 7kHz 'carrier' and nearly all noise are strongly removed by a LPF, leaving just the products. These are then expressed at % of the original, usually in rms terms.

With the AP *System One*, the HF tone can be swept against frequency ('SIF') from above 2.5kHz, or both tones can be swept against level ('SIY'), as shown in Figures 7.23 and 7.24 respectively. The ratio may also be changed. But in all cases, the intermod products caused are close to the HF tone and are liable to be masked. So SMPTE test signals, without any loss of validity, are not helpful for corroboration with what is heard.

SMPTE testing can be sensitive to high order non-linearity, namely some kinds of cross-over distortion, and LF effects such as signal transformer saturation, and thermal distortion. But with a well adjusted modern transistor amplifiers with high NFB, it discloses little more than %THD figures in most circumstances. But due to the superior discrimination against noise, in having a bandwidth of about 700Hz, it can yield a % distortion figure that is x2 (6dB) to x3 (10dB) lower, and closer to real %THD, assuming the associated %THD+N measurement is mainly noise-laden due to a 22kHz or greater bandwidth.

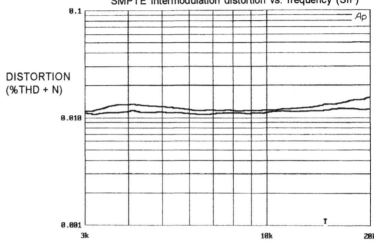

Figure 7.23

SMPTE intermodulation distortion swept against frequency. 3kHz is the minimum effective frequency. The almost flat curves are typical of linear, transformerless, electronic paths.

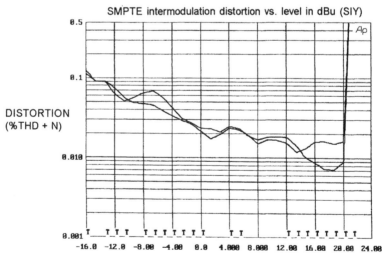

Figure 7.24

SMPTE intermodulation distortion swept against input level. This often manifests the 'V' shape for the same reason as dynamic harmonic distortion plots - the distortion-to-noise ratio, made visible by the low %THD of the AUT (as here) having respectably high linearity. The Timeout markers (T) confirm that the measured data was largely noise. The minimum may be taken to be the rated figure, i.e. about 0.01% SMPTE @ 1dB below clip.

CCIF method

In this test, originated by a French standards body, and useful for getting a high resolution handle on HF distortion, the test tones are the same size and closely spaced, e.g. 19kHz and 20kHz. Together, they generate any odd order IMD products at 1, 21, 39kHz, *et seq.*; and any even order products at 18kHz, 58kHz, etc.

Usually only the even order components are measured. In fact, the test may be simplified down to just measuring the 2nd, even order product (f2–f1), which may be standardised at 1kHz. The measurement can then benefit from the low noise in a 1kHz bandwidth. A steep low pass filter is used. *Note this simplified test will not pick up* symmetrical *non-linearities.* The (2f1–f2) and (2f2–f1) products may be more important. Alas, measuring the odd orders requires spectrum analysis. In all cases, at least the products are well removed from the test tones, thus they are not masked by the ear, and so in theory at least, CCIF %IMD test figures have a higher chance of corroborating with the listening. But in appearance and percentages, in the context of linear, transformerless signal paths, typical CCIF plots ('CIF' and 'CIY') results are not dissimilar to the immediately preceeding 'SIF' and 'SIY' plots.

3 tone method

In this far more sophisticated scheme [25], published by Ivor Brown in Wireless World (alias EW – see page XVI, *Publications*), the spectra of three mixed tones are viewed. The test begins with two relatively low level, close-spaced HF tones, e.g. at 4.8kHz and 4.6kHz. They are summed equal in amplitude, and have low %THD. They are then mixed with a large (–1 dBr say) 40Hz burst wave (Fig. 7.4). The 40Hz exercises the amplifier with the two tones at every level, while the burst limits dissipation. The frequency is chosen so the supply rail voltage is changed in each 50/60Hz cycle. The output is high-pass filtered to remove the 40Hz and its low order harmonics, then applied to a spectrum analyser. The products of interest are harmonics and IMD products clustered around the HF tone frequencies. Visual presentation is clear and the 3-tones can show products that are not apparent with a plain spectral analysis down to (for example) a –93 dBr noisefloor. There are rather more permutations in this procedure than with the simplistic two-tone tests.

7.7.5 Dynamic intermodulation (% DIM 30/100)

The DIM 30 and DIM 100 tests are two of the few of relatively recent origin (1976) to be embodied in widely-used test instruments. The DIM test, proposed by Shrock and Otala aims to test the error correction delay in high NFB amplifiers. To this end, the DIM test signal comprises a 15kHz sinewave mixed with a 3.18kHz square wave. The latter acts like a recurring fast HF transient. Before it reaches the AUT's input, the test signal is LP filtered at either 100kHz (DIM 100) or 30kHz (DIM 30). The test signals then have maximum slew rates of 0.32v/μS and 1v/μS *relative to the input level.* Rounded corners in the recovered signal indicate gross slewing. The intermod components have to be measured with a spectrum analyser, rms summed, and expressed as a % of the 15kHz sinewave. The subsequent analysis is messy and requires skilled eyes. A simplified version of the test allows a 'one number' readout while sensing both odd and even order non-linearities. As a feature on the Audio Precision test sets, this has become the *de facto* standard DIM test. Its limitation is that multidimensional information has been lost. So the causes of the %DIM cannot be inferred from the results.

7.7.6 Sundry intermodulation checks

Supply modulation may be examined. Colloms [11] runs the AUT at –4dBvr (re. 8Ω load) with a 37.5Hz signal, into a 4 ohm load. The loading is chosen to stress the power supply. The frequency chosen is an *in*harmonic of 50Hz. A spectral analysis is then made between say 30Hz and 500Hz, to separate the unwanted 50Hz harmonic components and assess the degree of supply noise breakthrough. In a good amplifier, the breakthrough should be at least –100dB at all audio frequencies.

Alternatively or additionally, spectra on the main DC supply rails may be read, each referred to the central 0v point between the main audio supply's reservoir capacitors, with the test signal as above. If the spectra on both + and –ve are identical, all is well. If either differs, and assuming a centre tapped DC supply, then the 'missing' signal component is flowing in the centre-tap (often 0v) wiring, and may be expected to degrade sonics.

7.7.7 Other distortion tests

Used in conjunction with a %THD+N analyser, suitably timed burst testing with a variety of low settings (i.e. from –6dBr, down to hard muted) can show up dynamic crossover distortion in BJT output stages, when the quiescent current is inadequate, in the period while the Vbe sensor is lagging thermally.

More sophisticated schemes for assessing distortion in audio power amplifiers have occasionally surfaced, but all employ relatively special or expensive instrumentation, and without being taken up in commercial test instruments, there is an insufficiently wide base of experience for there to be consensus on the value of such tests.

In 1975, the late Richard Heyser suggested (in amongst a list of radical tests) a way to make simple %THD measurements more meaningful, by using phase lock and *cisoid* mapping techniques to show if the phase of a particular harmonic varies with level [26]. If so, a change of tonal value with intensity and in timbre, is to be expected.

In 1978, Alan Belcher (then at the BBC) pioneered a test that used computer generated pseudo-random binary waveforms [27, 28]. These simulated music more closely than any other algorithmic (predictable) test signal. Comb filters and digital frequency-shifting were used to isolate the distortion products. Good agreement with what is heard, was reported, but there are no commercial instruments. In 1988, Jensen and Sokolich [29] proposed use of a comb-spectrum. This was of similar ilk, but more visual. Amidst the complex test signal spectra, added and subtracted harmonic and intermodulation components were able to be identified. Again, there are many permutations, but so far, no purchasable equipment and no standard procedure.

7.8 Power output (P$_o$)

For many users this defines '*how loud it goes*', and will be all they care about –
even though 'the wattage' can be surprisingly loosely related to the 'intensity'
(strictly, 'subjective loudness') actually experienced by the ears.

Definitions

Power is rated in watts. Meaningful, 'true' watts for audio are always *average*,
never 'peak' or 'instantaneous'. Power is 'Energy integrated over Time'. 1 watt of
power is the *outcome*, the work done, when energy is consumed at 1 Joule-per-
second. Over periods that are short relative to the rate of change, it is more appropriate to talk of *energy* (in Joules).

For the steady and repetitive sine wave signal, as is inevitably used for much testing, *average* power delivery is calculated by multiplying the *rms* volts and the *rms*
amperes flowing into the speaker:

$$P_{avg.} = (V_{rms} \times I_{rms})$$

Note that 'rms' is pronounced spelt out, ie. *Ar Em Es*.

Multiplying the two r.m.s quantities gives an *average* (avg.). There is no such animal as
'rms power'. *Every* mention of 'watts rms' always *really means* 'average watts'. Unfortunately, the term 'average' sounds 'average' to non-technical copywriters.

Scaling adjustments

To users and purchasers who do not know about ratiometrically- (or logarithmi-
cally-) scaled relationships, and about dynamics compression in both speakers and
amplifiers, the difference between notionally related amplifiers delivering varying
numbers of watts versus what actually emerges from speakers, is at least counter-
intuitive. For whereas a doubling of power costs something like twice as much in
parts, it has surprisingly little effect on SPL – at least at midrange frequencies.

For this reason, the IHF-A-202 standard proposed 'dBW' instead. This is an absolute dB scale of power, but scaled in voltage decibels, with 0dBW set at 1 watt
(2.83v rms). Thus 10dBW =10watts (8.9v rms), 20dBW = 100 watts (20v) and
30dBW = 1kW (89v). This way, the large wattage numbers are played down, but
still not far enough perhaps. This system has been long adopted by Martin Colloms
and also *Hi-Fi News*, for their power amplifier reviews.

Practicalities

With a speaker, *average power* delivery with a continuous test signal will likely
vary with frequency and history, because of the complexity of speaker impedance.
Also, most speakers (most particularly tweeters), are not rated for continuous signals at levels remotely approaching those that a potential partnering power amplifier has to be rated for, even though it should only reach these levels transiently in
practice. They quickly burn out. It is thus convenient and universal practice (though
strictly unrealistic) to use a well cooled, high power rated resistor as a 'dummy

speaker' or 'power load'. This allows *continuous average power* output capability to be determined without risk of damage.

Real rated power

The basic power capability definition above is not tied down adequately until waveform purity is defined. Without this, an amplifier's power rating can be made to seem large by driving some way into clip. A sign of this is when power is specified at '10% THD'. In honestly specified power amplifiers, rated power (at least in conventional high NFB amps) is practically defined either at the rated %THD+N (only possible for spot figures) or else at some %THD+N figure which unambiguously indicates the final rise on the way to clipping, often at 0.1% (see prior section on %THD+N testing).

As most audio power amps have unregulated power rails, the level for a given %THD, hence power rating, also relates to the main DC rail voltages, and by inference, and not forgetting mains supply distortion, a particular mains supply voltage. Assuming a conventional, unregulated supply, the DC level is in close proportion to the incoming AC's peak (FWR'd) value, which is readily measured compared to the HT, which would require wires hanging out of the AUT. Mains voltage adjustment is still required; see introductory section 7.3.1. Having established a stable supply, power rating is most easily and consistently tested on a variable-skill production line with a computer test set like AP *System One*, where the plot can be regulated ('locked') to 0.1% THD.

Optimistic powers

Power amplifier makers have long played number games with power ratings. These have included (and may have been applied in free combination):

* Adding together the ratings of the channels' outputs (in a 2 or more channel amplifier).

* Specifying power with 1, 5 or 10% THD, meaning an amplifier with NFB was being driven some way into clip, i.e. high or gross distortion.

* Entering the power equation with the peak values of voltages and currents - as in 'watts peak' (another quasi-existent quantity). This doubles the wattage figure. Or even using the peak-to-peak voltage values, to quadruple it. These approaches have been called 'Music Power' and even 'American Watts'.

* Misunderstandings between the designers and advert copywriters[30], resulting in a 2.5 watt (avg.). rated amplifier being advertised as 10 watts 'rms'. This problem was nicknamed in the UK 'Sinclair Watts' - though other makers across the world have been no less careless or naughty.

7.8.1 Output voltage capability (V_o rms, MOL)

This is simply the maximum output voltage (usually rms, and continuous, and either called or very close to 0dBr), at the threshold of acceptable distortion (e.g. 0.1% THD), on which the output power capability is based. See the prior section. If unmeasured, but the input drive at clip is specified, and the gain is known, then the output voltage (in dBu) is simply Vin (in dBu) + gain (in dB), e.g. (+4dBu + 34dB), Vo = +38dBu.

7.8.2 Dynamic output capability

In the 1966 IHF standard, the heightened power delivery for a short burst (up to clip) of 20ms was rated and called 'Music Power'. The period of 20ms was a compromise, chosen as it was brief enough to catch the unloaded supply voltage (before the rails began to droop), but long enough to cover the attack portion of a sound. Effectively, this test just measures how loosely regulated the power supply is. And as discussed in 7.8, there was at this time a period of highly misleading power figures. This lead the FTC to impose a 'steady-state testing only' rule in 1970.

Dynamic power

In the IHF's 1976 standard, A-202, the continuous delivery rating was kept, but it was also supplemented with an improved burst power test. It employs a 1kHz tone-burst gated on at half-second intervals, each burst lasting 20ms - less than a cycle at 41Hz. Drive is increased until clip. The equivalent power is then expressed in dBwr (or dBvr) relative to the rated continuous power. For example, if 300w is achieved in these bursts, over 150w, the situation is 'IHF dynamic headroom = +3dB' (dBvr). The test became EIA-490 in 1981 [1]. In practice, a great body of music has transients that last longer than 20ms, often up to 200ms or even half a second [31]. This is the justification for class G and H amplifiers (see chapter 4), as well as amplifiers with very high continuous ratings.

Real duration

In 1987, the late Peter Mitchell concluded that a proper primary test standard for audio power amplifiers is not the continuous test (or even those using AES/EIA pink noise). These are important secondary tests, but the primary one has to measure the dynamic power delivery with bursts of 200 to 300ms, corresponding to the typical duration of a musical note (a quarter note lasts 200ms at 75 beats/min), with a duty cycle 'on' period of about 20 to 30%. Note that in all these burst tests, the power delivery is nominally 1% in the 'off' period. Alternatively, the 1976 IHF test can be made far more realistic, indeed, worst case, by raising the 'off' drive level (between bursts) to −10dBvr, so there is full power delivery 4% of the time, and 10% of full power delivery during the remaining 96% of the time. The domestic THX spec of 1995 specifies 20ms of high signal, then 480ms low, or resting 24 times out of 25. This is less taxing but may be true of movie music.

Voltage solidity

With all those audio power amplifiers having 'stiff' main DC supplies, *or* Class A operation, the maximum level of a short burst voltage swing is no higher than the continuous, i.e. dynamic headroom is 0dB. This is not the problem it appears, as soon as the full power capability is recognised as an indefinitely sustainable head-room! The rationale of amplifiers providing higher swings for short bursts is that sustained power capability to the high levels required for ample headroom is too costly in currency, and often also weight and size, and with some sorts of music, a much cheaper, 'bouncy' power supply and 'pulse-rated' power stage will do the job, i.e. pass transients without clipping nor explicit compression. This can be valid enough with the HF end of live percussive music, where a 50 watts amplifier with voltage swing equivalent to '1000w' of headroom would do nicely, and for the full-range reproduction of purist recordings of orchestral works; but it is increasingly invalid and false economy with many rock/electronic recordings having highly com-pressed, high level bass components.

With low budget amplifiers 'tuned' so that small, cheap power supply transformer and reservoir capacitors give their all, burst voltage swing may be +3 up to excep-tionally +6dB greater than continuous power, as the reservoir capacitance is able to hold up some of the voltage in the short period of high current demand.

Practical testing

After synchronising the output waveform on the scope, and calibrating, the burst power is simply derived from the voltage (so burst power typically ranges x2 to x4 higher than the continuous) but drawing the line consistently can be hard, as it is not so easy to measure the % harmonic distortion of the discontinuous burst wave – since the reading has to settle, even if it can be read and comprehended faster by machine than by eye. *With practice, and a scope having a fine trace, the onset of clipping may be estimated by eye instead, to within 0.1dB.* This may be self-consis-tent, but likely shifts the %THD threshold up to 1 to 10%, where trace flattening becomes consistently visible.

Testing also requires an adjustable (or agreed, preset) burst signal, and a justifiable load condition.

7.8.3 Clipping symmetry

The peak or burst voltage capability may be checked independently in both direc-tions. For this, an asymmetric and polarity-reversible burst test signal is required. A function generator can readily produce the signal, and an audio gate the burst func-tion. Behaviour immediately after this test with symmetrical signals may be reveal-ing. The asymmetric test signals may additionally be driven into clip, to attempt to provoke misbehaviour, e.g. DC shifts in the output, or shutdown.

Hirata [32] uses an signal comprising asymmetric positive and negative pulses, arranged so there is (with a perfect amplifier) no nett DC component. Any DC at the output is then due to, and becomes a measure of, non-linearity.

7.8.4 Dynamic range

If the SNR is referred to full output (0dBr), and the amplifier hard clips *immediately* above this level, then the Dynamic Range is equal to the declared SNR. A difference of 0.5 to 1dB between 0dBr and hard clip may be safely ignored, as it is well within the tolerance of measurement finesse at both ends. The SNR figure must be unweighted. 'A'-weighting would give a falsely 5 to 10dB wider dynamic range figure.

In amplifiers which soft clip (ie. give a higher voltage than 0dBr with still acceptable distortion) above the chosen 0dBr, then if every dB counts, the 3 to 10dB over which this behaviour prevails before unacceptably gross distortion finally sets in, may be added.

7.9 Dynamic tests

A sinewave is 'dynamic'. It theoretically crosses absolute zero volts on its way to peak vales of either polarity that can be large signals. But it has discrete frequency and within a few cycles such a test signal gives no further information about an audio power amplifier. The sinewave is, moreover, nothing like as dynamic as a music waveform, which comprises many sinewaves, all different, and often shifting in frequency and level, while encompassing *n* other possibilities both simultaneously and sequentially. Despite these facts, dynamic testing of audio amplifiers is still mainly about response to pulses in time. This includes the 3-tone and DIM 30/100 test covered a few sections earlier. Before, and without computer control, non-sinusoidal ('rectanguloid') pulses were easier to generate, readily analysable by eye, and theoretically tell everything.

7.9.1 Rise time (small signal attack)

This is (or should be) a small-signal measurement. A squarewave or other clean-edged, repetitive pulse is used. Before testing, a new mindset from the world of RF and fast pulses must be worn [33]. Notably:

1. A 'scope used for risetime testing needs in itself to respond at least 5 and preferably ten or more times faster than the amp being tested. It is best to label the scope clearly with its risetime, so alarm bells are rung when a amplifier having a commensurate risetime are met.
2. The 'scope probe(s) must be aligned for a flat response at LF and HF. Usually there are two presets in the probe's body. For setting procedure, see the probe maker's instructions. A 10:1 setting (20dB attenuation) is the norm.
3. The risetime of the squarewave must be faster than the amplifier. If the test source is fixed, its risetime may also be usefully 'written large' on it, to ring alarm bells if approached, as above. If a clean-edged squarewave source is not to hand, consider the calibration squarewave(s) available on most modern scopes.
4. The signal should be small enough to not cause slew limiting. This will be obvious enough, as the risetime 'triangulates off' above a certain amplitude that is frequency dependent. Stay 20dB below the level where this first becomes visible.

7.9.2 Slew limit (slew rate, large signal attack)

The causes and nature of slew limiting have been looked at in depth in Chapter 4. Measurement is another matter. Caution! *Measurement methods described in this section may damage marginally designed amplifiers. Review electrical fire handling procedures!*

IHF standard

In the IHF-A-202 standard, slew limit is tested by applying a 1kHz and setting this for clip into a rated load. The frequency is then increased until the distortion (%THD+N) is 1%. The frequency at which this occurs, divided by 20kHz, is the *slew factor* (SF). Thus if 1% THD is reached at 10kHz, SF = 0.5. Likewise, if 1% THD is only reached at 100kHz, SF would be 10. Bigger here is safer. The specification requires the other rated impedances to be tested.

This method can give false results as it presumes that the cause of %THD at HF *is* slewing. If the %THD is non-monotonic and wavers around 1% (something only readily visible with hf swept plots), then only the highest frequency point can be valid. The standard makes no proviso for assessing amplifiers that do not reach 1% THD+N at frequencies at which a test set can reach.

de facto method

In practice, most engineers measure slew limit by driving the AUT with a square wave, and looking at the slope at the output. Because at high RF only small signals need get inside the circuitry, and because these may gain entry through gaps in the enclosure, or up the speaker cables, it is best to measure the equipment's real, bare slew limit, without any input RF filtering (LP filtering) in place. RC filtering can usually be disabled merely by lifting the relevant capacitor earth legs. This involves 'hacking about inside' and presumes knowledge of circuitry. For simplicity and worst case realism, the entire input stage may be bypassed by disconnecting the gain control, signal being injected after this point (assuming Fig. 3.7 type of signal path).

With a suitable scope, the input and output squarewaves may be displayed overlaid, so slope differences are evident. At low levels, the slopes may be almost the same. The difference is that caused by the amplifier's finite bandwidth, signal delay and the compensation that's necessary for stability. The square wave level is then increased until either the output slope ceases to change, or (in the case of a J-FET or MOSFET or else degenerated BJT, input stage) until the output signal peaks are at or just below the clip voltage. *An amplifier's clipping LED(s) cannot (always) be relied upon for this.* A scope with a delayed timebase is needed to make accurate measurement of the slope of (what remains of) the 'squarewave' at the output. An alternative [34] is to fit a differentiator circuit directly at the amp's output, *before* any inductor (Figure 7.25). For high slew, high power amp testing, the RC parts must have suitable values and appropriately high ratings. The slew rate can then be read off directly as the peak amplitude on the scope screen, e.g. with a scaling of 10mV per 1v/μS, a 500mV swing corresponds to 50v/μS.

Figure 7.25

Slew Testing Setup.

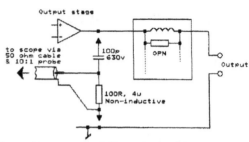

For accuracy, high speed scoping rules must be followed, notably to:

(i) ground the scope lead *at* the differentiator.

(ii) use a 10:1 scope probe with the correct 50Ω cable – not any old screened cable.

(iii) The differentiator parts should be close to each other, so the midpoint node is compact.

Restrictions

In practice, slew rate testing towards full output may not be practical nor safe. In many AUTs, even supposedly sturdy professional units, RF stability-providing Zobel and associated output networks are likely to fry. The amplifier may then become as a result self-unstable, causing further damage to itself and/or speakers, either immediately or later. For this reason, even if it continues to operate after a component burn-out, the AUT should be switched off immediately and repaired.

Another constraint is RF protection; highly specified professional AUTs have been known to wisely enter 'protect mode' shortly after the test signal is applied, having detected a rate of change that may damage or prove stressful to the output networks.

If the amplifier cannot be physically tested, slew testing can today be performed with some confidence by simulation [35]. Assuming the transistor models are reasonably specific, this can give results as accurate as or better, than the typically +/– 10% error, sum of the typical scope (3%) + cable (5%) + differentiator (2%) errors.

7.9.3 Transient response (impulse response)

Measurements of *transient* response are largely qualitative. They deal with an amplifier's behaviour over short periods of time. Both attack and also *settling* ('decay') behaviour may be seen.

Types of test

Conventionally, a repetitive squarewave is used for testing the transient behaviour, trigger synchronisation allowing momentary (but repeatable) events to appear fixed on a 'scope. But large squarewaves, particularly 10kHz ones, are not very representative of music signals. Sine-wave bursts may be used instead at three frequencies covering the band of interest. But accurate, low-noise capture – enough to allow high magnification – requires DSP techniques that are still relatively exotic [36]. By contrast, in SPICE-like computer simulation, transient tests are routinely commenced at T=0 with any waveform.

Setting up

In conventional squarewave testing, the amplitude is arranged for 'half power' in the load, being set so the squarewave's peaks are at the same level as a sinewave's would be, if set at –6dBr [37]. The input waveform is first viewed *in situ* with a 10:1 probe. Some small tilting is acceptable. If not otherwise intact and quite square, the input circuitry is affecting it and the source impedances should be checked. When ok, the scope probe is moved to the amp output. The 10:1 setting lessens the ill effects of cable capacitance.

Revelations

Assuming a square wave goes in, the squarewave response gives at-a-glance information on relative frequency response. Figure 7.26 shows some pathological conditions. Note that the appearance is relative, and will depends greatly on the squarewave's fundamental frequency and the timebase setting.

Figure 7.26

Squarewave Responses

LF Diagnosis
(i) LF rolloff with leading phase shift.
(ii) LF rise with lagging phase shift.
(iii) Controlled low-frequency rolloff.

MF Diagnosis
(iv) Narrow bandpass response.
(v) Poor LF response & peaking HF response.

HF Diagnosis
(vi) Heavy compensation, reduced HF response.
(vii) Heavy comp, slew limiting & reduced HF response.
(viii) Rising HF response.
(ix) Near perfect response - slight HF peaking and slight progressive LF rolloff.
(x) HF ringing.

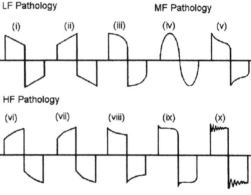

© Studio Sound

Figure 7.26 (x) shows ringing. This should not ever occur with a purely resistive load, but some ringing is to be expected into capacitative loads with most amplifiers, certainly with HF squarewaves. Provided the ringing has damped down by a nominal –20dB ($^1/_{10}$th) in as few cycles as possible, and well before the next transition, there is relatively little sonic ill. The ringing shown in Figure 7.26 (x) is prolonged enough for its effects to be audible.

Capacitative loads

The ringing implies a latent instability. If such a sharp edged transition as a 10kHz squarewave is needed to excite ringing, and even if that ringing is well damped, it seems that most music signals (even ultrasonic parts of) are unlikely to trigger RF oscillation. But other load conditions might. It is therefore prudent to check squarewave response into a range of capacitative loads. Values used will include something like 500pF (in practice, the test load lead capacitance), then commonly 470nF, 2μF and 10μF. To be safe for universal testing, load capacitors must be polypropylene, Teflon or polystyrene types (for low losses at HF), and rated at least 2.5 times the rms output of the largest amplifier to be tested, ie. 250v AC rms rated for testing up to 100v rms. This allows for the ESR's I^2R heating during any HF sine and squarewave tests and sweeps.

In the event of intractable or unacceptable stability problems caused by capacitative loads, the amp's output network (OPN) should be checked and possibly re-evaluated, preferably in conjunction with an overall Nyquist stability map. Today, this is readily achieved by simulation [38]. Compensation may also require re-tuning.

Inductive loads

Testing into inductive loads may also be carried out. This is unlikely to cause oscillations, even damped ones, in conventional high NFB amplifiers. But in some topologies and designs, inductive ill effects, such as current clipping or slewing may be observed.

Pulse alternative

Squarewave testing can upset an amplifier's DC conditions. If a test generator not made for audio is used, be sure to set the squarewave to be exactly centered on 0v, not offset. If the amp is direct coupled, check that DC offset is low enough (usually <2mV) at the levels and on the ranges being used. An amplifier that is upset by 'pure' squarewaves with 1:1 MSR can be driven by narrower pulse, ideally one with alternating polarity. This is closer to audio and less taxing on the AUT.

Hood's settling test

In 1975, John Linsley-Hood suggested that the settling time used by IC op-amp makers, could be a meaningful audio power amplifier specification [39]. With worst case reactive loading, his test required that a pulsed output voltage settles within about 0.5% of the final value, within a time that is short relative to the highest frequency signal that will be passed, e.g. 5μS was proposed.

7.9.4 Peak output current capability

Repetitive peak current capability can be computed from the continuous power output capability into reducing test load values, eg. 1250w (avg.) into 2 ohms =

$$\sqrt{W/R} = 25A \text{ rms}$$
$$= \sqrt{2} \times 25, \cong 35A \text{ peak}$$

But the kind of peak current capability referred to here is either the occasional, all-out burst; or regular but narrow spikes of current. In each case, either a high pulse current is drawn by using a series RC load (eg. 2 ohms + 10μF) ; or a difficult loudspeaker, known to draw high currents, may be used.

The 'one shot burst current' may be realistically tested using a defined portion of a loud, and known to be taxing music passage, possibly recorded onto DAT, together with a digital 'time domain' scope, likely DSP-based. Peak current is defined by capturing the peak voltage across a low value (typically $<<1\Omega$) current sensing resistor in line with the load. As the test signal is not continuous, a speaker may be used provided the repeatedly high acoustic output can be coped with.

Repeated short current bursts may be most easily tested using Baxandall's V-I test set (see chapter 5), together with an ordinary, analogue 'scope. Although the test is discontinuous, there is also a continuous, high level LF signal. So a speaker may not be comfortable as a test load.

References

1 *Standard methods of measurement for audio amplifiers*, EIA RS-490, 1981.

2 Details of the professional and domestic THX theatre sound system standards: Lucasfilm Ltd, THX Division, PO Box 2009, San Rafael, California 94912, USA.

3 Duncan, Ben, *Realistic amplifier testing: Procedures for professional touring music PA*, Proc.IOA, Vol.16, 1994. For details of the Professional Power Amplifier Test Standard, contact: BDR, Tattershall, Lincoln, LN4, UK.

4 (Nameless), *EIA/ANSI standard RS-426-A*, 1980. (later AES)

5 (Nameless), *Real load resistors*, Elektor, April '78.

6 Walker, Peter.J, *Audio amplifier load specification*, Wireless World, Dec 1975.

7 Stuart, J.R, *Psychoacoustic models for evaluating errors in audio systems*, Proc.IOA, Vol.13, part 7, 1991.

8 Woodgate, John, letter, p.41, Wireless World, Feb 1976.

9 Otala, M, & Huttdnen, *Peak current requirement of commercial LS, systems*, J.AES, June 1987. (This is just one of several papers. See chapter 2).

10 Scott, Jonathan, & Greg Lemon, *A non-linear load for power amp testing*, 5th Australian AES preprint B3-4027, 1995.

11 Colloms, Martin, *Technical introduction - Amplifiers and Stereo Tuners, Buyer's guide No.39*, Hi-Fi Choice, 1985.

12 Court, Stephen. J, *Damping factor*, Wireless World, July 1974.

13 Gerzon, Michael, letters, p.60-61, Wireless World, March 1976.

14 Jensen, Deane, *High frequency phase response specifications - useful or misleading?*, 81st AES convention, Nov 1986; reprinted with corrections, by Jensen transformers, 1988.

15 Lohstroh, Jan, and Matti Otala, *An audio power amplifier for ultimate quality requirements*, JProc.IEEE, Vol. Au-21, No.6, Dec 1973.

16 Green, I.W & J.P Maxfield, *Public address systems*, Transactions of AIEE, Feb 1923, vol.42.

17 Moir, James, *High Quality Sound Reproduction*, Chapman & Hall, 1961.

18 Nelson, Jeff, *Too many notes, Boulder notes No.1*, Boulder Amplifiers Inc., 1986.

19 McConnel, John.W, *Been there, done that!* (letter), Stereophile, Dec 1995.

20 Perrin, R., T. Charnley & J. dePont, *Normal modes of the modern English church bell*, Academic Press, 1983.

21 Lazenby, M, *How little distortion can we hear?*, Wireless World, Sept 1957.

22 Shorter, D.E.L, *The influence of high order products on non-linear distortion*, Electronic Engineering, vol 22, No.4, 1950.

23 Wigan, *New distortion criterion*, Electronic Technology, April/ May 1961.

24 Hilliard, John.K, *Distortion tests by the intermodulation method*, Proc. IRE , Dec 1941.

25 Brown, Ivor, *Trial by three tones*, EW&WW, Feb 1991.

26 Heyser, Richard. C, *Some new audio measurements*, 51st AES, preprint 1008-K4, 1975.

27 Belcher, R.Alan, *Audio non linearity: A comb filter method for measuring distortion*, BBC research dept. report, 1976/12.

28 Belcher, R.Alan, *A new distortion measurement*, Wireless World, May 1978.

29 Jensen, Deane, & Gary Sokolich, *Spectral contamination measurement*, AES preprint 272, November 1988.

30 Dale, Rodney, *The Sinclair Story*, Duckworth, 1985.

31 Mitchell, Peter.W, *A musically appropriate dynamic headroom test for power amplifiers*, AES preprint 2504-07, 1987.

32 Hirata, *Quantifying amplifier sound*, Wireless World, May 1981.

33 Williams, Jim, *High speed amplifier techniques*, AN-47-1, Aug '91, Linear Technology Corp. (USA).

34 Self, D.R.G, *High speed audio power*, EW+WW, Sept 1994.

35 Duncan, Ben, *Simulated attack on slew rates*, EW+WW, April 1995.

36 Duncan, Ben, *Loudspeaker cable differences, case proven*, Proc. IOA, Vol.17, part 7, November 1995. Also in Studio Sound, Dec 1995; and in Stereophile, Dec 1995.

37 King, Gordon.J, *The Audio Handbook*, Butterworths, 1975.

38 Duncan, Ben, *Viewing phase margin*, Spectrum News (USA), Summer 1995.

39 Linsley-Hood, John, *Settling time in audio amplifiers*, Wireless World, Jan 1975.

Further reading

40 Bowsher, J.M, A *new way of specifying amplifier outputs*, undated footer, JAES, 1974.

41 Cabot, Richard C, *Comparison of nonlinear distortion methods*, AES 11th international conference, 1992.

42 Duncan, Ben, *Harmonic convergence*, Stereophile, October 1992.

43 Metzler, Bob, *Audio Measurement Handbook*, Audio Precision, 1993.

44 Woodstock, Alexander, & Monte Davis, *Catastrophe Theory*, Pelican Books, 1980. ISBN 0-14-022250-2.

8

Real world testing – Rationale and procedures

8.1 Scope and why essential

This chapter details some of the real world tests that are never mentioned by, or are outside the remit of, the makers of audio 'test' instruments.

These real-world tests should be applied by designers, before amplifiers are made; and by makers, before products are shipped. They may also be employed by purchasers, before the money is handed over. The tests, like all sensible QA programs, are based on not accepting anything 'that *should* be able to be taken for granted'.

8.2 Listening

Skilled or even casual listening can (and does) reveal both blatant and subtle defects that routine (and even advanced) measurements are 'stone deaf' to.

Single presentation is the safest technique. You live with an amplifier for a week or a month. Daily (but wholly *ad-hoc*, unpressured) listening to a range of music is presumed. Then you try another, ideally one that is roughly matched in power delivery. You may not be sure of the differences straight after the changeover, but it is unlikely you will not hear some soon, and that later, you will be either glad to swop back, or else keep, the second amplifier; or if both are both flawed or good, you will at least have a feel for their respective advantages. On the other hand, there is a risk of getting used to changes. The fact that humans are highly adaptable to the imperfect or less good, is not an argument for ignoring the absolute difference. After years of experience, John Atkinson states "*It is an essential rule when judging hi-fi equipment, that the better component is one that lets you get into the music more easily, that opens your ears ... not the one that meets arbitrary standards ...*" [1]

Blind, double blind, and triple blind testing are ideas drawn from medicine. But unlike wines or drugs, the actual object is not being directly evaluated [2]. Also, in pharmaceuticals, the blind tester partakes of solely A or B, never both. And at the hands of hard-line objectivists, they have often developed (and without any deep consent from, or consultation with, listeners) into high pressure, and highly unnatural methods for most humans to make decisions about even large differences in musical emotional expression. In double blind testing, neither operator nor listener knows what is being heard (at the time). Then there is triple-blind testing. But does this stop pre-cognition? No.

Only a minority of individuals can 'perform well' when listening under these conditions. Pseudo-scientific types who mistakenly advocate only blind testing as the validator of audio differences, may be arrogantly ignoring the reality of human 'psi-fields', and their effects on other humans and equipment. in other words, *the pertubation of the measurement methods exceeds the effect to be observed.*

A-B and A-B-C testing and related schemes involve switching between one amp under test (A) and another (B) and maybe a third (C) and more (etc). At each switch setting, the amplifiers being listened to may be openly known (switch labelled), or the testing may be blind, double-blind, etc as above. If the latter, this type of test has the handicap and the danger described above.

It is well known that small differences in level can confuse inexperienced listeners, who hear a 'brighter' sound when one amplifier has a tad more gain or drive level. Levels must be very accurately matched. *At high SPLs and low frequencies, differences of as little as 0.1dB can be readily audible.* But matching is conventionally carried out at 1 to 3kHz, at least for a full-range amplifier. In an active system, the geometric mean centre of the frequency band would be optimal on this count, while the equal loudness contours demand that matching is performed at the more sensitive region of the band. For matching, precision equipment (AC millivoltmeter and high resolution, dB-reading DMM or other analyser) able to resolve below 0.1dB and with commensurately high accuracy, must be used. If detented knobs evade perfect setting, adhesive tape must be resorted to.

In a landmark paper [3] Martin Colloms cautions first that 5 to 10 repetitions are needed to thoroughly learn the music; else more is heard with each of the test stages, purely through the repetition. Second that music's lack of absolute consistency from bar to bar undermines the validity of switched A/B testing in some ways, to the point where random results are heard.

It is best used by designers or others working at depth, and listening should *not* be prolonged, else the brain begins to EQ the sound, and then fatigue sets in. Ideally, the duration of A-B testing should not exceed a few minutes in a day. Naturally enough, in any such tests, any financial or emotional interest in the test result is assumed to be suspended, if not absent.

In ABX switching, the listener switches at will between amplifiers A and B, and the centre position X. They then decide which X is and write this down. Ideally, what X really is, is then randomly changed. After twenty of these random, hidden changes in X, the result is 'statistically valid' if the assertion is correct sixteen or more times, i.e. > or = ($16/20$). Loudspeaker designer Mark Dodd describes the rigmarole as "Very objective, very insensitive."

Even if sighted, the abruptness of switching between units having small but highly complex differences tends to reduce the perception of differences, as the mind struggles to listen to a mainly non-repeating, dynamic, and emotionally-charged sequence of sounds, while making sense of these, and also offsetting minute (but ever present) tonal differences. As Wes Phillips puts it [4], *"Short term A/B'ing stresses evaluation over participation. You can't just study a situation and simultaneously respond to it."*

Moreover, few people have an acute aural memory that extends beyond a second or two. One expert listener, high-performance PA system designer Tony Andrews, tunes into the emotion, and notes how this changes at every switch over.

What about null results? John Atkinson reminds us that *"..a null result from a blind test does not mean that there isn't a difference, only that if there was one, it could not be detected under the specific conditions of that experiment."* [5]

8.3 Operable mains range

Relative to its rated supply, a high performance power amplifier, whether for professional or domestic use, should operate continuously without hiccups, with supply voltages 10% above (say) the rated 240v (ie. to 264v AC) and ideally down to 170v or lower (–30%) . This range (which would be *pro-rata* centered on 115v in the USA, and on 100v in Japan, etc) allows for normal, legal fluctuations in supply voltage of up to +/–6%, +6%,–10%, and +/–10% (depending on territory), and also provides an additional, small (–4%) allowance for (not all at once):

(i)   operation on reduced supplies (eg. 240v setting on 230v; or 120v setting on 110v) without the bother of tap changing,

(ii) the cyclic IR voltage drop in cabling, and

(iii) momentary (< 1sec), occasional sagging in the AC supply that is not registered as a lowered voltage by the supply utility or by a sluggish moving iron, panel voltmeter (analogue needle type).

Every power amplifier's operable AC supply voltage range should be stated *clearly* on the rear panel, as well as in the user's manual. Testing is performed with a Variac.

It is to be expected that a 'high supply' (e.g. up to 10% above 240v) will cause an amplifier to run hotter, and that eventual lifespan may be reduced accordingly, but even so, all internal parts should be within their safe ratings and there should be no early failure (in the next 12000 hours say) caused by several hours operation at the high end of the legal range.

With a 'low supply' condition (e.g. up to 30% below 240v), it is to be expected that the AUT's headroom and voltage swing will be reduced, and that fans may run slower. However, any regulated supplies that 'drop out' should not cause audible ripple (hum), nor damage; and cooling fans must not stall.

Ironically it is amplifiers using simple *regulated* supplies in their output stage that are potentially the most affected by voltage variations above a few %, as it can be expensive and wasteful for high current linear regulators to operate over the wide, ideal range.

With domestic amplifiers, the lower end of the operable AC supply range might be relaxed, but it is still helpful for amplifier performance to be maintained without muting, shutdown, or other hiccups, to considerably below the –10% that a civilised electricity supply is expected to stay within.

8.3.1 Inrush current

The inrush current at switch-on is worst when an amplifier is started 'cold' i.e. when the reservoir capacitors are wholly discharged. Under these conditions, and high mains voltage, switching on up to three amplifiers at a time with a one second period before the next group of three, etc., should not trip house breakers, nor blow house, nor mains plug (receptacle) fuses, nor disrupt other equipment's operation.

Miniature Circuit Breakers (MCBs) are classified by speed. Domestic amplifiers should ideally not trip type 1 breakers, the fastest tripping kind (Figure 8.1). But in large installations, with low impedance mains supply, it may be permissible to use slower rated (i.e. euro-types 2, B, C or D – or consult your local code) breakers in the event of nuisance tripping. *A qualified electrician should be consulted before changing circuit breaker types.*

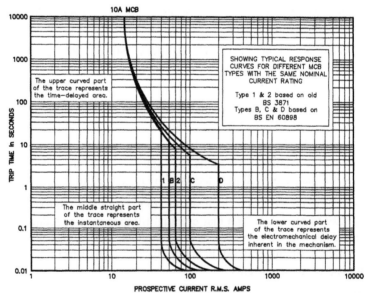

Figure 8.1

Time vs. current characteristics compared for a 1, 2, B, C & D types of 10 Ampere-rated MCB (miniature circuit breaker), according to BS 3971 and the newer BS EN6098 and also present IEE regulations (UK). Note that a type '1' breaker passing 40 Amperes takes anything between 45mS and 20 seconds to trip. The other types (2, B, C, D) will not trip on larger inrush currents of progressively extended duration. © Toby Hunt.

Inrush current may be captured and displayed with a digital time-domain scope, today using DSP (Figure 8.2). However, qualifying the effects that matter, on fuses and circuit breakers, remains more complicated and uncertain, than performing practical experiments.

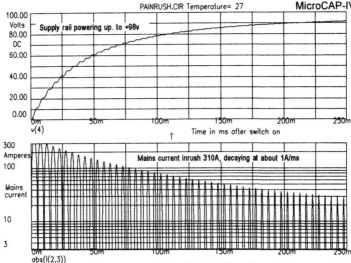

Figure 8.2

Mains inrush current Vs. *time. The lower plot shows how the current peaks and then decays immediately after switch on (lower), while the DC supply voltage rises (upper). The voltage wiggle is ripple.*

8.3.2 Soft start

The 'soft start' action of a power amplifier should be tested with a worst case speaker (or SLS) load. When testing, it is useful to connect a 250v AC meter or DVM across the soft-start resistor, so the cessation of soft-starting can be confirmed. The following tests (with the worst case load) will be instructive:

i)   Switch on then apply AES/EIA pink noise to input, set at clip. Switch off, then switch immediately back on. The soft-start should not stay latched-on.

ii) Set mains 30% low. Switch on. Again, the soft start should not stay on.

iii) Simulate high mains with high DC (even harmonic) content. The soft start switching device should not weld or burn-out when activated by repeated powering up.

Some audio power amplifiers employ *ntc* thermistors for soft starting. At the expense of a little added series resistance (quite allowable with an SMPS), this type is automatic and highly reliable. It takes several seconds to cool, however, and testing is different:

iv) Use a low impedance (e.g. 100 or greater Ampere rated) mains supply. Switch on, then apply AES/EIA pink noise to input, set at clip. After ten minutes, switch off, then switch immediately back on. The soft-starting ntc thermistor(s) should not be damaged, e.g. no fires or explosions. Also, the higher inrush current in this condition should not blow or stress internal power line fuses.

All these tests are far more meaningful when repeated hundreds of times on some sample amplifiers, by timer controlled relay. For the first two tests, the occurrence of a sustained voltage on the soft start resistor must be data-logged, or alarmed. Note: These tests may provoke a fire hazard – *review precautions*.

8.3.3 Mains current draw

The envelope of rms and even peak currents drawn by amplifiers is increasingly cited, at least by makers of professional power amplifiers that may be used *en masse* (e.g. Figure 8.3; see also the model specification in Chapter 7, Figure 7.1). Alas, the method used is highly specific to each maker, and the mains' gross waveform distortion is left un-mentioned.

Load (Ohms)	Max Power (midband)	AC Current Full power	AC Current 1/3 power	AC Current 1/8 power	AC Current Idle
8 + 8	800 x 2	25A	10.8A	5.7A	1A
4 + 4	1200 X 2	40A	17A	8.5A	1A
2 + 2	1500 X 2	58A	24A	12A	1A
8 + 8	550 X 2	19A	10A	5A	1A
4 + 4	825 X 2	31A	15A	7.5A	1A
2 + 2	1000 X 2	41A	22A	11A	1A

Max. sustained power w/ adequate supply — *UL.CSA.IEC rating standard*

(Rows 1–3: EX 4000; Rows 4–6: EX 2500)

QSC EX 400 and EX 2500 AC Power Consumption vs Load Impedance

Figure 8.3
One way of presenting AC power consumption information, from QSC. The Max Power column is in watts, average. The 1/3rd Power AC Current is for worst case, very dense (low PMR) music drive. The full power is for test sine waves only. The 1/8th power is more typical of full-range or HF music programme.
© QSC Inc.

Current draw can be measured and viewed with a 'scope connected to an *AC current clamp* (Figure 8.4) placed over the live conductor. A break-in adapter lead with a live wire that is 'split out' makes this easy. With a current transformer clamp, an AC milliamp meter must normally be connected in parallel, to yield the true waveform. Scaling is typically 100mV/A. The most basic scope will do, but the voltage (x) sensitivity of the set up must be calibrated, eg. with a test resistor across a 50/60Hz supply. An 11.35v rms 50Hz slate tone from a power amplifier connected across an 8 ohm, 16 watt resistor gives a 2 Ampere reference peak current.

Figure 8.4

Measuring the peak mains current drawn by the AUT.
The peak current drawn by amplifiers can be measured and viewed with a 'scope connected to a current clamp placed over either the live or neutral conductor. A break-in adapter lead with a 'split-out' conductor makes this easy.
Courtesy Studio Sound

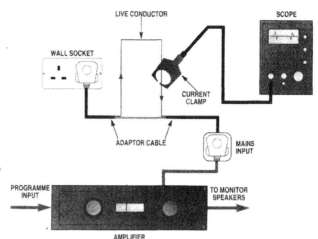

8.4 Signal present indication and metering

A 'signal present' LED should light at or within +/– 2dB of the level (in –dBr, or mv or mV) that should be cited in the specification. The specification should also explain whether the signal present shows

(i) the incoming signal or

(ii) the signal after any gain control; or

(iii) the signal available at the output ?

The importance of the signal present LED illuminating at a sufficiently low input level comes clear after considering that when driving a horn loaded speaker with (say) 120dB @ 1w @ 1m sensitivity, a 500w $\Rightarrow$ 8Ω power amp can generate a lot of SPL at just 50 milliwatts (which is –40dBr relative to the 500 watts). The threshold of a signal present LED is usually measured with a 1kHz sine wave. This threshold should be maintained within say 5dB, as the frequency is swept to the edges of the audible range.

Metering is comparatively rare on modern high performance amplifiers. Users rarely need to know the number of momentary 'watts'. As signal level meters, LED Bars are unforgiving; like digital recording, and unlike the meters on mixers and tape decks, as 0dB is the *hard* limit; going above means instant sonic nastiness. Professional metering has mostly shrunk down to the bare essential, which is the top, 0dB LED, marked 'Clip', etc.

If metering is fitted, whatever initial accuracy it claims should be marked on or below the metering. Then if anyone ever uses the metering for setting signal levels, they will know how far to take it. So, under many bar LEDs there might be printed: '+/– 2dB tol.'. This would remind meter readers of the possibility of up to a 4dB total reading difference at some levels.

8.4.1 Clip indication

Clip indicators and warning of overload should indicate within *at least* 0.5dB of (preferably below) the onset of clip. A steady test signal, usually a sine wave, is used.

The 'onset of clip' may be:

(i) where the %THD is beginning to rise above the *de facto* 0.1% or even 0.5% limit; or

(ii) where (on a scope) a spike first appears in the %THD residue; or

(iii) where the peak part of a fine 'scope trace first visibly flattens.

Testing should be carried out at the extremes of rated load, and with frequency swept over (and preferably beyond) the audible range. A late clip warning (> –0.5dB) at 30kHz may seem inconsequential but can prove fatal to a tweeter subjected to ultrasonic clipping. Testing should also be carried out at the extremes of the operable AC mains range.

8.5 DC at the input

If (or when, one day) the equipment feeding the power amplifier fails, and 'goes DC' and if it is direct coupled, or the DC blocking capacitors are wired the wrong way, then DC will be applied to the power amplifier's inputs. If these are direct-coupled, or again, if the DC blocking is selective about polarity, the input stage will be upset and may be damaged. If the amplifier's entire signal path is direct coupled, then the amplifier's own DC protection may act to protect the speaker, but even if the amplifier is partly or wholly powered down, it may still be open to damage.

DC voltages in consumer-grade signal source equipment are typically single rail, commonly +5 to +30v. In professional, and more refined domestic processors and sources, the DC voltages range from + or –12v to + or –24v. If an amplifier is not damaged when +30v, then –30v are applied, it will survive most catastrophes in preceding solid-state equipment. If the inputs are balanced, the two polarities of DC should be applied differentially, and common mode, and in the two unbalanced configurations, i.e. eight tests in all.

If a power amplifier either fails, or some major performance figure is degraded by the testing, then using a repaired or fresh unit, the DC may be re-applied at a lower voltage, increasing until damage or performance decay re-occurs. The ability to withstand up to a lesser +/–20v DC (say) applied in any permutation, would still confer useful protection, in this case against the majority of equipment.

8.5.1 RF at input

RF immunity is now tested for under EMC regulations that are compulsory in the UK and most of Western Europe. The same or similar tests (see section 8.9) may be used in other territories.

One aspect that is not necessarily tested for is occasional 'RF instability' (a euphemism, read: RF oscillation) of some types of input stages when a particular type and length of (usually stage returns multicore) cable is connected, and left dangling open at the source end.

8.5.2 Large signals at input

To withstand accidental mis-connections (particularly when XLRs are used at inputs *and* outputs), a well designed power amplifier should be able to withstand differential input voltages equal to the maximum, bridged output voltage without damage, eg. 90v rms for a 250w$\Rightarrow$8Ω or 45v AUT. Usually, input stages operate on low voltage (+/–15v) supplies and are limited to differential voltages in this range. Where protection exists, it is achieved with clamping schemes.

An ideal pro design would also withstand for short periods or suffer only limited damage, from the accidental mis-application of *common-mode* (not ordinary, differential, signal-like) voltages as high as the highest AC mains, e.g. 340v peak in the UK. Transformer coupled inputs have long provided this, *but in every case tested at Turbosound R&D, and elsewhere, it has otherwise been to the detriment of sonic quality.* Direct coupled, high common-mode voltage (hcmv) -withstanding input stages are also feasible, though rarely used so far in audio amplifiers.

Other than sine waves, program and pink noise, pulse signals as used for slew-rate, rise-time and V-I limits testing, may reveal weaknesses. *The relevance of these sorts of signals is that they are akin to some of the glitches and edges that emanate with switching from all kinds of mixing consoles, and preamplifiers.*

8.6 Output DC offset (output offset, V$_{oos}$)

There are no standards for DC offset (error) voltage at the output of an audio amplifier. A large offset, above say 1 volt (for 50w/8Ω), constitutes a *DC fault*. **This is a serious condition**, dealt with in chapter 5, sections 5.7.1 and 2. *Never plug an unprotected speaker into an amp suspected of having a DC fault.*

Checking

Checking for DC offset is straightforward, using an auto-polarity DVM able to resolve at least 1mV. Warning! If the AUT is faulty, moderately high voltages may be present (50v DC or more). The input may be shorted or left connected to a representative source. If the latter, there must be no signal.

Adjusting

Before adjusting anything, check the offset *is* in the output stage! If the amp has a gain control, and the offset increases when this control's setting is increased, then the offset is in the input stage *or* in the signal source. The same applies if the offset varies depending on whether the input is shorted or connected to a representative source.

Judging

Offsets below +/–1v are relatively harmless unless the amplifier is abused. However, they can cause loud clicks and thumps when sensitive speakers are connected or switched. This is disturbing though rarely harmful. On this basis well-designed amplifier should exhibit under +/– 50mV (0.05 volts) at most.

Excessive offsets may be caused by careless design, or manufacturing and set-up, including component value drift accelerated by high temperatures; or after an amplifier has been repaired with new but un-matched transistors in critical locations.

Testing

Establishing the DC offset voltage is usually carried out with the AUT warmed up, but monitoring from switch-on can be instructive. In direct-coupled power amplifiers, DC offset voltages soon vary from the peak value observed at switch-on, possibly changing polarity and eventually settling to a low value as the AUT warms up. This behaviour can provide fingerprint of the thermal and hence physical layout, and also of the semiconductors used, and may be used to uncover errors in production. It is acceptable for the offset at the instant of switch-on to be quite high, provided this does not cause protection or a loud bang when speakers are switched- or plugged-in

Trimming

In good designs, drift due to aging is foreseen, and offset can be trimmed. In the absence of data or instructions telling you what to expect, nulling DC to less than +/–10mV is a plausible aim.

DC offset nulling may be alternatively viewed as setting the midpoint between the output stage's push-pull or other halves. It follows that one way to adjust it (when you have a scope) is for symmetry when driving a load and clipping a pure (ly symmetrical) sine wave. *This argument is both partly nullified and partly reinforced, by music signals' noted asymmetry.*

When ok, nail varnish or something similar should be used to seal any adjusted presets.

8.6.1 RF at output

It is more common than believed, for audio power amplifiers to oscillate at (and radiate) RF [6]. The chances of capturing the odd circumstance(s) under which this might occur, are heightened by monitoring the AUT's output at all times with a 'scope. The scope should have as wide a bandwidth as is available, ideally over 100MHz. 10:1 probes should be used and the timebase should be set so RF > 1MHz is clearly visible as a fuzz on the test signal waveforms.

Alternatively, whether or not oscillation has occured may be logged digitally, or simply recorded with a LED clock, recording say up to 10 'RF events' within the detection window, and driven from a sensitive, wide-range-handling RF 'sniffer'.

8.6.2 Adverse loads

The following table sets these out.

The 1 to 10 of ADVERSE LOADs – all active types:

1. A legal, rated load.
2. Hard short across output terminals.
3. Hard short across far end of the speaker cabling.
4. Low loads - below amp's lowest rated impedance.
5. Heavily dipping load, eg. 'difficult' speaker with severe dips, sometimes as (3).
6. Highly capacitive load, e.g. certain driver/box combinations.
7. Wholly capacitive load, e.g. defectively wired crossover.
8. Intermittent or arcing connection to a legal load.
9. Intermittent or arcing connection to an illegal load, ie. any of 1-6 above.

Note: '*Hard Short*' below 0.1 ohms. '*Low load*' below 2 ohms, above 0.1Ω.

Transformer-coupled OPS only:

10. Open Circuit or far above rated load.

Note: '*Legal*' means '*rated as permissable*' by the maker.

8.6.3 Adverse load proving

Except where stated, all the following tests are most meaningfully carried out with the AUT driven by a 'continuous' AES/EIA (speaker testing) pink noise signal, to within 1dB of (below) visible clip, and with the AUT's output connected to a representative or worst case dummy speaker load.

8.6.4 Adverse loads, low loads and shorting

The short circuit protection capability of many a professional power amplifier has been theatrically demonstrated by holding a 6" nail across the output terminals, or inserting a shorting plug. While dramatic, this is not what happens in practice. In reality, the short is more likely to be :

i) a thin whisker at the terminals at either end, occasionally shorting or arcing.

ii) XLR or EP plug inserted cack-handedly or carelessly, momentarily shorting live pin(s) to chassis – if this leads back to amplifier ground.

iii) Damage *n* feet (metres) away, to the speaker cable insulation, and the conductors touch sometimes.

iv) Operation of a loudspeaker's protective crowbar circuit.

v) Connection of a 'repaired' speaker, miswired so it presents a short.

These conditions can be checked with three tests, with a relatively long, say 5m, speaker cable connected. To test worst case conditions, it should be the highest capacitance (per metre), lowest inductance cable, that the amplifier can stably handle with the cable connected to a representative or worst case dummy speaker load, and with the AUT driven with AES/EIA speaker testing pink noise.

Short circuits are then applied (i) at the output terminals, (ii) 1/10th way along the cable and (iii) at the end of the cable. Controlled testing may be carried out with a number of shorting devices, but for uniform timing, the choice reduces to a triac or other solid state 'crowbar' switch [7], or a relay. The former is realistic when speakers containing crowbar protection have to be driven. The speakers' crowbar circuit may be replicated as a test jig, and triggered by an external DC source. The triac's (or SCRs') rate of shorting, because there is no contact bounce, is much more incisive as well as faster than mechanically applied shorts. The back EMFs thus caused by triacs or SCRs may be destructive to a degree that is unlike ordinary shorts.

Figure 8.5 shows a simple test timer, using a monostable controlled relay to apply a short circuit for a preset period of 5 to 15 seconds. Shorting is initiated by the start button. A LED and buzzer indicate timeout. The relay must have low contact resistance while being rated to withstand and 'switch' up to 100 Amperes without welding.

Figure 8.5

Short-Circuit Testing Monostable.

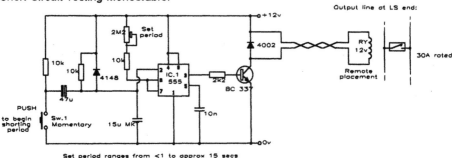

It is little recognised that *even when there is no music signal*, most solid-state power amplifiers can be stressed by (almost) *hard shorting* their output. This because in most designs, DC offset voltage provides a 'signal'. For example, if there is +50mV DC at the output, and the output's source impedance is 2mΩ, and the 'short' is 35mΩ some way down the wire, then

$$I = V/R = \{ {}^{50m}/_{37m} \} = 1.35A$$

flows. At the same time, one half of the OPS is sustaining its entire supply rail, less 0.05v, hence is left steadily dissipating (say)

$$(90v \times 1.35A) = 121.5w.$$

While a properly designed amplifier will survive this, it is clear that with even only moderately 'hard' shorts, not many millivolts of signal or DC in particular, is needed to tax the output stage in a most vulnerable zone, and test the amplifier's protection system. For example, if the DC offset crept up to 0.7v, and the short's resistance was 30 mΩ, then I = 23A. At 90v, this is getting serious. *It is therefore useful to independently prove the short protection at the highest level of DC offset that is reasonably expected*, or justifiable, due to drift, mis-adjustment, new parts, etc.

Professional power amplifiers, particularly any used for touring PA, are usually expected to survive all such tests.

8.6.5 Adverse loads, reactive

A 6.8µ or 10µF capacitor across the output is a stressful reactive loading. This situation can occur when a passive crossover or even active system speaker wiring is damaged or wrongly replaced, rewired or 'fixed'. A professional AUT would be expected to survive this condition for several minutes, or protect itself. It is a particularly stressful condition when the pink noise signal has substantial HF content - the AES/EIA test signal has some of this, but live sound *can have still more*. Alternatively, but with less realism, a 1, 10kHz or even higher frequency square wave may be used.

Another option is to drive with an adjustable sine wave. Then the *stability margin* may be evaluated using a phase meter to evaluate the excess phase shift at the output, i.e. the value of any phase shift above 0° or 180° relative to the input phase. It should be no more than 110° ahead, into a worst case reactive load. This is more conventionally expressed as 70°, when described in terms of Phase *Margin*.

Note: At 180°, the phase shift would has fully turned the NFB into *positive* feedback. Within most of the grey area, +/– 90° either side, instability is latent and dynamic response unstable.

8.6.6 Hard drive testing

This is essentially the warm-up procedure described in Chapter 7, under Pre-conditioning. However the load should be representative of the lowest rated, or a speaker load that is nominally within ratings. Any high quality audio amplifier should be able to withstand this kind of test. It may protect itself by shutting down, but if professional, this shouldn't happen too readily, i.e. shutdown shouldn't happen after the first five minutes.

8.7 Thermal protection and monitoring

Amplifiers overheat when they are pushed hard in an inadequately ventilated space; when the speakers and music are taxing; or when ventilation is grossly impaired by expired cooling fans, ducts blocked by dust, or carelessly dumped clothing.

Thermal protection is readily tested. In *blown* (air, not broken) power amplifiers, all outlet vents are taped-up or otherwise blocked. Another realistic approach, having a subtly different effect, is to disconnect the fan(s). If convection-cooled, a reasonably fire-retardant, heat insulating material (such as an up-ended, specially made wooden and glass box) covers the AUT. The glass allows LEDs and possible signs of smoke to be monitored.

A safe power amplifier will shutdown signal at the output. Some time before this happens, a warning is liked by professional users. This is sometimes provided, with an LED illuminating 10°c or so below the shutdown temperature. If the amplifier employs fans, it will ideally continue to cool itself whilst 'shut-down' A well designed power amplifier will also tell you why its service is temporarily suspended, and return automatically when it has cooled enough to safely recover. No intervention should be needed by the user, other than dealing with the cause of the overheating. A 'soft return' would be best, e.g. as pioneered by Crest.

Thermal monitoring is rare, but precious in touring PA, when intense stage heat and hard driving makes the backstage sound engineer fear that amps are about to start *thermalling out* like skittles. In the BSS EPC series, bargraph LEDs show either actual temperature (at some location like the air vents); or thermal *headroom*, i.e. °c left before shutdown.

8.8 Muting behaviour

If power up/down muting is effective, then when the amplifier is connected to a horn-loaded speaker having a sensitivity towards the high end of the scale, and the power is switched on, off, then on again (with varying timings) with no signal applied, any clicks, plops or 'blahtt' sounds should be quiet enough not to unduly shock or upset bystanding people. The peak SPL under these circumstances will be below 70dB. And with the majority of domestic speakers, it will be 10 to 30dB quieter – commensurate with the lower levels of impulsive noise that are acceptable in a domestic space.

If signal muting is fitted, and operated manually, remotely or automatically in the event of a fault being detected, the muting action should not leak (or at least make more than very dimly audible) full level, wideband programme, even when the amplifier is connected to a horn-loaded speaker having a sensitivity towards the high end of the scale.

8.8.1 Acoustic noise

Most power amplifiers made for domestic or some kinds of professional (or 'semi professional') use, are likely to be in the same room as the listener and must not detract from silence. Noise on standby is principally caused by either cooling fans (even temperature controlled fans); and by transformer buzz, which may be initiated by DC on the mains supply, and acoustically amplified by sheet steel casing.

THX's domestic specification suggests a distance (say 6 feet/2m) at which any noise will not be audible, with a plausible background noise level, say 30dBA$_{SPL}$.

There are less constraints on the noise levels of most professional power amplifiers. In recording studios, main monitoring amps are universally mounted in separate, sound-proofed cupboards. In PA systems, the noise they make is well below, or at least comparable to, the noise level of venue air conditioning plant, and the crowd. Problems may arise when a noisy amp rack is used for (say) a folk group, and has to be placed near to a quiet audience in an intimate (read: cramped) venue.

8.9 EMI and EMC

In the USA and Germany, national FCC and VDE regulations have long set limits on RF emissions from power amplifiers. For the most part, these regulations have not applied to audio power amplifiers with conventional 50/60Hz power supplies, while for units with HF switching power supplies, the regulations have not been enforced with much vigour, except by individual importers or specifiers. For example, the Domestic THX specification requires that power line conducted EMI exceeds FCC Class B specifications. The maximum allowable radiated hum field from the power transformer (s) is also specified (in magnetic field units, Gauss).

Recently compulsory EMC (Electro-Magnetic Compatibility) regulations in the European Community, and further, forthcoming regulations, are about to have a strong effect on the future design of audio power amplifiers. These pan-European regulations supersede VDE, while in the USA, the less stringent FCC regulations have been applied with more vigour. At the time of writing, the appropriate EMC test standard for most audio power amplifiers is the generic type. Still, professional help must be sought by makers or importers to determine which test standards must be used. New test standards may apply at future dates, or applicability may alter depending on the intended use. Tests which are inappropriate may be omitted provided their exclusion may be justified according to the intended application and environment of the amplifier. Generic testing, to EN-50081 and 50082 , may cover the following [9] respectively:

Testing for Emissions: (50081)

From the enclosure, over 30 to 1000 MHz – if the power amplifier contains a processing device clocked >9kHz.
Conducted onto the AC mains, over 0 Hz (DC) to 2kHz; then 150kHz to 30MHz.

Testing for Immunity to: (50082)

Ambient, RF fields, over 27MHz to 500MHz, at 3v/m, unmodulated.
* Ambient power frequency magnetic fields, 50Hz, 1 or 3A/m.
Electrostatic discharge, 4kV contact to case; 8kV air.
Fast transients up the in/out ports, common-mode, up to 500v, at 5kHz.
RF up the in/out ports, 80% modulated @ 1kHz, 0.15 to 80MHz at 3v rms, $Z_s = 150\Omega$.
Mains voltage fluctuation, 30% for 10ms; 50% for 100ms.
Mains power interruptions, for 5 secs.
Mains voltage sag or surge, +/–10%.
Mains borne transients, 2kV common mode, and 1kV differential mode.
Fast transients up the mains ports, common-mode, up to 1kV, at 5kHz.
Mains, RF, 80% amplitude modulated @ 1kHz, 0.15 to 80MHz at 3v rms, $Z_s = 150\Omega$.

* *This test is valid even though it is not part of the generic standard.*

At the time of writing, EMC and also LVD (Safety) testing are in a state of flux. *Up to date information should always be sought from, or verified by, experienced EMC and safety consultants, or test houses.*

References

1   Atkinson, John, *The puzzle of perception*, Stereophile, Feb 1992.
2   Atkinson, John, and Will Hammond, *Music, fractals and listening tests*, Stereophile, Nov 1990.
3   Colloms, Martin, *Some observations on the results of objective and subjective technical reviewing practice in high fidelity*, Proc.IOA, Vol.13, part 7, 1991.
4   Phillips, Wes, *Testing, testing, one-two, three*, Stereophile, Aug 1994.
5   Atkinson, John, *The blind, the double blind and the not-so-blind*, Stereophile, Aug 1994.
6   Duncan, Ben, *EMC: Testing times?*, Studio Sound, Dec 1995.
7   Duncan, Ben, *A quick crowbar*, HFN/RR, June 1982.
9   Marshman, Chris, *The Guide to the EMC directive 89/336/EEC*, EPA Press (UK), IEEE press (USA), 1992.

Further reading

10 Colloms, Martin, *The sound of amplifiers*, TAA, 3/1985, reprinted from HFN/RR, May 1985.
11 Duncan, Ben, *Realistic amplifier testing: Procedures for professional touring music PA*, proc. IOA, Vol.16, 1994.
12 Harley, Robert, *The listener's manifesto*, Stereophile, Jan 1992.
13 Harley, Robert, *Deeper meanings*, Stereophile, July 1990.
14 Heyser, Richard.C, *Hearing Vs. measurement*, Audio (USA), Mar 1978.

Choice, application installation and set-up

'On the big screen they showed us the sun,
But not as bright in life as the real one...'

Lyrics by Bernie Taupin.

9.1 Manufactured goods, a résumé

"A manufacturer not only has to design a worthy product; they have to be able to make it consistently and reliably. Each is important as the other, and both are important to users."   *John Atkinson [1]*

High performance sound systems can cost the individual or company alike more than their home, studio or warehouse. Now more than ever before, intelligent purchasers in Western countries are wary of purchasing pretentiously advertised, over-packaged, gleaming goods that will consume the disposable income or equipment budget for some months or years while having a high likelihood of failing to perform, or becoming boring, or ruinously expensive to maintain after a while.

The 'secret' is not just that most high performance audio equipment is built to last a lifetime or more, but that *it has a value quite unlike that of other material objects, as a tool that establishes a link or tunnel with higher realms.* After you have established the quality infrastructure, the idea is to spend the money on exploring what (musically) can come down that tunnel. Similarly upfront hi-tech tools giving access to higher consciousness include immersion chambers, spaceflight, and the gear of extreme physical sport.

That many manufactured items have a short life, or are unsuited to the purpose which their marketing image pretends they meet, is not a new problem. We can reasonably presume that Roman citizens complained about it two millennia ago. Certainly, over a century ago, one of Britain's arch-mentors of quality, wrote:

"There is hardly anything in the world that some man can't make just a little worse and sell just a little cheaper, and the people who buy on price alone are this man's lawful prey."   *John Ruskin, 1819-1900*

Fortunately, the dedicated users and the more intelligent reviewers of audio power amplifiers have become good at assessing quality, in all areas. So the frequent comment by those who are astute about quality in everyday life, "They don't make stuff like this any longer" when surveying surviving 40 year old artifacts against many modern goods, may be *un*true when it comes to power amps. The reverse may actually be true. The development of domestic high performance audio has undoubtedly benefited from the continuing, public scrutiny of product quality by experienced reviewers, in magazines.

One of the most important discoveries of the second part of the 20th century is that while quality is hard to define, humans have intrinsic quality sensors that precede intellect.

"Reality ... is the moment of vision before the intellectualisation takes place. There is no other reality."

And *"..what is good... and what is not good, need we ask anyone to tell us these things ?"* Robert Pirsig, in *Zen and the Art of Motorcycle Maintenance* [2]

Listening to equipment is an example of this skill being honed. But in any manufactured item, quality is more than this, and the skill of 'picking a good one' and 'by instinct', is there for the using.

9.1.1 Choosing the right power amp, domestic

Many domestic power amplifiers are of the 'integrated' (pre+power) type, but a minority of these may be described as high performance. A 'separate', ie. a discrete power amp, as used by dedicated listeners and professionals, reduces the number of opposing performance and feature compromises, and increases the scope for future changes.

How much power ?

Be aware that even if the cited watts are 'honest', the relation between wattage figures and loudness is counter-intuitive. Most notably, it is *under*-powered amplifiers that are more likely to burn-out HF drive units. Specifying the 'right wattage', a power rating that will satisfy the listener and also suit the speakers without risking 'blowing them up', is complicated by the uncertainty that 'apples are reallybeing compared with apples'.

Programme ratings

For example, most low-budget hi-fi speakers are already rated (in watts) for *'programme'* (music), so the power amplifier's, *average* watts per channel rating shouldn't range much above this. A power capability up to twice the speaker's ratings is likely to be harmless if programme material always has reasonably wide dynamics and no sustained high levels above 5kHz, i.e. avoid playing highly compressed sounds, and raw percussion and synths, at high levels. Then a 200 to 300w/ch amplifier would be the most for speakers rated at 100 watts to 150 watts programme.

With more honestly rated, possibly more expensive loudspeakers, the rating given may be a continuous type, based on the AES/EIA test. If so, programme ratings will be x3 to x10 greater. So a 60w AES rated enclosure will handle programme up to from 180 to 600 watts, provided it is not compressed, and that the amplifier is not driven into clip other than occasionally and momentarily.

Power rating truth

The amplifier wattage needed for a given loudness depends a little on how near you are to the speakers, and a lot more (often) on how sensitive the speakers are. *The loudness difference between the least and most sensitive Hi-Fi Speakers is up to at least 8 fold* for a given drive level. This represents a 1000 fold difference in power conversion efficiency, alias a 30dB difference in SPL. *Just changing to a 3dB more sensitive speaker* (e.g. from 90dB$_{SPL}$ @1w @1m to 93dB$_{SPL}$ @1w @1m) *is equal to doubling the amplifier power rating.*

Scaling

Increased watts give only inverse or anti-logarithmic (read: slack) increases in sound level, yet cost linear money. A 500 watt amp and speaker will both cost getting on for ten times more than 50 watts. But a 10 times more efficient speaker need not cost anything like 10 times more.

Power satisfaction

The easiest and most certain way to know what power rating is required is to forget about precise power ratings and listen with the prospective 'bride' speakers that the amplifiers under selection may be mated with.

The maximum useful loudness can be quite different to expectations, and may depend on the music being played, according to whether the amplifier has output protection, and whether this is being triggered by the speaker's dynamic load.

Active systems (bi or tri-amped) are well known for producing clean sound levels higher than the simple aggregation of the four or six (or more) amplifiers channels' power ratings would suggest.

High level defects

Another cause of inadequate or unsatisfactory loudness (e.g. where 100w of power amplifier Y sounds quieter than 30w of amplifier Z) occurs when the sonic quality degrades half a dB or more below clipping. A well-designed power amplifier using NFB maintains its sonic character up to just below clip, say within 0.2dB. But above clip, the distortion sets in extremely abruptly and if prolonged, soon reaches unbearably high levels. Some listeners prefer to accept 'gracefully decaying' sonic performance in the 10 to 3dB leading up to overload. Called 'soft clip', this is second best to adequate headroom. Yet even the high-end industry and the informed public have a historic and continuing avoidance of high headroom amplification, in part through fear of the high power numbers (*One Thousand* watts and above) that adequate headroom confers, and partly through a belief that such high powers (which actually have only up to an order higher voltage and current swing than conventional designs) must necessarily degrade sonic quality.

Blind specification

If buying without listening (or being able to listen) to how loud the amp will go with the speakers it will be partnered with, you will be reliant on a power figure. However notional this is, at least check what impedance a prospective amplifier's power is developed into. The issue is not that higher powers are always developed into lower impedances than your speaker has, but that the advertised or brochure's 'headline' power rating, *is* the one that is developed into the nominal impedance of your speakers. In other words, a specified '100 watts' won't be available into your 16 ohm speakers if the 100 watts is only developed into 2Ω. Only about 13 watts will be available. Misreading – or not reading – the maker's (or a good technical reviewers') specification, then not listening, could leave you paying dearly for hypothetical power and headroom.

Quality

Somewhere along the line, decor and aesthetics will ultimately matter to someone, and likely more so with physically bigger units, particularly if it (or they) will have to be positioned dominantly as a 'room sculpture'. Nothing short of a physical inspection and a relaxed period of hands-on use can allow quality to be perceived in an apposite perspective – and all the more so for astute and skilled purchasers.

Reliability must somehow be gauged. Parts inside most power amplifiers will be subjected to more stress than any other item of Hi-Fi equipment. Most domestic power amplifiers are still heavy, so the cost in carriage can be substantial if they have to be sent back for repair. But then an amplifier in a sturdy case will more likely be built electrically sturdily, so the likelihood of the need is less!

Rationalised controls

Domestic power amplifiers are usually even more bereft of unnecessary or tangential features than utilitarian pro-PA amplifiers, features that regularly appear on mass-market, Mid-Fi domestic and low-budget DJ-grade power amplifiers, most notably gain controls, 'swing needle' (or LED) meters indicating 'power' output , and any other kind of superfluous LED indication. The money saved can potentially be used to increase *real* quality. However this should not be taken too far.

Recommended features

A clip indicator LED is invaluable (with conventional solid-state amplifiers having high global NFB) for knowing for certain when you are exceeding limits that may quickly burn out part(s) of your speaker system. Even if you have enough headroom so you never see it light in the loudest passages, how could you be sure without it? The clipping of amplifiers employing valves and/or low NFB is less abrupt, far softer and far less likely to damage drive-units.

A *'we have a problem'* 'Error' or 'Protect' LED is most useful in the eventually probable event that something is (or goes) wrong, at least to save you wasting time and energy trying to work out why there is no sound.

A second set of parallel output terminals or sockets, for bi-wiring.

Headphone sockets are rarely seen on any modern power amplifier, excepting integrated types. For most users, they are superfluous. Fortunately, if required, usually for when you want to use medium-to-high impedance (rated at 100 ohms or greater) headphones, it is straightforward to add (or have added) a headphone socket. Switching of speakers is best avoided. An advantage of using the professional *Speakon* speaker connectors in that speakers can be disabled (and the plug contacts cleaned) without removal, by simply twisting. Low impedance (50 ohms or lower) headphones should never be plugged into an amplifier capable of more than a a few tens of watts, as both the headphones and the wearer's ears stand a chance of being traumatised if anything like full level is passed more than momentarily.

9.1.2 Choosing the right power amp, for pro users

In **touring PA**, reliability, low weight, a certain fashionable 'u' size for a certain power, and ability to 'kick ass', have often taken precedence over general sonic quality in any selection listing.

In **recording studios**, there is more chance of the sonic results being the priority, yet the amp may be 'chosen' by playing safe and buying on a name or untested recommendation alone; or the monitor speakers' maker or installer may specify what's to be used. The latter may be the best route, but the suitability of such a specification should be demonstrable by listening and comparison.

Power

The average power rating is commonly selected to be *higher* than the AES/EIA or 'rms' or 'average continuous' power ratings cited by most professional drive-unit makers. Provided clipping does not occur, or only happens occasionally and *very* briefly, both the speaker and in practice, the amplifier power ratings may safely be three to ten times this. For the speaker, this rating over the continuous is well known as *'Programme Power'*. This procedure naturally allows a given amount of hardware to be used to its fullest. This is important not just for operating costs, *but also for sonic quality, since it allows the number of sound sources to be minimised.* Unless there is severe compression (e.g. as might be caused by over-processing, or by clipping back at the mixer), the amplifier's unclipped maximum power delivery will be largely in reserve.

Bridging

With a cursory inspection, bridging sounds like a thoroughly good thing. It is after all *balanced* drive. But it's commonly overlooked that bridging halves any given output damping, doubles the amount of circuitry, internal connectors and solder joints in the signal path, and perhaps most important, almost wholly cancels all even-order harmonics created in the amplifier. As no amplifier has solely even harmonics, odd harmonics are left, and most are highly unpleasant and fatiguing [3].

Bridge frustrations

Usually, rated power is barely doubled. Amplifiers capable of about 1kW into 8 ohms when bridged are relatively commonplace and 'old hat'. Bridging makes a nonsense of precious power-to-size ratios: 2 channels of 1kW into 4 ohms become 1 channel of 1.2kW into 8 ohms, without any halving in size to show for it ! In its defence, bridging is fine if it does the job, but the potential cost in size, weight and sonics must be evaluated carefully.

On the other hand, bridging ideally reduces supply rail noise intrusion, and can be used to build more reliable amplifiers, where one half of the bridge continues with only a modest volume reduction, if the other side is sick or being repaired. *Overall, bridging two amplifier channels remains a mixture of good and bad in engineering and sonics, a frustration to the designer and the user alike.*

High power

Bridged or not, amplifiers that produce above 1kW/ch into 8 ohms/ch or above 2kW/bridged are increasingly available. In PA, appropriate use of these multi-kW per-channel amplifiers with the new class of drivers and suitable enclosures will reduce the number of sources, help simplify acoustic calculations and the related predictive accuracy, improve sonic quality, reduce rigging time, reduce costs and potentially increases system reliability, as there is simply less kit to go wrong.

Sonic chemistry

Some amplifiers sound good whatever they're used with. With others, reliability and sonics may be considerably influenced by the boxes they are driving. An amplifier that sounds poor or shuts-down driving one type of cabinet, may shine with a different kind – even one working over the same frequency band.

Amplifiers that measure badly can sound good to some users, but inaccurate to others. Pleasant amplifier distortion can mask unpleasant distortion elsewhere in the system, but if the whole system omits unpleasant distortion, the masking is unmasked. This means if you choose a given amplifier on the reasonable basis that it sounds best in your PA or studio, you may find that a completely different amplifier is preferred when or should you change the mixer or other major components.

Listening test caveats

How well a given amplifier sounds on the day you check it out may depend on whether it was warmed up, the quality of the mains supply, and the general health of the system. For example, if unbeknown to you, one channel of the mixer source happened to be marginally stable and putting RF at 1MHz onto the signal, then Amp

'B' with better RF filtering might win over Amp 'A', which has lesser filtering, and would otherwise sound better.

Reliability tips

While reliability can be estimated by seeking out the experiences of other owners, inspection of the construction, and asking the maker specific questions, there is no way to be absolutely certain. Production of any previously reliable amplifier model can prove unreliable in use at any point in the future if key factory staff are lost, or change, or if the design is 're-engineered' by accountants. And even a well-engineered, generally rock-solid design can fail if a particular user and their equipment accidentally uncovers out some weak spot.

Equally some users are happy with, and get reliable use from, designs that can be readily blown up when bench tested. An amplifier that blows up when driven continuously into a 2Ω load is a non-problem if the amplifier is only driven with music and driven into 8Ω speakers that do not exhibit significant impedance dipping.

Transistor reliability

With good protection, and otherwise solid design, it is the junction temperature and temperature cycling '*world line*' of the power transistors, that have the foremost effect on audio power amplifiers' reliability. Like aeroplanes, power devices have a finite number of cycles before metal fatigues, while every 10°c rise in temperature approximately halves a semiconductor's lifespan, through chemical diffusion.

Considering amplifier cooling, the large (and especially full depth) cooling tunnels seen in many high power amps are less potent than they seem. Frequently, when working hard, the temperature of the hottest device at the outlet end of the tunnel, *can be as much as 20°C hotter* than the one by the air inlet (Figure 9.1, overleaf). And this is just the surface temperature. Internal, junction temperature differences can be greater, especially if the transistor fixings differ in tightness. Such temperature differentials confuse power sharing and will lead to unreliability as the hotter devices blow prematurely.

An amplifier suffering significant temperature differentials between the output devices (whether this is caused by careless cooling design; or by poor mounting of devices) should be derated for reliability, to well below the potential power capability. Because the maker can only derate so far while meeting a salable price tag, some trade-off in reliability and sonics is inevitable.

A number of modern designs (e.g. specific models by ARX, BGW, C-Audio, Chevin, Crown and MC² Audio) employ a short but wide lateral heat-exchangers cooled by a plane airfront. With all the power devices exposed equally, they operate at near identical junction temperatures, sharing out the thermal stress equally (Figure 9.2). This approach allows every device to be safely used to its full rating, and fewer devices (and less stringent protection) may be needed as a result, for a given level of

Figure 9.1

The power devices on blown, tunnel heatsinks rarely operate at the same temperature. This degrades reliability and sonic quality and wastes precious silicon resource.
© Studio Sound

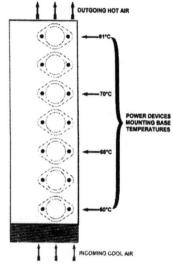

Figure 9.2

On lateral heatsinks, power device temperatures can be almost perfectly equal, so thermal stress is experienced equally.
© Studio Sound

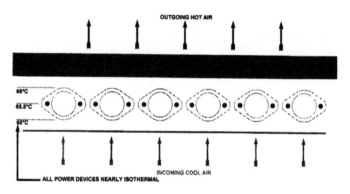

power handling and reliability. Suitably designed short cooling tunnels can also offer isothermal operation.

While one cannot really estimate the reliability of an amplifier by counting the number of devices alone, a broadside-cooled array of power transistors is a good point you can tell about an amplifier, just by taking the lid off.

Users can contribute greatly to reliability by shielding amplifiers from stressful occurrences, e.g. see section 9.7.

Mains universality

In territories where the AC mains differs from that in the maker's territory, users (or importers) should carefully evaluate and monitor the maker's success at the conversion. For example, the transformers in USA made equipment are designed for a nominal 60Hz and 110 to 115v AC. In the UK, the supply is *not* double as is often presumed, but 240 volts and often up to 256 volts (irrespective of theoretical EU normalization). Also, the frequency is 50Hz. Alone or together, the lower frequency and higher voltage will cause a transformer designed down to the last cent for 60Hz, and down to an inadequate primary voltage, to run hot, buzz loudly, and possibly fail at an early stage. Sonic quality may also be affected by line frequency, at bass frequencies in particular, as power supply energy replenishment is naturally less frequent at 50Hz (50 cycles per second), vs. 60Kz.

Users operating internationally should also check that the amplifier's power supply is suited to use on other mains supply voltages and frequencies, and conservatively rated, not forgetting that mains voltages outside of the UK and Western Europe are not so standardised and/or not nearly so tightly adhered to both in the '3rd world' and in the rural outback of spacious countries like the USA, Canada and Australia.

Listening

Relaxed listening is a good start (forget A-B tests unless you have an strong 'photographic' short-term memory of music). Make the repertoire extensive; listen over a few days, with lots of different source equipment, different speaker boxes and different people. If you care about checking in on finer differences, make a point of listening in stereo. This applies even for PA, and even if stereo content in concert PA has mostly so far been minimal. If you're looking for a bass amplifier (say), and one of the shortlisted amps you are trying is obviously not giving the performance you were after, it might be worth briefly trying it in other roles. This is the way great discoveries are made ! Equally, if you experience disappointing results, it's valuable to contact the dealer or maker. Getting them to supply another sample unit and making doubly sure that you're using it optimally may make no difference, but equally, the tables have been turned before on such occasions.

Mechanics

A mechanical layout that spreads the load can make a big difference to the comfort, safety and longevity of a touring power amplifier.

Examples of bad mechanical design include those amplifiers with:

(i)   Much or most of the weight focused in any one area away from the centre other than towards the front and symmetrically about the centre-line.

(ii) More than about 6"/150mm deep *without* rear rack supports.

(iii) Sharp (un-radiused) metal edges, liable to cut or injure installers.

Good designs may include such features as:

(i)   Recessing of vulnerable or critical controls or switches.

(ii) A single secure but easily removable cover giving access to all the breakable or wear-prone parts most requiring attention.

(iii) All air-spaced PCBs, allowing rapid drying out in the event of flooding.

Individual reality

Overall, power amplifiers are like other machinery that humans experience intimately: their '*world-line*' interaction with your system in your part of the universe is not predictable from any spec sheet. Even with in-depth technical understanding, no amount of technical data and bulleted, internal technical features can tell you anything definite or specific about the unmeasurable qualities. Therefore you may find it useful to do the listening tests *before* reading any spec sheets and technical papers, or dissecting or abuse-proving the unit under evaluation. If the amplifier sounds bad, you never have to wade that far.

9.2 Howlers

Here are some common and '*it really happened*' defects to look for in manufactured units. Based on hundreds of reviews of both domestic and professional power amplifiers costing from £150 up to £5000, or even more.

Illuminations

*   'Clip', 'Peak', 'Error' or 0dB bargraph LEDs that warn of overload more than ½dB early or far worse, more than ¼dB late.
*   'Clip', 'Peak', 'Error' or 0dB bargraph LEDs, where the accuracy of clip indication depends on the speaker loading and mains voltage.
*   Important panel LEDs that are invisible unless viewer's eyes are almost on-axis.

Noises

*   Transformers that buzz audibly with normal mains harmonic variations.
*   Overly noisy fans for domestic and some PA applications.
*   Hum at output with no input connection.

Anti-social behaviour

*   High inrush current at switch-on, which trips house power breakers, making use almost impossible.
*   Farting noises and sweeping whistles from speakers, when amplifier is switched off.
*   HF signal continues to pass when output is muted.
*   Binding posts deeply recessed between heat-exchanger fins.
*   Binding posts that are too slippery to be tightened, and/or undo themselves despite attempted hard tightening.
*   Sharp, un-radiused metal case corners.
*   Protection by a specially fast fuse with no spare(s) supplied. Fuse fails at worst moment through old age.
*   Manual intervention required to reset an amp – after just a momentary power down, or the briefest short at one output.

* Under-PCB insulating material that traps moisture despite hours of drying out in warm air flow. Important for outdoor use and even for parties.
* Thermal grease (especially anything silicone based) oozing around mounted parts, or smeared about.
* Unfiltered fans blowing into electronic areas; and un-louvred or shielded air vents, allowing dust and filth to accumulate on PCBs.
* Fans switch off when the amplifier gets too hot.
* A fault on one channel stops the other channel or the whole amplifier working, without specific indication.
* Confusing, complicated or illogical LEDs ('a green LED, marked 'OIC' flashes dimmer when headroom is exceeded') or unindicated events ('The power LED will continue to glow even though the amplifier has shut down').

Poor Q-A

* Supplied 'dead on arrival'.
* Something metal rattles inside the case.
* Uneven torquing-up of power device screws.
* Un-alignable lid screw holes.
* Random input polarity i.e. hot sometimes (XLR) pin 2, other times pin 3.

'Unroadworthy' construction

* Unlocked screws.
* Unsealed pots in an unsealed enclosure.
* Unpolarised connectors.
* Unretained, non-latching connectors.
* Unmarked connectors adjacent to others having identical pinout.
* Power devices, or audio or mains connectors or other wearable parts, hidden under an hour of metalwork dis-assembly.
* Sub-standard 4mm binding posts with short and/or slack receptacles
* Sub-standard 4mm binding posts with a slippery, ungrippable body.
* Flimsily mounted transformers that wrench free in transit, smashing other innards.
* Amplifier innards explode when outputs momentarily shorted.
* Amplifier emits smoke or fails when long speaker cable connected.

A small, worldwide amplifier interface quality catastrophe

A type of 4mm output binding post made in Japan and regrettably fitted to thousands of 'competitive' professional and perhaps millions of 'mid-priced' Hi-Fi amplifiers, mainly made in Japan and the USA, is diametrically at variance with real needs. To be of any use, the hexagon shaped posts, moulded from a slippery plastic that is certainly ungraspable if the grasper is sweating from any exertion, require an end-on hex driving tool, that is never seen or supplied. Second, the 4mm plug receptacle is foreshortened and has slack tolerances, so even the best sprung plugs are not gripped uniformly tightly, if at all. Third, the offending type is drilled to accept only thin wires, usually not above 0.75mm². Astute purchasers will insist on having these ill-conceived posts sent for non-audio metal recycling, and replaced by higher quality parts, which are hardly expensive. Cliff and HCK are the foremost makers.

In the following sections, first use, installation and sonic optimisation are methodically considered.

Figure 9.3

BGW's PS2, a compact, solidly-packaged I yet budget professional amplifier, showing (centre foreground) four of the sub-standard output 'binding posts' blindly employed by pro and domestic makers in the US and Japan. Note input jack sockets on the right, which should never be used for fixed installation. The remainder of this design gains high marks for ease of servicing and inspection. Most of the air exits at the sides – a type 'C' airflow. © BGW

9.3 AC mains voltage

New amplifiers should be factory set to the nominal mains (line) voltage for your country, and this voltage should be stated on the rear panel.

Across the world, there are many combinations of voltages and frequencies. Five AC mains voltage ranges cover most Western and many Pacific rim countries:

1. UK, Australia, New Zealand, Malaysia: **240v**
Note 1: Since 1.1.95 UK is technically 230v +10%, –6% by EU law but in practice 240v +/–6% is maintained. In the year 2001, AC voltage throughout the EU will be in theory standardized at 230v +/–10% but the UK power authorities will nonetheless maintain 240v + 6, –10%. Likewise, continental European 220v will not increase to 230v in the short run, in many territories.

Note 2: Western Australia ranges 240 to 260v.

2. Western Europe (continental)	**220v or 230v**
3. Singapore, Korea	**220v**
4. USA, Canada, Mexico, Taiwan	**115v**

Note 3: With regional and local variations from 110 to 120v.

5. Japan   **100v**
Note 4: With regional and local variations from 95v up to 110v.

Line frequency

Frequency, always either 50 or 60Hz, is usually of no concern for operability. But when 60Hz transformers are used in 50Hz territories, the 20% change down in frequency increases transformer losses. Tightly designed types may prove unsafe or unreliable in professional use, where the amplifier is working hard in a confined space; or too noisy in domestic use. The temperature rise may be ameliorated by extra cooling. At worst, problems can be overcome by having a purpose-made 50Hz transformer fitted.

Tap adjustment

The AC operating voltage of well designed power amplifiers made for international use can be adjusted (after disconnecting the supply) by removing the cover(s), and changing the transformer's voltage tapping, identified by labeling, colour coding and/or by diagrams in the manual. A minority of highly specified amplifiers (e.g. BSS Audio's *EPC* series) have automatic tap setting at switch on. The amplifier will set up for the mains voltage at this juncture. This type should *not* be 'brought up gently' using a Variac.

Headroom preservation

The significance of tap setting is that as the majority of power amplifiers' supply rails are unregulated, the rated or stated output power will only be developed if the mains voltage is closely matched to the transformer. Otherwise, continental Europeans get slightly fewer watts for their money from a UK made amplifier (without a 220v or 230v adjustment), since the continental 220 or 230 volt supply won't allow amplifiers designed for 240 volts to develop their full, rated power. Fortunately the fractional dB loss in maximum SPL will be barely noticeable – but remember the same dB loss is also part of a potentially cumulative loss of headroom, which correct tap setting can restore *without cost* or compromise. *The fraction of a dB could make all the difference as to whether a speaker is burnt out by accidental clipping.*

Operable range

Mains supply voltages are nominal. A 'solid' amplifier will be tolerant of wide voltage fluctuations that can occur in practice, and will continue to operate normally even if the incoming supply varies widely.

In the outback and rural areas of some countries, and even cities in some places e.g. in SE Asia, mains supply voltages can surge to well above the nominal. Over 310v has been measured on a nominal 240v supply, both in Thailand and Malaysia. In many amplifiers, the driver and output transistors are not (and perhaps cannot be) rated to meet this level of excess. The surest solution may be to reduce the voltage permanently. In the absence of a 280v tap, a suitable, communal step-down auto-transformer may be used; or as a temporary solution, a padlocked Variac.

9.3.1 Safety earthing

It is the norm in countries having high electrical safety standards, for substantially metal-cased equipment (or ideally, equipment with any exposed metal) operating from the AC mains, to have its casing bonded to earth, via a third wire, called the earth (or 'earth-ground' in the US; or CPC, the Circuit Protective Conductor in the British BS 7671 wiring safety standard). In a flexible mains cable (cord), the earth wire is coloured green/yellow by international convention, but in the USA and Japan, green is still used.

Raison d'être

Safety earthing is required to prevent or forestall the chance of electric shock, the former caused by any accessible metal or other conductive parts of an amplifier becoming live, through (for example) internal damage, or a loose wire or part. With safety earthing, and assuming good connections, and a heavy enough gauge, the voltage on the exposed 'en-livened' metalwork won't rise more than a few volts above ground, and fuses should immediately blow, alerting someone to the problem.

Exceptions

Mass produced consumer-grade goods with predominantly plastic enclosures, and/or where the mains connected components are doubly insulated, do not require a safety earth connection. The mains cable has just two wires. This method has its advantages, particularly if the audio connections are unbalanced. However, all but the smallest and most efficient amplifiers are sufficiently heavy to be unsafe to encase in the affordable but brittle plastics that are normally used. Shielding may also be questionable.

Amplifier cases made with glass loaded plastic are feasible, but would be more expensive and use more space, than the steel, aluminium, copper and related metals still used in 99% of high performance amplifiers – professional or domestic. Plastic cases are also rarely shielded other than with conductive paint or skimpy stick-on metal foil. More solid metal shielding would defeat the cost saving.

Side effects

Safety earthing is rarely electrically clean. Worse, it can readily form a loop in unbalanced systems, where the casework and signal common are one and the same, or are not wholly isolatable. One half of the loop is formed from 'low' side (usually the shield) of the signal connections. This 'earth loop' operates like a shorted turn, and high 50 or 60Hz and harmonic currents are induced, hence the names 'hum loop' for the condition, and 'earth buzz' for the sonic effect. Temporarily disconnecting the mains earth wire might prove the point, and is sadly practised by desperate or uneducated operators. But left off, it can put lives and property at risk. ***Removing the safety ground is dangerous,*** may invalidate all kinds of insurance, and should NEVER be practised. For hum solutions, look elsewhere (Section 3.3).

Ground solutions

Power amplifiers with correctly wired, balanced inputs are literally outside of the (earth) loop. In the event of suspected earth buzz, the input cable shield connection to chassis should be lifted – via a high-wattage-rated resistor or better, small $(0.1\mu F)$ capacitor; or opened. Some professional power amps have internal, hence tamper-proof headers or switches that allow several options to be tried.

Groundlifting

If inputs are *un*balanced, the signal ground should be isolatable from chassis and mains earth. This is the purpose of a *ground-lift* switch, or linking bar that can break the connection between the case or chassis (earth) to the signal ground (0v), on a power amplifier. If these facilities are absent, the internal grounding may be modified by a practitioner.

Groundlifting control may be accomplished with a resistor, an RC Zobel or even shunt network, or with back-to-back rectifier diodes. Unlike the former three, the latter provides high isolation until the difference in ground potentials exceeds about +/–0.6v, whereupon the diodes act almost like a hard connection.

Sound ground

In large systems, and wherever there is RFI 'smog', the performance of power amplifiers and all other audio electronics can be improved by establishing a low impedance connection with the real ground mass (soil, clay, sand, soft rock and water table) connected to audio system safety earthing (mains earth).

To be useful, audio earthing stakes must be away from others, as well as from buried power cables. If there are no accurate records of underground services, a power cable tracer should be used to select a location that is also safe to dig. The stake(s) must make a low resistance connection to the ground mass. To do this, the water table should be reached, which may be at least 4'/1.2m down, and possibly up to 25'/8m down. Electrical earthing rods are usually made of copper coated steel. Driving is made possible by the screw-together construction, like drain rods. The lowest, most RF-effective earth resistance is usually attained by driving-in several stakes over an area, or by excavating and burying earth mats, rods or wire grids. Water with a small amount of salt may be used to improve conductivity in temporary situations, but at the cost of rotting the copper and any exposed steel, as well as polluting the soil for many years. Urine has just enough salts and is the perfect eco-answer.

Misuse of a technical earth-ground may prejudice safety. Use may require special dispensation by the authorities. Employ a qualified practitioner.

RCCB backup

If for some reason it is impossible to get hum free results without breaking the mains safety earth connection to the amplifier casing, be doubly sure that a 'Residual Current Circuit Breaker (RCCB), alias 'safety power breaker', rated to trip at an imbalance of not more than 30mA, is used. If this trips out, it is trying to tell you that a potentially lethal current is flowing. Diagnosis is by unplugging everything then replugging one at a time. Often - but not always - the cause is the accumulation of small, individually harmless leakages spread across many items of gear, caused by capacitors to earth, in RF filters. This can be demonstrated if the equipment that appears to trip the RCCB is different after replugging in a different order. RCCB sensitivities vary also, between some 10 and 25mA for a maximum nominal of 30mA. The solution is to split the system between more RCCBs, each rated at 30mA. Don't be tempted to uprate to a 100mA RCCB, as it will not offer enough protection against shock.

Whenever power amplifiers are used outdoors, even if the only wetness for months in some regions is dew, and whether AC power is mains, or generated on site, an RCCB is strongly recommended for added protection against a 240v AC shock, which is far more likely to prove fatal in moist conditions.

9.3.2 Mains cabling

For the best performance in given circumstances, it's important that power amplifiers' peak current draw is not impeded by inadequate power cables. For, excepting the present minority of amplifiers with *power factor correction* (see next section), peak current draw can be far higher than the amplifier's nominal rated mains consumption in watts or VA, suggests.

Peak consideration

It is particularly important to consider the peak voltage drop when multiple arrays of amplifiers are in use, and/or when the incoming mains supply is more than a few yards (metres) from the amplifiers.

It is no less important to consider it in domestic and studio monitoring systems, even if the nominal amplifier power ratings are not especially high. For irrespective of rated power, the more the source impedance of an amplifier's conventional capacitor-smoothed power supply is reduced (as is done in many high-end power amplifiers and installations), the better it can sound, but also the more peak current it wants to draw. It would draw an infinite current for a moment, if it could ! A sign of low supply impedance is that the reservoir capacitors' recharging is bunched into *'more amperes in less time'*.

Well recorded or live, uncompressed percussive music is especially testing of power supply recharge capability. Using a particularly realistic percussion recording, domestic high-end researcher Russ Andrews cites a power amplifier with a modest 200VA rated transformer (enough for a nominal continuous output of at best 150 watts) where a peak mains current draw over 100 Amperes for 1 millisecond was recorded, on a low impedance mains supply wired with non-inductive ('fast') mains cabling, and where line fuses were replaced by non-intrusive protection.

Mains impedance

Irrespective of the power factor condition, the size (csa) of cable required depends on the cable length and the highest peak voltage drop that can be accepted. But the size (csa) of cable that it is *worth using* depends first on the mains supply impedance. *As the mains impedance is finite and rarely as low as imagined, there are limits to how much employing low resistance conductors can assist sonic performance.*

A 100 Amp rated supply, for example, should maintain the nominal supply with at most a 6% voltage drop at rated rms current. In turn, this implies a maximum source impedance (at 50 or 60Hz) of 144mΩ (0.144 ohms). A good starting point is to measure the mains supply impedance.

To do this:

(i) Read the mains AC voltage with a true rms voltmeter, at the incoming supply or a socket adjacent to it.

(ii) Decide on a known, *resistive* load (a heater is best) to be connected to the supply. Read the rating and voltage, e.g. 3kW, 240v.

(iii) Calculate the design resistance of the heater element, e.g.

$$V^2/W = \{ 240^2 / 3000w \} = 19.2\Omega$$

(iv) Observe the mains voltage fluctuation and if necessary, wait until the mains voltage is stable. Then connect (or switch in) the load, and observe the voltage drop. Note this. If there is any doubt about fluctuation, repeat the test: Stable, read off, then on, then off again.

(v) The impedance is then calculated as follows. If the rms voltage was 243.0v and reduces 2.7v to 240.3v when loaded by the test load, the supply impedance at 50Hz is

$$V/R = \{ ^{2.7} / _{19.2} \} = 140m\Omega$$

This would be about the maximum for a 100 Ampere rated, 240v supply, where the maximum legal drop is –6%.

Remedies

If you calculate that the supply impedance is enough to take the supply voltage below its legal tolerance limits at its rated official load, then consult the supply authority. There is likely a bad connection in their system. The public supplier must make the voltage at your premises fall within the legal tolerances at rated load, and this is tantamount to reducing the impedance.

If the impedance is within limits but a lower impedance is desired, ask your electricity supplier about the cost and availability of a higher rated supply, e.g. 500 Amperes. This will cost more to rent but the source impedance will be reduced by a useful factor.

Generator sonic limitations

If the power source (usually for outdoor events) is a typical site generator, even rated at 100kVA or more, the source impedance is likely not especially low, and rises with loading, as 'reactors' giving a choke-effect, protect the windings. Therefore source impedance should be measured using a substantial resistive load, e.g. *at least* 4kW *In any event, this largely prevents cables having lower resistance than that indicated by the rms current draw, from having any benefit.*

Cable rating

Once the supply impedance has been established, the connecting cable resistance can be computed. Based on not adding appreciably to the mains' finite impedance, it can usefully have a resistance that is down to a quarter of the supply's impedance. For example, the cable between the supply and the amplifier is worth reducing to about 35mΩ, if the supply impedance measures 140mΩ (say). This ratio might be taken further (e.g. to a tenth) in high-end domestic installations.

The table below shows the DC resistance for some common cable sizes x lengths. Note that due to skin effect, the effective resistance at 50/60Hz is slightly higher.

POWER CABLE RESISTANCE – loop value

Length	2.5mm²	4.0 mm²	6mm²	10mm²
1m/3'	13 mΩ	8.5 mΩ	5.6 mΩ	3.5 mΩ
5m/16'	65 mΩ	42.5 mΩ	28 mΩ	17.5mΩ
10m/32'	130 mΩ	85 mΩ	56 mΩ	35 mΩ

9.3.3 Power factor correction

The peak current draw or demand just described is known to power engineers as a 'bad power factor' – in part because the network capacity is being under, then over-utilised. *It is pertinent to note that peak current draw is markedly reduced in all cases if the power supply is an HF (switching) type with power factor correction* (PFC).

In EU countries, EMC legislation is (since Jan 1996) setting compulsory, and increasingly tight, limits on the peak current draw of most mains powered equipment sold in the EC. In legislation that is forthcoming, PFC will be mandatory for all new audio power amplifier designs rated above a few hundred watts overall. As power factor corrected current draw is purely sinusoidal (varying in phase with the mains voltage waveform, as a resistive load behaves), mains cables can be thinner, and if long, simply and conservatively specified for a resistance that yields a reasonable voltage drop (say 4v peak re. 240v rms) at the nominal, maximum rms current draw cited by the maker, with pink noise drive.

In touring PA, power factor correction will close the sonic difference between hearing one small section of a large system (sounds good) and hearing the whole system (louder, but never sounds quite so good). Without it, the interaction, and crosstalk through the mains supply, of many power amplifiers all connected to a common main riser or generator with typical supply source impedances, can be immense. In theory, PFC could be 'bolted onto' an existing multi-kW rig but this would be percieved as prohibitively expensive, weighty and bulky. Its use *en masse* will have to wait for overall re-equipping with power amplifiers incorporating PFC. This is a far less expensive and radical change for those amplifiers already employing HF switching power supplies.

Figure 9.4

Crest Audio's goliath model 10001 cannot be safely lifted by most individuals without aid ! It contains four channels of class G, each capable of over 1kW of power delivery, and as such, is relatively light and compact. The total potential continuous mains power draw, particularly on 100/115v supplies, is so great, in excess of standard power outlet ratings, that individual hosepipe-sized power cords are fitted per pair of channels - hence the two power switch/breakers seen here. Notice the central tunnel cooling, and the general symmetry of the weight disposition, with end-on toroids up front, and reservoir capacitors behind. © Crest Audio Inc.

9.3.4 Mains fuses and breakers

Fuses are purposefully weak links that will burn away safely and cut further current flow. In most power amplifiers, a *cartridge* fuse or fuses either located on the front or (better) on the rear panel or inside, protect(s) the amplifier and the building from fire, in the event of a major fault, or short from live to the casing.

When installing or servicing, check the tightness of the fuseholder cap, if screw fitting. If installing a second-hand amplifier, it is worth checking the fuse in situ is the type specified in the manual, or close. If the value or speed rating are higher, the fuse(s) may fail to protect on the rare occasion when needed. Insurance may be invalidated.

If fuses are tarnished, they may be best cleaned with a super-fine abrasive, then degreased with Hurd's *No.1 Grimewash* comprising 5% by volume of household ammonia in isopropyl alcohol; or even better, a proprietary and qualified electrical contact enhancer, e.g. from Caig Labs, USA.

Fuses age, and with time, *can fail at any time.* As with lightbulbs, weakened ones usually expired at switch on, when the surge current heats and stretches them most. You could avoid this by changing all fuses annually. But this would be a waste. For near uninterrupted use, either keep spares at all times, *or* fit a circuit breaker.

If the mains fuse has blown or blows repeatedly, the amplifier is being misused, is wrongly set up (are you really sure '*there is no mains volts setting*'?), or is faulty. To save blowing expensive, scarce or hard-to-access fuses each time the amp is powered up after the fault is believed to have been cleared, professional troubleshooters employ a simple jig which places a low to medium wattage mains lightbulb in line with the AC supply. If the fault persists, the lamp will light and the fuse will be saved. Do not expect to pass a high level signal into a load in this condition.

Ideally only ever replace a fuse with another of the same type and current rating. A temporary current uprating, to twice at most, is just passable so long as the correct fuse is sourced and installed ASAP. Mark the equipment:

'Temporary Fuse! Date:_ /_ /_ Replace ASAP in 24 hours'.

For more information, refer to chapter 6.

Circuit breakers (thermal–magnetic and/or pneumatic 'breakers') are alternatives to fuses. In touring PA, common industrial types double as panel power switches, give effective protection, clear fault indication and few problems, and commonly last out the equipment life. If not fitted originally in a given amplifier, breakers may be installed externally in the amp rack, and the amp's interior fuse value(s) may be increased (with due care) so the breaker trips out instead, at the correct current.

Circuit breakers of the magnetic or pneumatic type are to be preferred *in lieu* of fuses for *all* high performance power amplifiers. First, they are solidly connectable by machine screws or soldering, unlike most fuses. Second, suitable types have a lower resistance than a like rated fuse, and also a relatively invariant resistance, with temperature and over time. The contact resistance of the average, miniature cartridge fuse when fitted in the low quality holders that are commonly used, is potentially high enough to swamp the benefits of a heavy-duty supply and cabling. Third, in conventional supplies (without PFC), fuses that are correctly rated to protect the power supply parts from a fire will be appreciably heating up on current peaks under hard drive conditions. The significance is not just fatigue but also modulation, since a cartridge fuse's resistance will approximately double before the wire melts.

9.3.5 AC mains connectors, amplifier end

In consideration of the high peak currents that can be drawn by high performance power amplifiers, and the fact that most connectors have (or can develop) as much resistance in their contacts as a metre or more of the cable they are on, superfluous mains connectors are best avoided, however convenient.

If the AC mains cable you are fitting *has to* plug into the amplifier, at least make sure the fit is secure and tight. For those plugs where contacts are clamped by screws (*cf.* soldered), a regular tightening programme is required. Even if the screws do not loosen (due to the relaxation and creepage of the soft metals that are best for electrical contacts), vibration and/or thermal expansion cause 'fretting', where the microscopic clean contact areas are eventually exposed to the air, and/or over which oxide 'crud' is spread. This applies to *all* non-soldered contacts [4].

Professional operators will usually replace the socket with a fixed cable, secured by a gland. Well designed pro power amplifiers have accessible mains terminations and even optional panel plates, to facilitate such changes. If a connector *must* be used, whether on the rear of every amp, or on the rear of each amp rack, the order (for ruggedness *and* security of contact) is approximately as follows:

⇐Solid					Flaky⇒
CEE,125A	EP3 20A*			IEC 16/20A	
CEE,63A		EP LNE#	Neutrik *Powercon*		
CEE,32A				Bulgin 15A	
CEE,16A				IEC 10A	
		XLR LNE #		Bulgin 5A	IEC 6A

A = nominal rating in Amperes. # Officially not for domestic use.
* Used in older professional systems. Not legal. Must not be used !
CEE = CEE-form type, round pin industrial connectors to BS EN 60309-2:1992, previously BS 4343.

Figure 9.5

The IEC 6 Amp connector is unlatching, has quite loose contacts, and is readily broken. © Canford Audio.

Figure 9.6

The EP mains connector. It is rugged, compact and makes good contact, but is not legal for domestic use. © Canford Audio.

Figure 9.7

CEE-form *AC mains connectors, a development of the 1930s British round-pin mains plug, available in ratings up to 125A rms, are the preferred choice for*

touring power distribution, and for advanced high-end amplifier installations, where the size and garish plastic are not a problem. © Canford Audio.

Figure 9.8

Neutrik's new Powercon *AC power connector is based on their* Speakon. *Seen here are the blade connections on the* Powercon *receptacle inside an amplifier case. © Canford Audio.*

9.3.6 AC mains connectors, the power-end

Where feasible, the best sonic performance is most likely to be achieved when the AC mains is 'hard-wired' throughout. In fixed installations and high-end domestic setups, this is possible (at a cost) using 60 Ampere or greater, metal-cased 'industrial' switch-fuse boxes (e.g. MEM), which contain heavy-duty, solid brass terminals with double screws.

Failing this, seek the heaviest connector that maintains the lowest contact resistance in practical use. The UK's 13A 'square pin' (actually rectangular) and pre-WW-II 15A and 5A 'round-pin' power plugs and sockets are the world's chunkiest domestic AC mains connectors, and the better quality examples are increasingly used by Audiophiles and serious audio facilities throughout the world, where mains connector quality is mostly of a far lower order. Whatever connector is used, clean the contacts regularly. In fused plugs, clean the fuse contacts and fuse also.

Caution: Copper cables 'creep', causing brass or other terminal screws to slacken in time. In mains connectors, *this can ultimately cause arcing and fire. All mains conductor clamping screws* must be regularly checked, shortly after installation, then annually at least.

9.3.7 Mains wiring practice, domestic and studio

Note: In some countries, some of the recommended techniques may require special dispensation by the electrical supply or safety authorities.

Ring and Spur

In many installations, but only in countries adopting the custom, the mains power for domestic and light commercial buildings is wired in a ring (Figure 9.9). This saves copper. The idea was originated in the UK in 1939, when the import of vital copper ore was for the second time in under 25 years at risk from Germany's U-boats. However, AC power for high performance audio power amplifiers is best fed from a dedicated radial spur cable, taken directly from the incoming supply, where impedance is lowest (Figure 9.10).

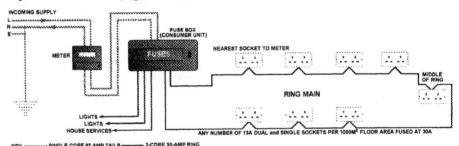

Figure 9.9

The 30 Ampere diversity ring supply that is universal in the UK for domestic and light industrial premises. However, it is unsuited to powering audio equipment, especially power amplifiers. © Studio Sound

Figure 9.10

In a dedicated listening or monitoring installation, every power amplifier and all other major audio equipment is supplied by an individual spur noded at the incoming supply.
© Studio Sound

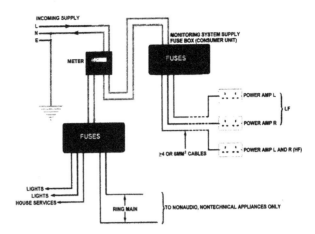

Star system

This technique of running individual cables to equipment from the lowest imped-ance, nodal point is analogous to 'star earthing' and virtual earth mixing. Like these, it helps each source reject external influences, and reduces the sources' own influ-ence on all the other circuits. This means: (i) sag (including music-modulated stuff) caused by the amplifier(s) and experienced by equipment on other circuits will be attenuated, and (ii) the amplifiers' reception of interference from noisy appliances on other circuits will be attenuated, and these noises quite possibly will be done away with altogether.

Rejection

Typically, a spur's interference-cum-disturbance attenuation ranges from a modest (but still welcome) –8dB, up to –40dB at audio (10Hz-20kHz), and more at radio frequencies [5]. By contrast, ordinary mains interference suppression filters are pre-dominantly focused on RF: they provide *no* protection below 200kHz, and only come in strongly above, in the MHz realms.

Sonics

For these reasons, radial spurs are the norm for the technical, audio equipment sup-ply in most competently wired studios and dedicated listening rooms. If you have mains problems, it pays to check, particularly in installations where despite using all the right equipment, the sonic quality is never quite right or changes daily.

Installation example

Begin by measuring the mains supply impedance (section 9.3.2), remembering that the amplifiers' power cabling will be based on this. If not considered low enough, a higher current supply may be installed; this may require special dispensation from the supply authority – as you will not be buying more electricity commensurate with the enlarged supply capability.

With a multi-amped monitoring system, powering the amplifiers from individual, exclusive spur cables, one per amplifier, is worthwhile for the closest approach to a clean supply. Ideal cable size is determined according to distance, and the mains impedance: see the discussion and table in section 9.3.2. In installations where the power cabling is to be buried in the structure, it's best to err on the side of generosity with the cable size, as the mains impedance might be reduced, and even 10mm^2 conductors won't incur a great deal of added expense alongside the cost of typical building work.

Low inductance cables improve sonics and reduce 50/60Hz fields and harmonic radiation, but types made for speaker wiring may not be safety approved for permanent wiring use in some territories. *In lieu* of using purpose-made low inductance cables, place Live and Neutral (return) conductors in 'close parallel' configuration, so they are adjacent throughout the run, i.e. no centre earth conductor. This is achieved easily enough with single conductors. As the conductors are likely overrated for current, the proximity should not normally require the circuit's current rating to be derated. But note that close send/return placement is also subject to proximity effect, which causes a dynamic resistance increase at high currents.

When planning a new room or a revamp, if the mains supply impedance is low enough for it to be worthwhile, it's useful to organise the house's incoming supply and the amplifier placement to be adjacent, for the shortest power cable run. This may mean moving the supply or the amplifier placement, or both. Even in domestic installations there is a great deal to be said for having amplifiers in a room that is separate from the music, to lessen the effects of vibration on sonic quality. This 'other room' might be also where the house mains supply enters. *At the same time, the length of the speaker cables must also be considered.* The location and positioning of the speakers will almost certainly have to take precedence, but if a room is created afresh, the incoming power and the amplifier room (if one) would be best placed behind the wall(s) nearest the speakers.

Mains ecology

Dedicated low impedance power connections to power amplifiers resolve (or at least greatly reduce) the interference problems within the building. Outside is another matter. Most AC power has a high harmonic content (from 5 to 20% harmonic distortion). On a scope, it appears flat-topped (Figure 9.11). In those mainly urban and industrial areas where the incoming AC power is excessively 'brown', interference ridden, or has erratic voltage and harmonic fluctuations, and if the supply authorities are unable to improve matters, you will need to either install a power conditioning plant which may range up to a dedicated substation, or else generate clean AC on site [6]. This will involve common-mode chokes which must be rated at high currents commensurate with the supply's nominal capability. This is for the same reason of not adding unnecessarily to source impedance, as well as not having the choke's core saturate, so its value abruptly changes.

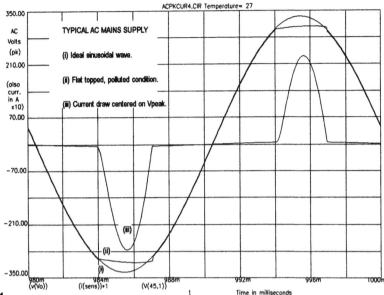

Figure 9.11

If the mains were only loaded with lightbulbs and heaters, the voltage would be a pure sinewave (i). Today, the supply voltage is nearly always flat-topped (and flat bottomed) (ii) which implies high harmonic content. The flat top/bottom is caused by thousands of audio and other capacitatively smoothed supplies drawing current in a peak (iii) only about the peak voltage point.

Failing this, AC mains pollution commonly varies throughout the day and the days of the week. Outside of heavy industrial areas where power-polluting processes (e.g. steelworks) operate every day around-the-clock, most listeners find that the cleanest power and best sonic quality is experienced after midnight, through to the early hours; and at weekends. Peak voltage is also higher.

Conclusions

All power amplifiers affect and are affected by the mains supply. Class A amplifiers are least affected, but can still cause disruption to other equipment sharing the supply. By increasing the potential for high peak currents, dedicated, low resistance spurs accelerate the recharge of class B, A-B, G and H's amplifier's reservoir capacitors after power-hungry transients, and class A amplifier's reservoir capacitors after every dip in the mains supply. A limited number of modern amplifiers have regulated supplies. They're largely immune to performance variations caused by supply voltage and harmonic content (seen in the non-sinusoidal waveshape) [7], and while this isn't automatically conducive to good sonics, it helps a great deal, the more the mains supply conditions are bad.

Overall, the benefits of a low impedance, spur or other diversified, clean*er* supply are clearly evident to the ear at higher playback levels on even modest setups, but are all the more apparent in high performance systems.

9.4 Input connections

i) Phono, unbalanced (RCA connector)

With domestic power amplifiers, even high-end models where makers should know better, input connections are still almost universally based on 'Phono' plugs and sockets. They were originally developed by RCA (Radio Corp. of America) as an early constant-impedance coaxial connector, for VHF. Audio use was through historic accident. Quality and price ranges enormously. For a list of reasons, phonos are frankly unsuited to high performance audio. In the past decade, high-end makers and accessory suppliers have worked on solving or ameliorating some of the defects of the Phono plug and socket, namely the plug's poor cable support and the need to solder down a hollow tube; the way in which the live contact (when plugging-in) is made before the earthy side (causing loud bursts); the tendency of the inner and outer contacts on both plugs and sockets to slacken; and sockets' isolation from the chassis – to mention a few. On the plus side, metal bodied Phono connectors can do a good job of shielding the inner connector at RF. As they have only two connections, Phono connectors are usually limited to connecting unbalanced signals, as follows:

(i) Outer to shield.

(ii) Inner to signal hot.

On a few amplifiers, there is a balanced input via phono sockets, which requires connection as follows:

(i) Outer to signal cold (note socket body is isolated from chassis).

(ii) Inner to signal hot.

(iii) Shield soldered to fly-wire. Connect to chassis.

It does not matter whether the source is balanced or unbalanced, but if the latter, do not connect the shield at source end.

Figure 9.12

An improved phono connector.
© Canford Audio.

ii) XLR, balanced ('Cannon')

In high performance, professional power amplifiers, *XLR* connectors are almost universally used, usually the 3 pin type. Whilst not perfect, they have proven to be intrinsically rugged and reliable. The XLR system, originated by Cannon (US) in the early 1960s, comprises matching, latchable male and females in both line (cable) and chassis mount 'plugs' and 'sockets'. XLRs are also used by some high-end domestic equipment makers, usually when the equipment has balanced connections. But some of these makers appear isolationist and ignorant of their professional colleagues' long established pin connection and gender conventions.

Figure 9.13

XLR connections

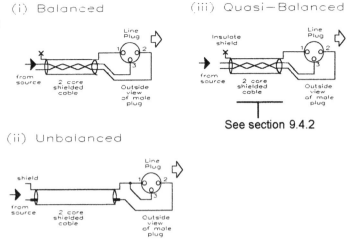

(i) Balanced

(ii) Unbalanced

(iii) Quasi−Balanced

See section 9.4.2

Female XLRs (sockets or receptacles) are almost universally used as the chassis mounted input on power amplifiers. There are at least two ways of wiring the incoming male cable plug, assuming one is used per channel.

For operation from a balanced source, a 3 pin XLR plug is wired as follows, using 2 core (1 pair) shielded cable: (Figure 9.13, i)

Pin 1 – Ground or shield

Pin 2 – Hot   or   +

Pin 3 – Cold   or   –

Today, this is an international convention (e.g. AES14-1992, and within IEC.268). It was however first adopted in the '60s by (for example) RCA in the US and the BBC in the UK. Yet many makers in the USA have been slow to revert from an alternative standard that developed, where the polarity is reversed, i.e. Pin 2 = cold and Pin 3 = hot. This can cause problems with polarity 'matching' when equipment wired to opposite conventions is used together. See section 9.4.1, *Balanced Polarity and Shielding*. This section also considers shield connection options.

Unbalanced XLR

For short cable runs (ideally under 10'/3m) from unbalanced sources, ideally only in a rural, domestic environment, unbalanced connections at the amplifier end may be satisfactory. This arrangement requires only single core shielded cable, but has no hum and noise rejection, so *it* should be rejected unless it proves satisfactory. The following connection (Figure 9.13, (ii)) applies to (i) using a balanced input from an unbalanced source with an unbalanced connection, and (ii) in a wholly unbalanced system.

Pin 1 – Cable screen/signal return

Pin 2 – Live input (Hot or +)

Pin 3 – Link to pin 1 (if input balanced) or leave open (if unbalanced).

357

Figure 9.14

Neutrik's XLR connector, the world standard for professional line connections, and many professional speaker connections. © Canford Audio.

iv) Jacks, 1/4", balanced and unbalanced (Phone Jacks)

Mono (2 pole) and 'A' gauge ('stereo' or 3 pole) jack sockets are used on low quality and 'contractor' equipment. Originated by the British Post Office nearly a century ago as telephone jacks, they make reliable contacts when regularly inserted and withdrawn, but not when installed and left. For this reason, *jack connectors should never be used for fixed, high performance audio.* The sole exception is in systems that are moved *and replugged* almost daily, or where labour is available to polish the connectors weekly – using a proprietary micro-abrasive or other suitable cleaner. Otherwise, in equipment having solely jack connectors, or requiring jack connectors for practical application, the connectors are best changed.

The balanced connection is like XLRs, where:

Sleeve = pin 1 = shield

Ring = pin 3 = cold (–)

Tip = pin 2 = hot (+)

See also section 9.4.1, *Balanced Polarity and Shielding.*

v) DIN

DIN connectors, dreamt-up in an austere post-WW2 Germany, are the most fiddly connectors to wire, are wholly unsuited to normal wire gauges (anything fatter than 34 swg), and the cable clamping and assembly quality is commonly mediocre to poor. Pin contact can however be surprisingly reliable. The tinned contacts do not look pretty, but sonics can be good. While most commonly encountered in the 3 or 5 pin 180° format, there are versions with 4, 6 or more pins, and some types span 240°. While rarely found on the inputs of few (if any) modern high performance power amplifiers, they have played a part in several classic designs. In many instances where power amplifiers are designed as part of bespoke, esoteric or early systems approaches, DINs have been used to carry DC power, control signals, etc., sometimes along with line level audio. Higher quality types made by Preh have robust latching.

The DIN wiring standard is prolific, but also idiosyncratic and outdated, e.g. there is nothing for balanced power amp inputs. The central pin (on 180° plugs) is always 0v (shield) and is always numbered (as pin 2), regardless of the number of *and* numbering of, the surrounding pins.

358

vi) Terminal strips (barrier strips, screw terminals)

These are found on the back of low budget amplifiers intended for (or as an option for) fixed installation. They can give good results provided the metal is substantial enough to allow positive screw tightening without fatigue or thread stripping, and provided the metal and plating are both appropriate to high quality sound. The correct eyelet or ring terminal must also be used on the wire, which must be either soldered, or crimped correctly with the proper tool. Support of the delicate cable conductors is usually nil, and may be difficult to arrange. And the nuisance of disconnection can create a psychological barrier to regular amplifier inspection. If connections are balanced, see also section 9.4.1, *Balanced Polarity and Shielding*.

vii) Proprietary input connectors

In both high-end power amps, and professional touring amplifier racks, unconventional input connectors may be used for good reasons. Mostly these are shielded, multipin types, often with a latching interface, specially plated, with specially low resistance contacts. Most are made for arduous use in professional, industrial, military or aerospace equipment. Examples include Camac, Lemo and Lytton. Wiring conventions are naturally *ad-hoc*; for connections consult the amplifier or rack maker.

9.4.1 Balanced polarity and shielding

With balanced connections, signal polarity can be one of two ways. With amplifiers, variations apply mostly (but not always just to) inputs. It is not unknown for the inputs on samples of a given model of amplifier (one intended for use in the US say; or previously rewired under the hood when hired to work with other gear) to be wired differently from today's 'Pin 2 hot' norm. *Random or indeterminate signal polarity can cause anything but the simplest loudspeaker system to perform below its potential.* Some highly specified professional power amplifiers have featured PCB links, headers or other internal switches allowing XLR (or other balanced connector) polarity to be set for 'pin 2' or 3 'hot' (or equivalent thereof).

It is worth noting that if all the connections are balanced, then if *every* amplifier in a given system is wired for the 'wrong' polarity, then only the absolute polarity of all outputs is flipped. This would have a relatively subtle effect on sonics, and could be simply corrected by flipping the polarity of the inputs to the active crossover. Otherwise, if the balanced (XLR or whatever) input sockets in the amplifiers in a system have opposing hot and cold conventions, system polarity should be made 'all one way' by rewiring the amplifiers' sockets internally. In large systems, nothing should ever be taken for granted: one should double confirm polarity with a tester. This way, amplifiers that have been carelessly serviced, so the polarity is flipped inside by wires being replaced wrongly, show up. When power amplifiers or sources with non-standard polarity wiring cause polarity errors in a pro-touring system, temporary 'phase-changer' *tails* (short leads) may be employed as a fix. Ordinary XLR-to-XLR signal cables should *never* be cross-wired.

The optimum connection of the shield in a balanced system is less fussy than in unbalanced arrangements, yet there are options and permutations that require consideration and often, practical evaluation. The starting point for shield connections in all-balanced connections is simply tying it to ground at *each* end. In practice, this usually means the chassis, or whatever else the mains safety earth wire (CPC) is connected to. The problem is that this almost inevitably creates a loop down which surprisingly high currents can flow, which can cause or aggravate hum in other parts of the system, if not directly.

In the following choices, it is assumed that pin 1 at each end is *not* connected to 0v (it is the wrong place to connect a primary EMI shield to), but solely to chassis/earth-ground, as just discussed.

The shield connection options are:

(i) Connect the shield to pin 1 at amplifier input only, as illustrated in Figure 9.13 (i). This connection, *uniquely important to power amplifiers*, ensures the shield is tied to ground as long as the cable is plugged into an amplifier's input. Without this, there is a risk that damage might be caused to the amplifier and/or speakers.

(ii) Connect shield instead to pin 1 at the source end only. This system seems no different on the surface, but it offers superior at RF noise rejection. This 'tie back the shield' connection cancels the effect of conductor-to-shield capacitance imbalance, which otherwise reduces CMR at RF, where it is most needed. *It is thus the optimal connection for most audio connections, where the previous consideration does not apply*. But this connection is risky as it stands for amplifier connection in large systems, as by just dislodging the returns multicore connector at the mix position end, n kW of amplifiers would be left dangling with 100m or so of unshielded cable, and in the words of Steve Dove, would be likely 'Screaming off into the ionosphere'.

(iii) Connect as (ii), but also connect as (i); but via a small capacitor. This may be fitted between pin 1 and chassis inside each amplifier. This has no appreciable effect at audio frequencies, but ties the shield increasingly to ground (chassis) at RF, at the amplifier end, safeguard if the other end is disconnected.

9.4.2 Quasi balanced (unbal-to-bal)

Even if the source equipment doesn't have balanced outputs, a power amplifier with balanced inputs can still be usefully connected in *Quasi-balanced* configuration. This arrangement, while offering less protection than good overall balancing, can nonetheless greatly reduce hum and RFI pickup over simple unbalanced connections. Two core shielded cable is required. Alternatively, twisted unshielded conductors may be used, at the owners' discretion.

At the amp end, the connections are conventional, provided that pin 1 on the chassis-mounted socket is connected (or can be re-wired to connect) to chassis, instead of signal 0v. At the source end, 'cold' goes to the unbalanced output's earthy '0v' side. The shield is not normally connected at this end. See Figure 9.13, (iii).

9.4.3 Input cabling

Input cabling to power amplifiers should have above all low *and balanced* capacitance (unless very short), low microphony, and good screening.

Capacitance

Low capacitance avoids placing any unnecessary loading on the source equipment at HF – particularly when the cabling is long, as in PA applications. 'Low' means below (say) 100pF/metre. 600pF/metre would be considered high, and 200 to 300pF/m medium. The pF/ figures can be looked up in catalogues or measured. In balanced cables, the pF/ figure used should be 1.5x the capacitance between the cores, as the capacitance from each core to shield adds half again to the total. *But if the cable shield is 'tied back' to the source end, the core-to-shield capacitances no longer appear as loading (at least in balanced systems), leaving only the core-to-core capacitance, as in twisted but unshielded cables.*

Microphony

Low microphony is achieved firstly, with a low equipment source impedance. If the source impedance is necessarily rather high (anything above 1kΩ), microphony can be reduced by suitable physical design that isolates or quickly damps or counteracts vibration between the cable's insulators and conductors. It is an important consideration in systems where the load and (particularly) the source impedances are higher than the norm, usually when the source is valve equipment. It may be instructive to listen to a sound system, after suspending the amplifier's input cabling (e.g. on string hangers) – so it's no longer in direct contact with the room surfaces. In PA systems, return multicores are frequently already hung or at least slung, for practical reasons.

Shielding

Cable shields have two jobs. They can totally reject electrostatic fields that might cause hum or signal breakthrough. They can also intercept RF, and depending on its field strength, at least partially drain it away. But note they have little or no improving effect on magnetically induced hums or buzzes. Good screening is achieved by choosing the right cable for the job, as well as by correct screen termination. Correct termination is always direct to chassis (via pin 1 if XLR) – at least at RF [8]. In unbalanced systems, this *must* be performed at the power amplifier end. In balanced systems, the connection can be either source or power amp end. If the chassis grounding of the shield conflicts with another grounding scheme, then connection may be made via a capacitor that maintains a very low to relatively low impedance across RF, e.g. 100nF, 400v rated, in parallel with say, 100pF.

At RF, the best shielding is achieved with aluminium foil, or with multi-layered braided (knitted) shields [9]. The latter is expensive but may be needed if RF pollution levels dictate it. Single braided cables are next best. Good braided shields are highly resistant to damage by repeated coiling, twisting and trampling. Without this factor, single twisted 'lap' shielding would also be rated, but random gaps develop or existing gaps enlarge as soon as it is just bent around a corner. This makes single lap shielding unsuitable for shielding high performance audio connections unless ambient RF levels are known to be low. The double lapped version is better, but still inferior to a good braid. Likewise, but far less readily, the foil in foil-shielded cables will tear apart if over-handled. While degrading RF shielding, gaps in a shield have little or no effect on intercepting hum and buzz frequencies.

Summary

Foil- and lap-shielded cables are best kept for fixed professional and domestic installations, and internal wiring inside equipment that contains no RF, digital or switching psu circuitry.

For touring PA, the best and longest-lived RF immunity will be had with braided and even conductive plastic shielded cables.

In domestic installations with low levels of RF, and E-M fields (e.g. by spacious layout and/or use of low inductance mains and speaker cabling), *unshielded* cables may exceptionally be used, particularly if twisted or maximally mutually–coupled, and connected to a balanced input.

9.5 Output connections

Power amp output connections are unlike most others in audio, in terms of the current magnitude (commonly peaking in the order of low tens of Amperes yet well over 100 Amperes in some cases), the peak voltages (potentially enough to cause shock), and the consequent size and weight of conductors and insulators that have to be accommodated.

These stress the need for a stably low resistance, together with good enough insulation. Excessive safety legislation at the connector is pointless when the speaker cables could just as easily be abraded or bitten into. Now that audio amplifiers exist with output voltages equal to the highest single-phase mains voltages, it seems inevitable that sometime in the future, safety protection using some form of current-balance monitoring (as in an RCCB) will be required on the speaker circuit.

Contact care

If unplugged when programme is present, current at the 'moment' of separation could be at any value between zero and many Amperes. Whenever a substantial current flow is abruptly ceased by breaking connections, inductance in the broken circuitry sets up a *back EMF* or 'kickback voltage' that is high enough to cause arcing. This draws out the contact period. Even if it only extends it for milliseconds, any arcing (just like welding) will inevitably vaporise some part of the connector surface. It will also create oxidation products that can corrode or corrupt the remaining good surface metal. *It follows that in high performance audio systems, power amplifier output connections ideally shouldn't ever be broken or remade when large signals are present.* With small signals, typically below 1v peak, there is little current flow, so arcing doesn't occur or has insufficient energy to more than heat the metal surface. Connection under these circumstances is safe. With some types of connector, e.g. 4mm, arcing will with usually occur at a location (in this case the tip of the plug and socket) that isn't needed for conduction.

Practicalities

If there is any choice in the connectors used in a system or installation, consider that misconnections are likely on adjacent amplifiers or channels which use input and output connectors which are either compatible (eg. unbalanced XLR-F in and XLR-M out); or where there are converter leads in use. For example, the outcome of a *Speakon* to male XLR cable could be easily (in some large systems) plugged into an amplifier input, *particularly if speaker cables are not always readily distinguishable from input cables.*

i) 4mm binding posts (4mm terminals)

Binding posts are amongst the oldest means of connecting electrical apparatus. Traditionally, bared wires were threaded through the hole in the post, then wrapped around one or more times, then clamped down tightly under the knob part. But unless the binding post can be torqued down tight enough to crush the wires solid, this method provides little area of gas–tight contact, or at least can leave a lot of conductor bare to the air, so sonic performance quickly decays with the oxidation and pollution of the conductors' skin.

In modern practice, the procedure has been simplified to threading the bared ends of speaker wires straight through the post, then clamping them down securely. This way, the post and conductive shoulder makes immediate gas tight contact with the single thickness of conductor – all that is needed. By crimping *bootlace ferrules* onto them, professionals may prevent conductor tails, if stranded, from getting frayed and scrunched-up.

Modern binding posts also allow a 4mm plug ('banana jack') to be inserted from the rear. Alternatively, in the better designs, the conductor hole is 4mm (dia) or slightly larger, so both a 12mm^2 conductor and a 4mm plug can be clamped down. This latter best suits the poorly-sprung plugs that most need clamping. High grade plugs with low contact resistance have 'leaf' sprung contacts on all sides, and localised hard clamping will damage them unless the torque is judged carefully.

The binding *posts* should be turnable with ease with the fingers. Slippery plastics are out *(see 'Howlers' section 9.2)*. Knurling, fluting or surface texturing are best. To be of any use, the commonly encountered hexagonal shape would require a suitable end-on driving tool – that is never seen. The best designs (Cliff in UK) give good access for threading fingers, and have plain, large screwdriver slots end-on, for tightening or merely checking tightness, when they are deeply recessed or otherwise out of easy hand reach. They can even be tightened with a coin. No special tools are needed. Spacing is conventionally 19mm (once 3/4"), to allow double 4mm-jacks to be used.

Traditional and modern binding posts as well as 4mm plugs and jacks, single and double, may not meet modern safety standards, e.g. LVD in EU countries, as bare conductors are usually exposed. This may signal their demise in some territories. Alternatively, added insulating shrouding, firmly clipped-on after wiring, may enable them to meet safety requirements. At the time of writing, some makers are fitting a removable, shrouding 'end bung'.

Perfection in the external details and utility of binding-posts is fruitless without suitable terminations behind the panel. These are nearly always solderable, but also threaded, allowing wires to be connected via eyelets or tags which can be securely nutted down. The eyelets may even convert to push-on type blades. These options mean that sockets can be changed without soldering. The significance is that 4mm binding posts are proud components and usually made from brittle brass, which is occasionally snapped or bent in transit, installation or professional tour rigging. So the need for an occasional replacement *is to be expected*. The benefit of *not* soldering is that more than one wire of more than one gauge should come together at output sockets; and that it is almost impossible to desolder heavy, PVC insulated wire without melting-back and damaging the insulation.

Notice: Copper-based cables, particularly those with a small number of relatively thick strands 'creep', causing brass or other terminal screws to slacken in time. To arrest degradation in sonic quality, all connectors based on gas-tight conductor clamping must be regularly checked, shortly after installation, then annually at least.

ii) 4mm sockets

Sockets (receptacles) are notionally 'safer' than binding posts, as live parts are inaccessible with plugs fully home, and are more deeply recessed when plugs are absent. Special high voltage types are available with an extra deep recess. But unless the plugs used with them have high voltage sprung shrouds, live terminals can still be exposed by partial withdrawal. Some types can offer switching when a plug is inserted. Contact rating is typically up to 32 Amperes for quality types. Connections may always be soldered. The better types have optional threading for nutted eyelets, or push-on *Lucar* blades.

The preceding connectors (i) and (ii) are single pole. This gives ample flexibility, and the cost is low, but it increases the risk of wrong connections, and the time taken to make connections, in large 'active' and multi-amped systems. Multipole connectors with 4, 6, 8 or more pins are therefore de rigueur *for the external rack connections to speakers, in touring systems.*

iii) XLR

See prior section 9.4 for a general description. XLRs have long been employed as output connectors on professional amplifiers. But with modern safety legislation, and also increasing maximum output voltages, they are viewed as unsafe, because of the long established convention that 'signal emerges from males' (i.e. on pins). This is fine at line levels, but not at anything above 40v, the voltage at the output of a relatively modest amplifier. Some professional users, including the BBC, have overcome the problem, by simply reversing the gender on output connections, so the output is on a female receptacle.

Another snag for use for output connections, is that few XLRs can comfortably manage the diameter of sheathing on 2 core PVC cables above 1.25mm^2.

XLR *output connections* are usually wired as follows:

Pin 1 – Black or –

Pin 2 – Red or +

Pin 3 – not connected.

XLRs were originally expensive. Today, thanks to the efforts of Neutrik (*cf.* the earlier US makers), utility has been enhanced and high overall quality maintained, while the real price has halved in the past decade. *Completely plastic* XLRs have been made available by other makers. These cost a little less, but most of the quality is lost, viz. EMI shielding, ruggedness and inter-plugability tolerancing, for a start. They are not recommended.

In some systems, particularly theatrical PA, 4 pin and other XLRs have been used for bi- and tri-amped or other multiple speaker connections. Such wiring is *ad hoc*; consult the maker or installer. Use is inadvisable in high power systems, above about 50w to 100w (re. 8Ω), as the current rating of the 'higher-pinned' XLRs is greatly reduced.

iv) Speakon

The *Speakon* was expressly designed in recent years by Neutrik (Switzerland/UK) to satisfy the modern requirements of professional and even domestic power amplifier users (where others had failed), and to meet tighter safety legislation (the IEC 65 'finger test', and LVD in EU countries).

Figure 9.15

Neutrik's dedicated Speakon *speaker sockets, 8 and 4 pin types.*
© Canford Audio.

There are two types, 4 and 8 pin, called NL4 and NL8 respectively. Presently, both sizes are plastic housed. While a tough plastic has been used, the experience of several touring PA operators is that they are more easily broken than older, metal-bodied connectors (e.g. EP, XLR-LNE series). Due to inadequate shoulder flaring, they are also considered unwieldy by touring professionals, being particularly prone to snag when long cables are tugged while de-rigging. Another annoyance is the locking ring, later changed to a latch, which has fooled many a sound engineer. None of these will bother a domestic or studio user. But then there are the fiddly screw connections, which will demand regular inspection and maintenance as there is no option for soldering. Lastly, when wiring *Speakons*, the pin numbering can be almost impossible to read. These foibles preclude this well-intentioned connector from universal adoption and professional endorsement in the UK.

On the plus side, *Speakons* are realistically designed to accept conductors up to 6mm^2 and cables up to 15mm^2, have high torquing hex (Allen) socket terminal screws, are latching, and have solid cable clamping. Contacts are designed to withstand attempted damage by high inrush/outrush currents and arcing.

Conventional wiring is as follows (subject to amendment).

Speakon connection (provisional IEC Standard)

System category:	*Speakon* type:	Pin numbers: 1–	1+	2–	2+	3–	3+	4–	4+
Full range	NL4	F–	F+	(F–	F+): these may be paralleled				
Bi-wired or bi-amped	NL4	L–	L+	H–	H+	♦	♦	♦	♦
	NL8	L–	L+	♦	♦	♦	♦	H–	H+
Triamped	NL8	L–	L+	♦	♦	M–	M+	H–	H+
Quadamped	NL8	L–	L+	– L	M+	– H	M+	H–	H+

Note: L = LF. LM = Low Mid. M = MF. HM = High Mid. H= HF.

   ♦ = no connection.

Attention: Copper cables, particularly those with a small number of relatively thick strands 'creep', causing terminal screws to slacken in time. To arrest degradation in sonic quality, and avoid risk of fire in high power systems, all connectors (such as the Speakon) based on gas-tight conductor clamping must be regularly checked, shortly after installation, then annually at least.

Figure 9.16

Speakon *plugs. Note the newer latch on the 4 pin type (left); and the older twist-collar on the 8 pin version (right).* © Canford Audio.

v) EP series (originally made by Cannon)

The EP series have been widely used for multi-driver connections in touring PA since the early '80s. They are the 1950s predecessor of the XLR, larger and stouter, and current ratings per pin are proportionately considerably higher.

A *de facto* pin configuration, as used to connect Turbosound and other professional, 3 and 4 band actively-driven enclosures, uses the EP3 as follows:

Pin:	Connection:
1	LF −
2	LF +
3	Mid −
4	Mid +
5	HF −
6	HF +

EP4 and EP8 are also used. For wiring, consult the box maker.

vi) The PDN speaker connector

This type was a competing precursor of the *Speakon*, that addressed the safety concerns that were raised with traditional terminals and XLRs. While it appears in catalogues, and offers an adequate electrical connection, it is not known to be fitted to any commercially manufactured amplifiers and is not recommended by many professional operators in the UK.

9.5.1 Speaker cabling

A loudspeaker cable's foremost requirement is to act as if it has little (ideally nil) length; it has to connect the amplifier's low source impedance as tightly as possible to the drive unit's terminals (assuming voltage control). To do this, it should have very low resistance *and* very low series inductance. This is tantamount to saying that the cable should have maximum *mutual* inductance (magnetic linkage between send and return), and maximum capacitance, these being the dual (reciprocal) elements.

A loudspeaker cable that has low enough resistance, but where series inductance is moderate, as is the case with ordinary AC power cable commonly used for speaker connections, will (unless very short, under 12"/0.3m say) *degrade the ability of a high performance amplifier to damp the speaker on musical transients associated with all audio frequencies*. Figure 9.17 (overleaf) shows how cables' damping may be tested, using burst signals and a double differential analyser [10].

Loudspeaker cable that has low inductance but too much resistance will suffer a power loss. Long before this has any effect on loudness, it will cause degraded damping at most frequencies. The same series resistance will cause a voltage drop across the cable that follows the speaker's impedance *vs.* frequency plot, thereby altering tonal balance as well.

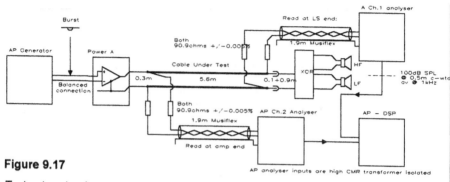

Figure 9.17

Test set-up to abstract error signals from a typical length of loudspeaker cable, under realistic conditions.

The following table specifies minimum or ideal cable c.s.a., for a notional damping factor of 100, based on 10mΩ per ohm of nominal speaker impedance. In other words, all the cable lengths for the sizes listed will be under 80mΩ for 8 ohms, and under 160mΩ for 16 (or 15 ohm) speakers. For 4 ohm speakers, simply step up the gauge by one or two column(s). The cable gauges required show why it is advisable to feed tandem drive units (as used in many bass enclosures) individually.

Never run speaker cables alongside input cables. This might cause positive feedback at RF, hence RF oscillation, which is readily destructive if the amplifier employs BJTs in its output stage. In an ideal system layout, all input cables approach the amplifier at right angles (90°) to speaker leads. Don't forget that wires running up and down are at right angles to wires running in either horizontal plane.

Speaker Cable Gauge

Nominal speaker impedance	Min Cable size (mm²) for distance:					
	8' 2.5m	16' 5m	25' 7.5m	32' 10m	65' 20m	(feet) (metres)
8Ω	1.25	2.5	4.0	4.0	10	mm²
16Ω	0.75	1.25	2.5	2.5	6	mm²

Inductance

Low resistance conductors have greater cross-sectional area (csa) than thin ones. They also have slightly lower inductance, per unit length, but it is essential to recognise that the inductance that counts is the series inductance when the cable conductors are looped. This 'loop inductance' is much smaller than the inductance of an individual conductor, provided the send and return conductors are close to each other. Other than reducing the loop inductance (LL), the proximity reduces radiation and confers EMI shielding. Makers of very low inductance cables include Goertz and Kimber (US) and Jenving (Sweden). These types can improve sonics by helping the amplifier's NFB get a faster grip, and by receiving less RF hash.

Spaced cables

Loudspeaker cables with spaced conductors are purely cosmetic. The low capacitance they exhibit is not ideal, as the wide spacing needed to gain it also increases loop inductance, causing smearing in high resolution speakers, particularly those where the drive-units are passively crossed-over and/or don't have conjugate networks. *Such spaced cables are also far more open to pickup EMI, and in multi-driver/multi-channel systems, crosstalk.*

R-to-L balance (DC resistance / loop inductance)

For each cable construction, there is an optimum c.s.a., where increased inductance is balanced against reduced resistance. The higher the speaker impedance (e.g. 15Ω rather than 6Ω say) the less conflict there is between the two, in some disproportion. This is the benefit in connecting paired drive-units via individual cables, rather than one cable equal to the sum of the individual csa's.

Other effects

Skin effect (see also Chapter 5) is a frequency-dependent conductor strangling effect, acting like inductance, but less severe. Its measured and sonic effects come into play significantly, and at increasingly lower frequencies, as loop inductance is reduced. A cable with low skin-effect is not useful unless loop inductance is first low.

Proximity effect is another cause of constriction in the cable, so effective c.s.a. shrinks at high currents. *It causes heavy gauge cables to under perform at the time they are most required.* It can even cause cable resistance modulation, as the temperature rises in the small area of conductor into which the current is confined. If current-hungry speakers are in use, cable designs which combat proximity effect are worthwhile if they also have low loop inductance.

Dielectric quality in speaker cables is in many ways far less important than it is in line-level cables, since the cable's shunt capacitance is highly damped by the amplifier's output impedance, if not the drive-unit. This is one of the benefits of a high damping factor.

Properly researched audio grade speaker cables may also offer purer or specially alloyed or plated conductors; the damping of the mechanical effect of proximity effect and/or air-borne vibration; and lower losses up to low RF, then higher losses.

Shortcomings

Purchasers should be wary that gold and copper form bad-sounding oxides, and that super-pure conductor metals oxidise very readily on their skins as soon as the cable is manufactured. For a start, the common types of plastic insulation covering speaker cables releases reactive plasticisers as soon as it is made, that attack the conductor's surface. The better sounding cables employ conductors that are plated with tin, or made of solid silver or other metals or materials even. They also are clad in a relatively inert insulation, e.g. special PVC with low chloride emission, PTFE or silicone rubber.

Makers of 'monocrystalline' copper cables do not mention that the crystalline structure is brittle, and that except with very careful installation, the properties will be lost.

Split options

The quality of loudspeaker cable is most significant in passively crossed-over speaker systems. Bi-wiring is mentioned in chapter 2. There may be different optimum types and sizes of cable for the individual frequency band. Connecting each drive-unit via its own cable also reduces each cable's 'workload'.

In his studio control room designs, the monitoring systems quality researcher (and historic Virgin record producer) Philip Newell takes this one step further. Noting that in speakers with passive crossovers, the power amplifier to passive crossover connection commonly exhibited more sensitivity to speaker cable quality, he moves the crossover to the power amplifier, so the cable length in this section is reduced to a few inches [11]. Again, specific cables might be used for the different frequency bands.

Other hints

In large systems, and where XLRs are used for input and outputs, be sure to employ a standard form of speaker cable that is readily distinguishable (by colour difference, markings, size, tactility) from input (line level) cables, in dim lighting, and by people of limited perception.

9.5.2 Impedance setting

A minority of transistor amplifiers (e.g. APT, BSS Audio), but most valve amplifiers, have a switch that optimises the amplifier to the nominal speaker impedance. The switch is thus marked usually 16/8/4 ohms, or somesuch. In most cases, taps on either an output- (if the output stage is valve) or power supply- (if transistor output stage) -transformer, are being changed. Mis-setting this switch may temporarily degrade headroom or other sonic qualities, but it should not cause damage. Usually, it is inadvisable to change such switches when programme is playing, certainly at high levels. If a speaker is 'hard to drive' (has bad impedance dips), it is possible for sonic quality to be improved with a setting lower than the nominal speaker impedance, eg. 2 ohms for a difficult '8' ohm speaker.

9.5.3 Output polarity

Pay particular attention to amplifiers set to operate in bridge-mode. The Channel 1 red or '+' output terminal is usually '+' in Bridge-mode, but be prepared to check the maker's manual. In absolute terms, the lower channel No's + is taken to be the '*real*' +, i.e. polarity refers back without inversion to the hot (or pin 2) *input*, while the higher channel's output is usually the one that is inverted.

9.6 Placement and fitment

Care ! Many high power or highly specified audiophile power amplifiers are heavy enough to damage or stress connective tissue in your spine or other places, if they are not lifted with due care. Always bend your knees and lift by unbending here first, transferring the weight gradually to the spine. Reverse the process when letting down, ending up with bent knees. Unless you are a weightlifter or PA rigger, quite a few highly rated power amps are too heavy to lift comfortably on your own – at least two lifters will be needed. In warehouses, on stages and in demo rooms, handling aids, such as trolleys, wheelboards and carts are used. These are urged for moving weighty power amplifiers in the home also.

When installing amplifiers into racks (usually for touring), the racks are normally faced up, and are often on a wheelboard, which is locked. Rigging gloves (with part open fingers) are worn to protect fingers when letting down a heavy beast into the rack. With heavy amplifiers, and whenever weight distribution is rearwards, rear rack fixings are essential. These should be supplied by the maker.

In domestic and studio installations, the sonic performance of power amplifiers may be improved by avoidance, damping and/or 'grounding' of air- and structure-borne vibrational energy. For perfectionists, this means

(i) evaluate sonics in different locations, with different types of music, at different sound levels. Moving just 12"/0.3m away from a corner or from a wall, can be significant.

(ii) then evaluate damping and conductive aids, e.g. *'Blutak'*, foams, dampers, mats, spikes, etc.

(iii) then re-evaluate the new, best position.

9.6.1 Cooling airflow conventions

Most domestic amplifiers are cooled solely by convection. Fan cooling is only permissible if the fan noise is extremely low at the listening position. The further ear-to-amp spacing possible in a large room helps. In the highest quality designs, a considerable volume of cooling air is moved with much lower SPLs than is the norm, using a *centrifugal* fan. The more refined professional amplifiers have also long had continuously variable 'servo controlled' fans, which with a class B, A-B, G or H amplifier, only speed up when there is substantial current, generally when the programme level is quite high. The snag is that high speeds may persist for some time after a loud passage is followed by a quiet one, or silence, and the ears gradually notice the fan noise. At least any changes can be gradual. Some designs have a slow setting and enough heatsinking capacity to handle occasional loud passages, while keeping fan nose uniformly low. A third kind have two fan speeds, viz. slow, switching up to fast if heat exchanger temperature gets into emergency realms. This type causes abrupt changes in fan noise, making them highly intrusive in the listening room.

Any amplifier's cooling can be defeated by simply racking it with other amplifier, because the direction of forced air isn't standardised. In one standard, call it 'A', air drawn from the rear emerges from the front. This helps users know that the fans are working, and by assessing the waste air temperature, one can gauge some idea of the amplifier's work condition. Also, the back of the rack may be kept cool.

Other makers employ standard 'B': they draw air in from the front, and exhaust from the rear. If the rear of the rack is inadequately ventilated, so the main ambient air temperature is lower, this yields cooler running. On the minus side, it stresses *or can prove fatal* to other amplifiers in the same rack employing the reverse airflow direction, and also any equipment that lacks forced cooling, especially if it runs hot and depends on convection cooling.

A third type of amplifier, 'C' has no front venting. Hot air is exhausted or drawn in from the sides and lid, and vice-versa from the rear panel. When racked, the rack sides should ideally be vented (as a chimney or intake) at top and bottom. If not, the rear of the rack should have a substantial volume of cool air or draught.

Figure 9.18

The modular heart of BGW's PS2 budget professional amplifier.
The BJTs (not visible) are cooled with enhanced reliability by the lateral, extruded-aliminium heat-exchanger.
A small fan blows air at the centre of the fins. Most of the air exits at the sides - a type 'C' airflow.
© BGW

Figure 9.19

The Rauch DVT 250s , another example of type C airflow, but with a rear intake. Note bi-polarity true error indicator LEDs above the knobs, covered in chapter 5.
© MS&L.

The temperature of the waste air from the small number of very high quality amplifiers employing centrifugal (alias tangential) fans, which move air slowly but at high pressure, can be hot enough to burn users. When hot air exits from the rear of a rack containing such amplifiers, take care that there are no flammable materials (newspapers, old polishing rags) present. Any PVC cables passing up the rack must be dressed to one side of the hot air path.

9.6.2 Cooling and air pollution

Fan cooling concentrates air-borne dirt inside amplifiers. A filter should be used but if the air is dirty, this soon chokes the air flow. This in turn stresses the common axial type of fan, reducing its lifespan, while thermal headroom suffers. Cleaning a filter is easy enough; gently wring out in hot water laced with a spot of detergent. When many amplifiers are in use, it is expedient to filter the incoming air, if this can be hermetically connected. The large filter will be easier to maintain and the individual amplifiers' filters will need checking and cleaning less often.

Fan filters are sometimes removed, often by those who have never had to clean out an amplifier. The parts that then get dirty eventually have to be cleaned, and are *far* harder to clean, and prone to be damaged. For example, air also contains some moisture, which is liable to be acidic or corrosive both in heavy industrial and urban areas, and within 10 miles of the sea. Also, since 1980, smoke machines have become increasingly *de rigueur* in many musical and performing art circles. Some of the oils have proved quickly and fatally corrosive to amplifier electronics.

Fan cooling exacerbates the deposition of corrosive substances, and as most PA amplifiers have to be fan cooled, any working near to smoke machines for prolonged periods can end up costing the user or owner a new set of PCBs and a major decontamination exercise. Air intake filters provide little protection against smoke and corrosive chemicals. Spraying an elastic protective ('conformal') coating over all the parts can provide some protection, but unless the coating is known to have sealed every crevice in every corrodable part inside, amplifiers have to be kept away from the worst of the smoke. Otherwise, the highest quality fan-cooled amplifiers, particularly those made for the stage and installation in other arduous environments (e.g. by the sea, or in desert areas) have air passages that solely pass through anodized aluminium, stainless steel, or some other chemically inert surface.

Convection cooled amplifiers used in most domestic systems and some professional setups can have internal or external heatsinks. If external, they may be casually fan cooled if desired (for a loud session) without worries about filtering the air, by simply directing a suitable fan at the fins. If the heatsinking is internal with louvres, or particularly top venting, there is again a dust and chemical pollution problem unless the cooling chimney is substantially separate and sealed from the remainder of the circuitry. The solution is to clean the air in the room, and avoid opening windows when outside air quality is low. This is good for the listeners' health too.

With convection cooled amplifiers, the user can choose how well they are cooled. Cooler running will increase reliability. To achieve this, arrange the siting or mounting so air can freely circulate around the case on most of all sides. If stacking, put substantial spacer blocks or bricks between amplifiers.

9.7 The 1 to 5 of prudent amplifier use

1. Turn on audio power amplifiers *last* – and the remainder of the system *first*. In high resolution systems, preferably only connect amplifiers inputs after the source system has settled, e.g. after a few seconds for well designed transistor equipment.

2. After 1, and only then, connect the speakers after the amplifiers and everything preceding them, has been switched on, and all have settled. This is a wise precaution when valuable loudspeakers are connected to an amplifier which lacks DC protection and turn-on muting. It will prevent the transmission of loud thumps or *bhlatts*. Don't use the power up/down switch on an amplifier or preceding equipment as a mute button, unless previously proven, as poorly designed units have been known to produce wild or embarrassing sounds up to several minutes after being switched off.

3. Keep as cool running as possible. If fan cooled, check and clean filters regularly, according to need.

4. Clipping damages or harms human hearing and loudspeaker drive units alike. Reduce levels immediately.

5. Before turning off the system, disconnect the speakers or set the amplifier to mute. Then switch off the source equipment. After 2 minutes settling time (see Section 3.5), the amplifiers may be switched off.

[Note: Steps 2 and 5 are not practical nor often necessary in large and professional systems and will be waived in such cases].

9.7.1 System back-end troubleshooting

Buzzing or humming

Check input connections. Try other cables. Does the buzzing stop when the source equipment is disconnected ? What does this tell you ? Try to use balanced connections whenever possible, *even if your source is not balanced.*

Coloured sound

(In active systems): Check cables, amplifiers and drive units in missing or amiss frequency range(s).

DC fault

When an amplifier's power or driver transistors fail short-circuit, to cause a large DC fault, the first symptom is a raspy buzz, as the DC supply's ripple is fed directly to the speaker. At the same time, the cone will move largely in or out. If this condition persists for long enough for you to realise it, i.e. above one second, then either the amplifier omits a DC protection relay, power shutdown, or other means of stopping the DC power source, or the protection isn't working very well.

Either way, the speaker cables need disconnecting *immediately.* Then switch off the amp and leave it to discharge. If the amplifier is 100% direct-coupled, do not be too hasty to blame it, as the DC fault might just be in the preceding gear.

Fuzz

When heard on decaying notes, and during quiet passages, may be (i) crossover distortion or (ii) a rubbing voice coil. A second speaker proves which.

Hiss

Not much should be audible when the amplifier's input is disconnected and shorted. If prominent, it may signify a sick semiconductor or RF oscillation.

Intermittent sound

Look for a mechanically loose connection. First, try exchanging cables. If this fails, *disconnect the power*, allow time to discharge, then try turning the amplifier on its sides, and over, listening for loose items. **If this fails, look inside.**

Noise

If really a hum or a buzz, rather than hiss, see "Buzzing or humming".

Overheating

If an amplifier overheats (and shuts down), or the casing becomes hotter than usual, then check:

(i) the speaker impedance – is this within ratings?

(ii) the connectors and cabling – is there a partial, possibly intermittent short?

(iii) the heatsink cleanliness. If inside the casing, with a chimney vent, or fan, when was the dust last blown out? Is fan airflow enough?

(iv) signal – is this clipped or unusually compressed?

(v) stability, especially when the amplifier isn't being driven hard. Use a 'scope or RF sniffer to check. If the amplifier overheats with no signal, then RF is a most likely culprit. Overbiasing of a class B or A-B output is also possible.

Polarity

If polarity testing reveal inconsistencies, first check (by eye, and with the help of a continuity 'bleep tester'):

(i) the output cable and connector wiring. Does the brown (or whatever colour) wire follow through from the amplifier's + terminal to the speaker's + terminal without transposition?

(ii) the amplifiers' input wiring. Polarity transposition is most likely when different amplifier models are in use, particularly if they have balanced inputs and/or different XLR connection standards, i.e. is pin 2 or 3 hot?

(iii) perhaps the amplifier is inverting. This is reare, but quite possible.

If no inconsistencies are found, check elsewhere in the system wiring. Are the speakers consistently wired ? A polarity checker is often best used, as at least one loudspeaker maker, JBL, has an inconsistent definition of the '+' terminal. Don't ignore the possibility of nested or other multiple, interactive faults. Before settling on rewiring, be aware that the polarity (or 'phase reversal') switching on an active crossover can solve (or cause!) polarity errors at the push of one button.

Weak sound

Check LEDs indicating shutdown or malfunction/abuse have not illuminated.

Check the gain control settings, if any.

Check the power up/down muting system has un-muted. Try re-powering.

If the signal present LED is dim or out, check the input connections and cabling. If lit, check output connections and cabling.

Weak, distorted and intermittent sound

Check for whiskers of wire shorting inside plugs or between damaged conductors.

When tracing faults, be thankful for the redundancy of the typical audio system. Most sound systems have two or more amplifier channels, and two of almost everything else, invaluable for comparison. Work by tracing the signal, or absence of signal, from one, the other, or both ends of the faulty part(s) of the sound system. In a large system, time will be saved by checking in the middle too.

Don't forget that faults can be 'nested', i.e. two or more independent events have been known to occur. When this happens, a sure cause (like a bad connection) is found and rectified, but the problem persists.

References

1 Atkinson, John, *When things go wrong it hurts me too*, Stereophile, Feb 1989.
2 Pirsig, Robert., *Zen and the Art of Motorcycle Maintenance*, Hodder & Stoughton, 1976.
3 Duncan, Ben, *Ultra high power amplifiers: the rationale*, Proc. I.O.A., Reproduced Sound conference, Nov 1993.
4 Bacon, Mark, *Contact basics*, Speaker Builder, June 1990.
5 Duncan, Ben, *The mains issue*, Hi-Fi News, Dec 1991.
6 Duncan, Ben, *Black Box, about mains ecology*, 7 parts, Hi-Fi News, Nov & Dec 1995 & Jan to May 1996.
7 Duncan, Ben, *PSU regulation boosts audio performance*, Electronics World, Oct 1992.
8 Giddings, Philip, *The proper use of grounding and shielding*, S&VC, Sept 1995.
9 Lampen, Stephen.H, *Shielding the information*, S&VC, Sept 1995.
10 Duncan, Ben, *Loudspeaker cable differences, case proven*, Proc. IOA, Vol.17, part 7, Nov 1995. Also in Studio Sound, Dec 1995 and Stereophile, Dec 1995.
11 Newell, Philip, *Monitoring systems*, Pt.9, Studio Sound, Jan 1991.

Further reading

12 Don & Carolyn Davis, *Sound System Engineering*, Sams, 1975. Appendix III, *Recommended wiring practices*.

Maintenance and surgery

10.1 Classifying failures

When an amplifier needs repair, the cause is usually one of the following:

Electrical abuse damage

This is the most common cause when the less rugged kind of solid-state, mainly audiophile-grade domestic power amplifier, which stands little abuse, needs repair. The most common victims are output and driver power transistors and associated, fry-able parts, usually resistors. Such equipment commonly omits high temperature warning, thermal protection, effective DC protection, discriminative fusing, power up/down muting and other stress-relieving or protective mechanisms. When high sound quality is being provided at low cost, this understandable. Alas, the maker often neglects to remind the user of this. For 95% of careful owners, this has no consequence. However, if an amplifier is driven into difficult speakers, and backed up to a radiator; or driven relentlessly for several hours for a party, and then over-heats, there is no safety net, and it duly expires. This saves *n*-thousand careful or lucky users (say) £9.37 on purchase, but costs say 50 unlucky users *each* £120+, in repairs and shipping.

In some cases of electrical damage, it is endlessly debatable whether the maker or the user is at fault – irrespective of the local 'consumer' law.

Physical wearout

Excepting valves, little wearout will usually occur until an amplifier has been in use for over 5 to 10 years. The parts most likely to actually wear-out in a physically obvious way are pots, panel legends, switches and poor quality or roughly-used connector contacts. Less apparent, the most wearable parts inside include main reservoir electrolytics, (surviving) output semiconductors; and zener diodes and resistors that routinely run quite warm or hot to touch.

To parts that become hot, *cycling*, where equipment is turned on and off every day, causes wearout faster than continuous operation. For this reason, prudent owners of installed (*cf.* touring) systems leave their amplifiers switched on at all times. Sonic quality may improve as a result, but standby energy consumption also becomes an issue.

As an amplifier grows older, certain birthdays count. If you do not know the date of the amplifier's birthday, look for date codes on electrolytic capacitors, e.g. 3194 is a capacitor made in week 31 in 1994. So the unit won't be older than this (unless the capacitor is a later repair), but it might be a year or two younger, with parts lying in stock.

Weekly or monthly

Check and clean fan filters (mainly touring).

1st birthday and annually

If casework open, blow out dirt. If fitted, check and clean air filters. Check all connectors and switches. Check electrical and mechanical screws and nuts all 'firm tight'.

7th and 14th and 21st birthdays

Replace main reservoir electrolytic capacitors. Check other elcaps and fans. Replace fuses. All output stage and driver transistors may be replaced if they are BJTs, and have been cycled and/or hard driven. Expect some random failures.

Mechanical damage

All but a few power amplifiers are heavy for their size. At least they seem it in comparison to most other audio equipment, excepting some speakers and reel-to-reel tape machines. Even with all the weight-saving advantages of transistors, some professional and high-end power amps are amongst the heaviest electronic equipment packed into 19" rack-sized boxes, requiring two strong people to lift move them very far without handling aids.

By contrast, most electrical and electronic panel components are delicate. Even supposedly 'heavy-duty' controls and connectors can be snapped, distorted or otherwise trashed when a heavy metal box is dropped onto a solid surface, or just awkwardly handled. With a lightweight unit, if somebody accidentally knocks or kicks a switch or other panel object the unit will probably move back and the force will be dissipated, so quite delicate switches give no problems; but on a heavyweight amplifier, the box isn't going to move, so the component takes all the strain.

For domestic amplifiers, mounted on a deep and stable shelf, or otherwise secure *in situ*, and excepting little fingers, the worst physical abuse is likely to be arising from inadequate packing and transport by a careless carrier, in the event of moving house; or having to send the equipment away for fixing. *This shouldn't happen if the original packing is saved and used.*

10.2 Problem solving procedures

If the fault is mechanical, what needs remedying will usually be obvious enough. Otherwise the problem is almost certainly damaged electronics, and invisible as such. As with testing any audio electronics, power amplifier troubleshooting may begin with some combination of:

(1) Trying out and listening with different sources, speakers, cables, etc. Section 9.7.1 gives some examples of symptoms.

(2) Visual inspection.

(3) Critical component testing.

(4) Inter-channel or inter-unit measurement comparisons. In depth testing is ideal (see chapter 7) but is often not essential, if skilled listening can detect no problem.

(5) Provocation and monitoring. See chapter 8.

Procedures 2 and 3 are covered in the remainder of this chapter. They will deal with the majority of power amplifier faults, which involve outright component failures, rather than subtle defects.

10.3 Universal repair procedures

In the following sections, the repair of principally solid-state audio power amplifiers is covered. It is assumed that there are no schematics or other service manuals. If so, they take priority, but the following will still prove informative.

A suitable, creative merging of the information by the reader is expected, since in reality the active components may have diametric roles to the ones described (e.g. a MOSFET front-end to a conventional BJT output stage), giving more permutations than there is the space to deal with individually.

New, specialist items mentioned in the following procedures are described in the subsequent sections 10.4.1 - 10.5.3.

If an amplifier is said to be 'dead', begin by disconnecting the mains power and removing the cover(s). Allow 5 minutes to discharge. Check DC rails (across all large capacitors) and discharge with *bleeder* if above +½v. It's assumed for now, for simplicity, that all active devices are BJTs. If not, see also section 10.5.2.

Observations

In a naturally well lit room or using adjustable bench lighting or a small hand torch, make a careful physical inspection, looking for any of the following. This may require several separate passes:

i) Blown fuses. If glass bodied, eye inspection is often enough. If ceramic, and to be sure in all cases, check to see if there are any DC volts across them. If nil volts, *only then* check with a continuity or ohms tester. If a voltage is present, discharge it with a bleeder resistor – see end of section 10.4.2.

ii) Damaged or completely missing PCB copper tracks. Look for the telltale 'shadow', and the torn or melted copper edge at one or more 'ends'.

iii) Discoloured, browned or charred parts. Also a *nasty mess*. This could be signs of chemical emission from electrolytic capacitors and wound components; a decaying mouse; various kinds of 'biological waste'; or simply the remains of a spilled, sugary cocktail drink that has collected several years' dust.

iv) Any flux residue across tracks that is darker than the majority. If you can see any, gently scrape or pick it off.

v) Loose screws or nuts. Also shake the casing, looking out for free bits of wire, or any other loose parts, usually screws, offcuts of wire, hairpins and plectrums.

vi) Loose feeling connectors, switches, pots, etc. Wiggle all large PCB mounted parts including elcaps. If the soldered pins or leads seem loose, the joint is probably dry. Remove solder and remake.

vii) When archaic, sub-standard, fragile 'Paxolin' PCBs are used (as employed in nearly all mass-market products made in the Far East), look for cracks, particularly radiating from hard fixing points. Test potentially affected PCB tracks for continuity, repeating while flexing or tapping the PCB.

Decisions

If an obvious, minor cause has been found and sorted (such as the later items *v* to *vii*) and nothing catastrophic appears to have happened to the amplifier, pass along to '*First repowering*' (page 387). Otherwise, assuming signs *i - iv* are discovered, surgery will be needed.

Preparations

The first task is to remove the PCB(s). *At this point, it's important to make notes and sketch plans of any wiring that has to be unsoldered or looms that have to be unplugged.* Sometimes self-adhesive labels or a variety of different coloured spirit-based felt-tips can be used to code connections. Beware of washing such marks away later with a cleaning agent. Also take care to retain nuts and bolts removed during dismantling; placing them in a labelled, resealable plastic bag. Once the sick PCB(s) have been removed, the chassis can be put to one side for a while.

A de-soldering 'gun' will be needed, to help remove soldered parts with a reduced risk of damage to the PCB. Desoldering braid is an alternative that can cause less damage in some circumstances. When desoldering proves troublesome, a sharp *rap* can help disgorge deeply entrenched solder. If the copper track is lifted by the desolder gun's suction, try reducing the iron temperature, or use a smaller gun.

Cleaning-up

'Burnout' in an amplifier made of bipolar transistors is often literal, involving a component fire. Such fires are usually self-extinguishing but even when very localised, they can make, like building fires, a disproportionate mess. If there is charring and soot, the first job is to remove this. If done gently by brushing while the PCB is nearly vertical with an aqueous aerosol foam, or else an alcohol-ammonia mixture (i.e. *Hurd's No.1 Grimewash* – see section 9.35), this reduces the risk of loosing clues as to the value or part number of the charred parts – most helpful if no service data is available.

Burns surgery

Where the PCB is fibreglass and has caught fire, the hard surface is lost and the core turns into a mass of spongy, carbonised fibres. These are conductive, shorting all the connections they contact. *For safety, a burnt PCB should be scrapped.* Such a board should be used only for parts, alias 'cannibalisation'.

If the burnt area is small, the price of a new PCB (with parts mounted) is uneconomic or a new PCB unobtainable, or in an emergency, the board may be safely reused, *provided the carbonised area can be cordoned off, by scraping away all the loose carbon, and cutting away tracks that enter the zone.* A mini-fretsaw might be used to cut around the carbonised fibres. These tracks must be re-routed with suitable insulated wires, and any components will need reconnecting in mid-air, and securing with silicone compound, or more securely by mounting via fresh holes, drilled on an unburnt area.

Lost values

If the amplifier has more than one channel, then obliterated parts may be ascertainable from the other channels, with due care being taken if the layout is in *any* way different, even just symmetric. Failing this, don't guess. Contact the maker, importer or their agent for advice and information on service information and spares; or place an advert in a sympathetic magazine, newsletter or bulletin board, looking for either service information, or else sight of a working example from which values can be ascertained.

Rs and Cs

If for some reason spare resistors and capacitors are not available from the maker or agent, then not being able to obtain exactly the parts used is not an insuperable problem. If the unit is of audiophile quality, consult specialist audiograde part suppliers as to equivalents or upgrades on the parts used. Generally, you can replace a given wattage or voltage rating with parts having higher wattage or voltage ratings, so long as the % tolerance and quality of the part is also similar.

In most cases electrolytic capacitor values (in μF) can be considerably increased without concern about the limit, again provided quality is approximately comparable. Other, mainly non-electrolytic-type capacitor values and some resistor values in conventional power amplifiers are relatively critical, some (not all) within +/–10% or less. These should not be varied unless there is good reason. Some resistor values are more likely to be critical in amplifiers with low or nil NFB.

To be safe, resistor wattages and capacitor voltage ratings should never be reduced, unless it is known that the maker had over-rated the original part for example, to overcome a sourcing difficulty. It is unlikely that a maker will fit anything over-rated except under the shadow of a prior bad experience.

Output Network

The OPN (OutPut Network) is a perennial weak spot in semiconductor amplifiers. The resistors here can smoke, burst into flames or even explode if driven with sufficient HF/VHF level. It's easily done by accident when playing a test CD, or sweeping a test generator. Or RF may be present on the incoming signal.

In amplifier servicing, discovering slightly singed zobel resistors is not unusual. They are best looked at, whenever amplifiers are overhauled; and also after anyone discovers that a faulty console (or preamp or other equipment) has been driving RF up the system. Thoughtful makers will place the Zobel resistors where they can be readily inspected and replaced. Ironically, once the Zobel network is at all damaged, the stabilising resistance value usually increases, *thus the amplifier is all the more likely to go unstable, generate its own RF output, and finish off the burnout.*

Transistor testing

After re-instatement of damaged passive components, the next step is to test the transistors. As the destruction generally begins in the output stage, it is most productive to test by working backwards from there.

Testing bipolar transistors is most reliably carried out with them isolated from the circuit, i.e. removed. *This does not mean it is essential, just preferred.* If removal is impossible, some testing can still be accomplished. In either case, basic *Go/No-Go* testing is often all that is needed. See section 10.5.1.

With TO3-cased power transistors, be aware that even with the base and emitter wiring removed or unplugged, a low resistance can occasionally occur if the leadouts are unsleeved and therefore capable of touching the thickness of heatsink metal through which they pass; or if solder or wire slivers held in place (and possibly hidden) by thermal grease, are making devious contact.

Other active tests

Diodes, zeners and op-amps may need testing also. Op-amps are rarely implicated except as prime causes (very rare with good, modern ICs), or input overdrive or fire victims. If there is no evidence of an IC having been harmed, and especially if the PCB has plated-through holes, then it is best left alone until everything around it that is more readily tested has been proven good. If replacement is necessary, always replace with a high quality IC socket so the problem of desoldering does not recur.

On a D/S (double-sided) PTH (plated through holes) PCB, *do not attempt* to de-solder an IC unless you have the proper vacuum tool. Instead, nip the legs flush to the board, very gently. Then there is a slim chance of being able to use the IC again. The amputated legs can then be desoldered and removed individually.

As two legged parts, zeners and diodes are easily tested in isolation by desoldering and extracting one leg alone. See section 10.5.3.

Spares sourcing

Resistors, capacitors, diodes and zeners occur in a limited number of internationally agreed values (e.g. 400mW, 500mW, 1.3w and 5w for most zeners) and have simple specifications which make substitution anywhere in the world straightforward if the need arises.

Transistors have far more parameters. There are no standards but due to the use of negative feedback, and other error-correction schemes many are 100% interchangeable. Appendix 3 lists the key data on many widely used audio BJTs. For more information, the standard reference is Towers [1].

Except where designs have been pushed to the limits, or where very close performance matching is essential, replacement transistors of different makes and types having notionally similar specifications often show little if any audible or measurable deviation in performance. Differences may be practically indefinable – except possibly over a large population. This is a useful benefit of amplifiers which use some global NFB. Then satisfactory replacements or equivalents can be obtained from the various specialist audio and electronics parts suppliers.

Special semiconductors

For a number of professional and high-end designs, where the maker has selected special parts, and even gives them special numbers, replacement transistors are best first sourced from the makers or agents. Special selection allows high performance to be attained consistently, despite low or zero global NFB. The selection process raises costs though.

Imagine 1000 transistors are tested for 6 things. Imagine 10% are rejected in each test. That leaves 53%. The price each has now nearly doubled, for any given % profit. Now imagine one additional test, the yield is a far from unreasonable 5% (of the 53% or 531 remaining). Now the price for the 26 items would need to be thirty eight times the original cost. The upshot is that High-End semiconductor spares are special and can cost plenty. Provided the maker can convince you that he has done the selection testing, and that the out-of-spec devices cannot be sold to amortise the selected part prices, then very high prices are a statement mainly about the wide tolerances of semiconductors and the benefit of NFB. These same high prices, and the ease with which even experts can occasionally vaporise silicon by 'one inadvertent mistake' makes it safer for all parties' feelings, that such repair work is carried out by the maker or retained agent.

A few makers have abused the system. The transistor numbers on one well known British PA amplifier were re-marked with the company founder's car registration number, but the transistors were quite ordinary.

Allowable equivs

The key voltage rating is V_{ceo}. This should be at the very least 10% higher than the sum of the supply rails, e.g. +/-45v rails sum to 90v, so a V_{ceo} of 100v would be (just) safe. A higher voltage is preferable and any higher rating is ok on this basis, but going too high (more than x1.5..x2 the minimum) may compromise other parameters. V_{cbo} is a common voltage rating condition that may appear if V_{ceo} is not shown. It can often be 20% to 30% higher than V_{ceo}, so derate accordingly, e.g. If V_{cbo} is 120v, V_{ceo} may be under 90v.

The main current rating (in mA or Amps) is I_C. If many transistors are paralleled for redundancy, a slightly lesser I_C rating may be passable. Otherwise it must be the same or higher. Again, too much higher (say 1.5 to 2x or more) will likely conflict with other parameters. If there is any doubt, the SOA curves of the original and replacement parts should be located and compared.

F_T (transition frequency) is an indication of open loop frequency response and 'speed'. It is a relatively loose parameter. If the f_T of a replacement part is more than 50% lower or higher, RF stability and/or HF distortion may be affected. This might be latent until a particular combination of load and temperature occurs. So picking a part with a similar f_T (within 50%,150%) is advisable. If not, choosing a higher f_T will also avoid increased distortion or limited power gain at high audio and ultrasonic frequencies.

H_{FE} (or H_{fe}) is similarly loose, +/−50%. If comparing the values, note the test current, which must be roughly similar (in A or mA) to make the comparison. It is more useful to match any new parts that are being paralleled to each other and to the H_{fe} of any surviving original parts, rather than meeting the relatively arbitrary textbook specification.

If a transistor type is unobtainable, unknown, or far too expensive for no apparently good reason, a knowledge of the circuit can allow a standard type to be designed-in. Otherwise, spares may be had from cannibalised equipment. However, if the equipment is over 10 years old, then with output transistors (and other wearable parts like elcaps), there is a risk of fitting parts that are entering their wear-out phase.

Reinstatement

Before replacing power devices, clean down their mounting positions with *Grimewash* (page 349). Also ensure each mounting area is smooth and level. Any roughness must be gently ground or rubbed away. Old insulating micas or rubbers are best thrown away. If re-use is the sole option, clean and inspect carefully. Rubber insulator pads shouldn't show signs of stretching or tearing. Micas should be cleaned by shaking in a jar of solvent and inspected for tiny holes and splittage. Unless these occur on the protruding edge, the mica is unsuitable for re-use.

Until recently, white zinc-oxide-loaded thermal conductivity compound or 'grease' was an essential, messy nuisance in any high performance output stage assembly. It is needed to fill in the microscopic hills and valleys, between the mating surfaces, to greatly improve heat conduction between surfaces that appear flat to the human eye. If it isn't available, silicone grease or even car-grease can be used in emergencies as inferior substitutes. But as Zinc Oxide compounds are inexpensive, substitutes are not recommended.

In recent years, rubber-based insulator washers have been developed which don't employ or require grease. The best types can (at the time of writing) slightly outperform greased mica. These types are highly recommended, since Zinc Oxide grease is unpleasant to handle (though of low toxicity) and a regular cause of further fatality – as it can hide particles that prevent proper seating , and wire slivers, that can cause shorts at a later stage. For these same reasons, if grease *has* to be used, don't be tempted to 'economise' and re-use what is around by re-spreading it. In any event, spread it *thinly* exactly where it is needed. A flexible, glue-spreading spatula is best. Only a few thousandths of an inch is needed. Any excess – which then exudes from the edges and spreads all over – must be cleaned away.

Power device insulation also depends on nylon bushes. Replace them if they are crushed or damaged in any way that reduces their insulating thickness or integrity. When heatsink-mounted transistors are re-mounted, correct tightening is essential. A torque wrench should ideally be used, set to between 60 and 90cN×m, or 6 to 8in-lbs or else tighten by hand so the tightness is 'slight firm', i.e. not particularly tight. Be aware that initial tightness relaxes after a few hours. Preferably after a few hours, go over all the screws to check the tightness is about the same. Nail varnish may be used to lock after final test. Ideally, correctly sized Belleville washers should be used on all power devices to preserve the optimum torque band.

When replacing the remaining, usually PCB-mounted semiconductors, take care not to stress them. The leadouts of metal can parts are not generally strain-relieved inside. They are therefore advisedly mounted on plastic (and stood-off by) spacers. Be sure to re-use these. Plastic cased transistors are more rugged in this area; but all transistor leadouts are easily fatigued if bent hard (>45°) or more than once or twice: look for tiny cracks if this happens. Such a fatigued part is good only for scrap. Some PCB mounted parts are mounted on or bear small heatsinks. Often, this class of heatsinking is un critical. If in any doubt however, treat as being as critical as the output stage, use heatsink compound. Particularly if the parts are plastic, torque correctly.

Other checks

When bipolar output stages require repair, certain associated resistors are commonly implicated. It is possible for these to appear undamaged when they are actually open circuit or have 'gone high resistance'. Check the following:

(i) Low value (0.1 to 1 ohm) OPS sharing resistors, typically 1 to 20 watt rated, and usually wirewound, in line with emitter or collector terminals (R_e in Fig 4.14). Usually mounted close to the output devices.

(ii) Low value (10 to 220 ohm) 'flushout' resistors between OPS base and emitter (Fig. 4.6). May be close to the output devices or on the driver PCB.

(iii) The resistor(s) in the output Zobel network(s) (Fig. 5.32). Usually 5 to 20 ohms, 1 to 20 watts, may be carbon composition, metal or carbon film, or wirewound.

PSU pre-testing

In most high performance amplifiers, damaged power supplies are remarkably rare and *non-coincident* with blown output stages. Still, before proceeding, prudence dictates that the PSU should be checked if it is easily disconnected from the repaired amplifier circuitry. Often this is easy, because the two, 'hot' DC supply wires (usually red and black – but not always) are either pluggable (e.g. at the bridge rectifier) or can be un-clipped or un-screwed (e.g. at the elcaps). Be sure to insulate all bare, loose wires!

If there is no other means of isolation and test connection, there may be DC rail fuses. These may be removed for isolation. Section 10.4.2 outlines gentle powering up techniques, including use of a Variac. If when testing the power supply, the Variac'd AC supply sags, or the series bulb lights, turn off and check for:

(i) correct AC voltage setting.

(ii) healthy reservoir capacitors - no signs of chemical leakage. If they have been disconnected, are the wires replaced in 100% the correct places and to the correct polarity terminals (irrespective of physical orientation) ?

(iii) healthy rectifier bridge. It is assumed the DC feed out from the bridge has been disconnected. Then if the bridge is internally shorting, the high current will cease when the AC input to the bridge is disconnected. In most cases, this means isolating *both* feeds.

If the problem is not found in the above, but persists, the mains and PSU circuitry should be isolated and systematically checked with a DOM (or other ohm-meter) for shorts or insulation breakdown.

When using a Variac, this is a good instance to determine and note the minimum supply voltage at which the amplifier will turn-on.

All is well if the power supply (after being isolated) can be powered up and delivers a DC voltage that is in proportion to the AC input. If a lightbulb is used, the AC input voltage on the bulb's output side will need monitoring. If the supply is say 190v, then relative to 240v, the DC rails would be about 79% of normal, eg. +/–35v, if normally +/–45v.

After testing, discharge the supply (see end of 10.4.2) and reconnect it to the output stage and/or reservoir capacitors.

More caveats

'If it ain't broke - don't fix it'

In a stereo or multi-channel power amplifier, one or more channels may be unaffected by damage. To protect these from unnecessary stress and disturbance during testing, you may choose to isolate the DC power to this/these, until after the repair is proven. If powered by wholly separate power transformers, it may be simpler to disconnect their AC input feeds.

Bias presetting

Power amplifier biasing is covered in chapter 4, notably 4.2.5, 4.3.3, and 4.6. Class G and H amplifiers are biased as class B,A-B or A. Under-biasing isn't harmful but sonic quality will be degraded, most at low SPLs, and at HF. Over-biasing can be fatal to amplifiers with BJT output stages. However, with cautious setting, there will be no problem. Turn preset slowly and pay attention.

Before proceeding, the biasing preset should be identified. This will be present in most Class B and A-B -based designs, and absent in most Class A designs. If there is no information, the preset is the one nearly always coupled to a small signal, NPN bipolar transistor, connected as a V_{be} *multiplier* or V_{bex}, which is usually connected to the main heatsink and electrically between the driver transistors – or other stage, immediately after the VAS. See section 4.3.3.

If there are a number of presets *do not guess*. Either get the maker's information; or work out which does what by tracing-out the surrounding circuitry.

Optimum biasing of bipolar transistors requires a harmonic distortion (%THD+N) analyser. The bias for a new set of transistors will almost certainly require resetting. Failing this or any other distortion test equipment, it is possible to set biasing by iteration with just a milli-ampere meter. First, turn the preset so bias is at minimum. In most cases, this means turn the preset 'down', i.e. CCW. This is not always true. If unsure, and there are no diagrams or PCB legend giving this information, set halfway, soft power the amplifier (via a lightbulb) and immediately establish at switch-on whether current is reduced by turning CW or CCW, by watching the bulb's intensity.

Biasing of amplifiers with MOSFET output stages is usually class A-B or A rather than B. In practice, this means at least 25mA and up to 100mA for *each pair* of MOSFETs. As biasing is relatively non-critical, there may be no preset. There is then nothing to worry about so long as the heatsink on the repaired channel runs (when quiescent) at a similar temperature to another, intact channel. 15°c above ambient is usual.

First re-powering

In the following initial tests, it is assumed that no signal is being applied and there is no load. If a bulb alone is used (page 391), current can be crudely monitored by noting its brightness. Apply the power. If after a brief build-up to brightness, the bulb dies back to a dull glow, if any at all, all is probably well.

If a Variac is use, bring up the voltage slowly to the minimum established (above). If the series bulb again initially lights, but dies back down, all is well. If the bulb stays lit, turning the bias preset down in the previously determined direction, should reduce the brightness. If it does not alter it, there is a fault. Turn off immediately and recheck the repairs and any re-made connections. Likely causes are a reversed connection, or a missing or wrongly placed insulator.

Safe bias tactic

Once (with AC power applied) bulb brightness can be controlled by varying the bias preset, accurate metering is worthwhile carrying out. Switch off and allow the DC rails to discharge. Disconnect the +ve wire to one amplifier channel and insert a mA meter, or else a low resistance (<1Ω) current sensing resistor. Rating should be at least 10 watts. Connect a voltmeter across it. If the resistor is 0.1Ω, the scaling will be 0.1v per Ampere. Make sure connections are firm. In either case, the bias current can now be viewed.

In the absence of a harmonic distortion meter, power up again and adjust the bias preset for about 30mA. This is a typical, safe value for many Class A-B amplifiers. *Now leave the amplifier to warm up for 30 minutes.* If the bias hasn't drifted more than 10% (ie. not <27mA nor >33mA) you can bypass the lightbulb, and re-power. If the bias has drifted up, the lightbulb will light to restrain it. If this happens, or the heatsink is hotter than usual, turn down the bias to say 20mA, and leave for another 20 minutes.

If one side of the amplifier is working, a suitable thermometer, or your finger (take care !) may be used to compare the heatsink temperatures. Assuming separate and symmetrical heatsinks, it is reasonable to set the bias so the heatsink temperature in the repaired channel matches the other channel after say 30 minutes of settling.

If all is still well, and the current hasn't increased more than a percent with the extra voltage, power-down and disconnect the mA meter before testing into any kind of load ! The amplifier can now be tested with a realistic load, and at high drive levels, including with music into a speaker. If a small (<<1Ω), suitably rated resistor (current shunt) has been used to read the current, it need not be disconnected immediately (in case a final bias check is needed), as it will not be harmed.

DC nulling

Many power amplifiers, even if not direct-coupled, have a DC offset nulling preset per channel. With no signal, test for DC offset (see section 8.6). If more than +/– 50mV, or if it causes an unacceptably loud click each time the amplifier is turned on whilst connected to the speakers, identify the DC offset nulling preset. If there is no diagram, the required preset is nearly always located near to, and associated with, the input differential pair(s) (LTP: see 4.3.2), and symmetrically connected.

If new parts have been fitted, allow as many hours stabilisation as you can afford before final trim of DC offset, as the rate of drift will slow appreciably with 'bedding in' time, measured in tens if not hundreds of hours.

Some power amplifiers employ DC servos (3.4.1). Their *raison-d'etre* is that they do not require adjustment. Most employ an IC op-amp which may have been selected for low offset voltage (V_{oos}). If official spares are unavailable, buy several ungraded parts and try each, while measuring the output offset.

Soak test

To complete, re-assemble and connect the amplifier to the normal mains supply, i.e. Variac set to local mains voltage and lightbulb switched out. Then arrange a music signal that peaks just below clipping, continuing for at least 12, and preferably 24 or more hours. An appropriately tuned radio receiver is the easiest source. If a suitably high power rated dummy test load isn't available, use enough electric fire elements or high wattage AC mains light-bulbs in parallel, with suitable, safe heat-radiating arrangements, to load the output with about 8, 4 or 2 ohms. Or, drive into a heavy duty old loudspeaker, packed into a padded, fireproof box.

Limits

With only basic test equipment (handtools and one or two multimeters) this is about as far as we can go. *The most important additional instrument is a scope. Without it, RF stability cannot be positively identified.* See section 10.6.

10.4 Repair tools and equipment

The equipment described in this section is of the immediate practical kind needed to get a power amplifier apart, to make good damage, fit new parts, and prove basic operation. Some of the more specialist equipment required for fuller testing and in-depth characterisation is described in chapters 7 and 8.

10.4.1 Useful tools

These tools (and quite possibly others) will be useful, although not always essential:

1 Screwdrivers, flat bladed, a wide selection, from 1.5mm wide jeweller's to 15mm wide engineer's.

2 Screwdrivers, *Pozidrive* (often abbreviated 'pozi') sizes 00 to 3. May use a ratchet, electric or pneumatic driver for rapid lid unfasten/refastening.

3 Screwdrivers, stubby types, bladed and Pozi, ratchet useful.

4 Wire cutters, various suited to diameters ranging up from 0.5mm² PCB precision, up to ½" (12mm). Mini and standard hacksaws and some hole punches and drill bits (to cut control shafts, modify metalwork). Hammer and set of small chisels (to remove rivets and wrecked screws).

5 Pliers, ranging from miniature snipe-nosed, long-nosed and square jaws, up to standard mechanics' type.

6 Spanners, ranging from M2 and 8BA to M12 and 0BA, and a few sizes above. Ring, open and ratchet types. Also nutspinners and nut drivers to fit onto ratchet driver. Ditto, onto torque driver. Alternatively, a suitable ¼" drive socket set can cover all the requirements in one go.

7 Clampers, ranging from miniature crocodile clips, surgical seizures and miniature, long-nosed 'mole-grips', up to engineers' bench vice in various sizes.

8 Soldering and desoldering equipment. Temperature controlled iron of about 40 to 60watts with a range of bit (tip) sizes. Ideally all low voltage (12 or 24v) AC powered. 22 and 18swg (gauge) 60/40 (tin/lead) solder. Also HMP solder for test loads, and LMP solder for SM PCB repairs. Large and small desolder guns and spare ends. PCB poking, scraping and track slicing tools.

9 A miniature torque driver, for setting power semiconductor screws. A trim-tool (or very thin bladed screwdriver) for presets.

10 A 4"/100mm paintbrush to clear debris away. Old toothbrushes. Hurd's Grimewash (9.3.4). A pressured airline, *which must be clean and free from water and without an oiler.* Blasts dust off PCBs and out of heat exchangers. Blasts broken glass (care!) out of jammed-up fuseholders.

11 A collection of rubber and heat-shrinkable and other insulating sleeves, tapes and boots.

12 A collection of common fuses.

13 Collection of spare screws and associated items for lids and power device mounting.

14 One or more meters, multimeters or DMMs, and selection of leads with croc clip ends and probes.

15 A scope. At least 10MHz bandwidth, dual trace.

16 A signal source, at least a sine generator capable of 0 to 5v rms, 20Hz to 50kHz, and square wave and pink noise signals also, if possible.

17 A mains 'quick connector' or 'rat-trap', for rapid and easy connections for amplifiers which have bare-ended mains cables.

18 A Variac. (10.4.2)

19 A series-connected mains lightbulb. (10.4.2) Typically 60w up to 500 watts.

20 A DC test supply. (10.4.2)

21 A test load. (7.3.3 and end of 10.3)

22 BJT go/no go tester (10.5.1). MOSFET go/no go tester (10.5.2).

() indicates section where described..

10.4.2 Test powering tools

Initially applying full mains voltage to a tentatively repaired power amplifier isn't recommended. It is possible that damage or crucial faults have been missed. Then, within the time it takes for a fuse to blow, the mains supply can deliver *kilo*joules of destructive energy, and hours of repair work can be vaporised before anyone can react.

Using a Variac

To greatly lessen the risk of damage, the ideal approach is to use a *Variac*, while monitoring the supplies. Most power amplifier main rails are unregulated. So rail voltages will be *pro-rata* for input voltages lower than the line voltage that is set. If the line voltage is set at 240v, then for a 48v AC input, the rails will be 20% of their norm, ie. 15v if normally 75v. This 'creep up slowly' approach works provided there are no auxiliary and soft-start circuit parts (eg. relays) that need higher rails, as a minimum to turn-on.

In any event, when the mains voltage into the amplifier under test is slowly raised from zero, any problem causing an abnormally high current flow will be immediately evident with a little experience, as the Variac suddenly or progressively buzzes more loudly than usual, the DC rail voltage (which should be monitored) stops

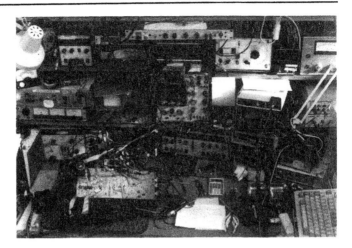

Figure 10.1

The author's No.3 test workstation, with an amplifier PCB under investigation. Equipment on the top shelf, from left to right comprises (on the right of the lamp) an RF signal generator; a lab power supply used to test zener diodes; stacked frequency meter & function generators; HCMV, balanced scope-probe amplifier balanced on top of a DMM; auxiliary audio generator; 20dB line amplifier; and part of an auxiliary % THD meter. On the lower shelf are a loudspeaker voice-coil thermometric test set, on top of DC test supply, with 2 voltage and one current meter visible; three 'scopes; a roll of cleaning tissue; another DMM; and part of an Audio Precision System One test set. On the bench are (left) two Variacs with a 'rat-trap' quick connector block, four lab power supplies, two DMMs stacked, soldering station with AES pink noise generator below, and far right, the ubiquitous PC keyboard hovers. Also visible are three bench lamps, hooks to keep cables almost tidy, anti-static rubber bench mat, carpet, notepads, calculator and hand tools.

rising, and faint arcing can be heard or felt, as the voltage setting is varied. If this happens, the Variac can be immediately backed off and usually, no damage will be found to have been caused.

The Variac must be adequately rated and fused; repairing a Variac is not the aim.

A minority of highly specified amplifiers (e.g. BSS Audio's *EPC series*) have automatic tap setting at switch on. The amplifier will set up for the mains voltage at this instant. This type should *not* be 'brought up gently' using a Variac.

With a light bulb

Whether or not a Variac is available, a series-connected mains light bulb may be used to limit excess current (Figure 10.2). With 240v AC for example, an everyday 100 watt bulb looks like 576Ω when 'driven' at 100w. Cold resistance at first switch

Figure 10.2

Test Lamp Wiring.

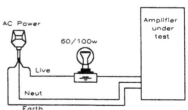

on is about ten times lower, enough to have a useful surge-limiting effect, being about as much as a soft start resistor. After that it increases rapidly to the hot value, as substantial current passes (over say 100mA). The resistance variation with current is complex. Suffice to say that a bulb of this or adjacent rating will in practice light vividly and nearly always prevent damage, if a high current is drawn, with amplifiers rated from say 40w⟹8Ω up to 600w⟹4Ω.

This connection is *not* suited to testing class A amplifiers other than of very low power; or driving test signals at high or even medium levels into real or dummy loads. Otherwise, full swing testing of Class B, A-B, G and H amplifiers, is feasible.

In some test conditions, the lamp may be on for a while and will thus run at normal high temperatures. It is also vulnerable. It should therefore be caged or boxed to prevent meltable or flammable items touching it, or breakage.

DC test supply

This comprises an unregulated, usually dual, HT power supply (typically +/–40v to +/–160v), possibly part of a old power amplifier, with enough voltage and current rating, to power whatever is being tested. This supply is used to first power a tentatively repaired amplifier. Unlike the repaired amplifier's own power supply, to minimise potential damage, the test supply will have small reservoir capacitors (e.g. 1000µF) and relatively low value, high wattage bleeder resistors(e.g. 1kΩ). DC power may be applied through a high current switch, initially via 100 ohm, 50 watt series resistors. Analogue voltage metering after this point clearly shows whether too high a current is suddenly being drawn, by dipping.

Bleeder

Unwanted energy ('voltage') stored in capacitors is discharged rapidly with a high wattage rated (e.g. 200w) resistor with a value of a few ohms, fitted for safety with a series 'T' fuse, say 5A rated. The resistor may be mounted on a heatsink and has long leads with crocodile clip on one end and a probe on the other.

10.4.3 Audio test tools

Test load

16, 8, 4 and 2Ω nominal loads for soaking and go/no go power output tests may comprise adapted heating elements, rated at or above the highest power test capability likely to be required in the lifetime of the load, e.g. 750w for domestic amplifiers; 5kW for pro-amplifiers, per channel. Precision loads are considered in section 7.3.3.

Even a 'rough and ready' test load must be safe for prolonged and possibly unattended use, by provision of caging, firebrick surrounds, overheat cutouts and alarms, and suitable hot air venting. If cooled and monitored by fans and electronics, a mains-power sensing *no-volt relay* should be used to prevent connection of test signals when the fan cooling is not able to operate.

10.5 Testing components

10.5.1 Testing BJTs (bipolar transistors)

Testing Bipolar transistors is most reliably carried out with them isolated from the circuit, i.e. removed. In nearly every case, basic *Go/No-Go* testing is all that is needed. This can be accomplished most easily with the simple tester circuit shown in Figure 10.3.

Figure 10.3

Bipolar Transistor Go/No-Go Tester.

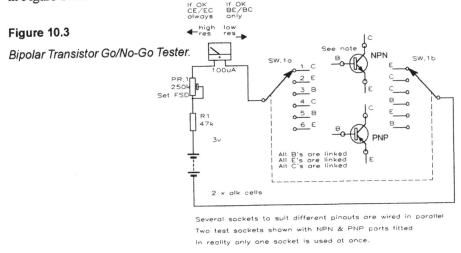

Go/no-go testing

For good transistors, two of the lower four of the six switch positions, either B+E– and B+,C– (for NPN) and B–E+ and B–C+(for PNP) give a low resistance i.e. FSD type reading. *All other positions should show high resistance* (i.e. no or very little needle movement), unless the transistors are germanium types, in which case rather more leakage deflection is expected in one or other C/E test positions.

Meter choice

Alternatively you can test with a DMM (or DOM) on a low ohms or diode test range. *If you haven't done this before, use another voltmeter (ie. DVM) to check that there is at least 1v but less than 5v on the meter prods when switched to the intended range. Some meters output over 5v or 6v on some ohms or diode test ranges. This is most useful, but could damage or degrade BJTs by zenering their B-E junction - even if it is harmless to diodes and most other parts. Preferably mark such ranges and avoid their use for BJT testing!*

The following table shows the readings you should get. '+' means the lead that is positive and usually red. On most analogue (mechanical movement) meters and older DMMs, this is the –ve (black) terminal's side. In this case, it may be best to swop the leads, so the lead to this *effective* +ve socket is the red one. 'Low' means conduction.

Out-of-circuit readings for healthy BJT transistors

Applied Polarity	Reading - Ω (ohms)	
	NPN	PNP
B/E: + base, − emitter	low	high
E/B: − base, + emitter	high	low
B/C: + base, − collector	low	high
B/E: − base, + collector	high	low
C/E: + emitter, − collector	high	high
E/C: − emitter, + collector	high	high

'High' means at least 2MΩ for most silicon BJTs.

Tips

If the leadouts aren't known, it's reassuring to know that if a device only reads low in only two directions with respect to one terminal, then it is almost always ok, and that terminal is the base. Discovering which leg is emitter or collector is more tricky unless you know, or can work out, whether the part should be npn or pnp, from its circuit position.

Blown devices will nearly always show low CE/EC resistances. But other patterns are possible. Audio power amplifier transistors have been exclusively silicon for over 25 years, and any leakage enough to reduce 'high' resistance to below 1MΩ is suspicious, but check first that the transistor and socket is clear of grease and dirt. *Any touching of probes with bare flesh will give worryingly high leakage.* A hot soldering iron placed against the part will cause true, semiconductor-caused (*cf.* external contamination-caused) leakage to shoot up. As the transistor cost is out of proportion to the price of further damage, it is best to secure a replacement if testing proves leakage. If this too proves leaky, it may be either an acceptable characteristic or else a bad batch − consult the maker.

In circuit go/no-go

Transistors *can* be tested using the low ohms or diode test range of a DMM (or DOM), without removal. But this route can be more uncertain. Power is assumed to be off and all reservoir capacitors drained - allow 5 minutes for this. There will be considerable leakage through other parts where a high resistance reading would otherwise be obtained. So it is best to concentrate on looking for low CE *and* EC resistances, i.e. low in both directions. Important: See 'Meter Choice' above.

Active probing

If an amplifier shows no signs of catastrophic damage, or it is believed to have been fixed, it can be actively probed. This means checking the voltage across each BJT's B-E terminals.

First, arrange the amplifier *so mains and other high voltages exposed by removing the covers are suitably shielded with insulating materials and warning notices*. Soft powering (see 10.4.2) is assumed. Bring up the mains supply to a fraction, but at least 10 to 15% of the norm and more, if necessary to give at least +/-15v DC rails. Read both DC supply rails. When they have settled, read the three inter-terminal voltages for each transistor. In every case, one should lie between 350mV (for large power transistors running at low current) and 800mV (for a small signal transistor heading into saturation). These are extremes. The active, forward V_{be} range for most BJTs in power amplifiers at around 25°c is 450mV to 700mV, and usually lower with high power-rated devices.

H_{fe} matching

Figure 10.4 shows an elegant, simple scheme for matching of BJT pairs for H_{fe} [2]. The bane of simple H_{fe} testing at moderate to high current levels is junction temperature rise, which alters the figure as it is being read. This tester overcomes most of the bother this causes.

Figure 10.4

Beta Matching for BJT Complementary Pairs.

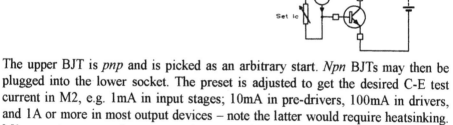

The upper BJT is *pnp* and is picked as an arbitrary start. *Npn* BJTs may then be plugged into the lower socket. The preset is adjusted to get the desired C-E test current in M2, e.g. 1mA in input stages; 10mA in pre-drivers, 100mA in drivers, and 1A or more in most output devices – note the latter would require heatsinking. M1, a centre zero analogue meter (or auto-polarity DMM) then shows the state of balance. If H_{fe}s are matched, M1 will read 0v.

The preset sets the base current by programming a current-source (or mirror) connected between the bases. The meters will require over-range protection if analogue. The batteries may be replaced by lab-type supplies, of adequate current rating, with adjustable current limiting.

10.5.2 Testing MOSFETs

MOSFETs are assumed to be L-MOS power types. The testing of D-MOS types and most small signal types is identical, except that the on-resistances will generally be lower, and the gate threshold higher (4v minimum).

Output stage MOSFETs are usually employed in parallel. Testing can be *in situ*. Allow time for full discharge and ideally disconnect any gate drive wires.

Test 1: Gate-Source integrity

After shorting the gate to drain for a second (with a jump lead) to positively remove residual charge, the resistance between the source and gate terminals should read nearly infinite, that's >>1M. Less than 1M (sometimes called a 'soft' or blown gate) usually implies one or more dead devices.

Care! If the heat-exchanger assembly is damp, or the amplifier has been recently removed from a cold and/or humid atmosphere, high resistance checks may give misleading results. Use a hair-drier to dry off. Another source of indeterminate or intermediate resistance readings is the sooty and partially conducting residue, possibly deposited on the underside of the PCB, after a fire in adjacent parts. Check carefully for evidence of soot. After cleaning with *Hurd's Grimewash* [9.3.4], again dry off before testing.

If a low G-S resistance is confirmed, snip the gate leg of the offending part to take it out of circuit, to permit continued *in situ* testing of the remaining MOSFETs. Assuming the gates have series resistors the guilty part(s) will exhibit the lowest of the low resistance readings. If the gates are not individually isolated by gate resistors, all but one gate lead in each array will have to be 'lifted', i.e. desoldered and isolated from the PCB, to uncover the culprit.

Two Ohmmeters

For the subsequent tests, two different ohmmeters are required, as seen in Fig 10.5.

Ohmmeter 1 (M1) has to source ('put out') a voltage of at least 1½v (up to at least 4v for D-MOS)*. Voltages in this range are commonly available from analogue test-meters switched to the 'low ohms' ranges, though the lead polarity is often reversed, relative to the Volt and Amp ranges, as described in section 10.5.1. Digital test-meters also source a suitable voltage when switched to either the 'Diode test' or even 'Continuity' function, and also on some 'low ohm' ranges. With digital multimeters, the voltage source polarity is usually identical to that implied by the leads, ie. red lead is +ve. If in doubt, use a second meter (switched to DC Volts) to check the polarity. **Note: The very first power MOSFETs (Yamaha's audio type) required up to 25v on their gates to turn them on [3].*

Ohmmeter 2 (M2) can source a voltage, but it's not essential to do so. Ideally, it should be wide-ranging, and auto-ranging, from <<1 ohm up to >1M. Failing this, you will need to switch between these ohm ranges for each test. Also, if an analogue test-meter is used, the leadout polarity of the low ohms range should be confirmed. Otherwise the internal reverse diode may conduct (fig.4) and cause confusion.

Test 2: Discharge

Connect M2 between drain and source (D-S).

If n-channel (2SK), connect end that will be +ve on the low ohms range, to the drain.

If p-channel (2SJ), connect end that will be +ve on the low ohms range, to the source.

If the gates have been discharged, the MOSFETs are biased off and resistance should be high, >>1M. In a humid or static atmosphere, the resistance may creep downwards after a few seconds or minutes if the gate/source are not kept shorted. A typical reading is between 5MΩ and 50MΩ.

If the humidity is low and D-S resistance reads below 2MΩ, despite attempts to obviate leakage paths, it's likely that one or more MOSFETs are defective. Again, the individual devices can be investigated by unhooking the gate resistors from their feed rail.

Test 3: Charge

With the MOSFETs still discharged, leave M2 in place and connect M1 to gate and source (G-S) to turn on the gates:

If n-channel (2SK), connect end that is +ve on the low ohms range, to the gate.

If p-channel (2SJ), connect end that is +ve on the low ohms range, to the source.

The MOSFETs should now turn on, and the D-S reading on M2 at room temperature should drop to below about 1 ohms per n-ch FET (if LMOS single die) and 0.2 ohm or less per n-ch FET, if D-MOS. Thus 4 x L-MOSFETs should read about 0.25Ω, give or take 50%. Commonly, p-channel FETs exhibit up to double the resistance of their n-ch complements.

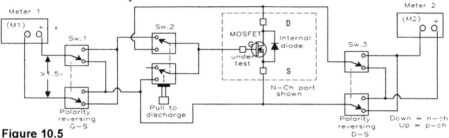

Figure 10.5

MOSFET Go/No-Go Tester.

Figure 10.5 shows a simple MOSFET go/no-go circuit which performs tests 1,2,3, with the addition that the gates can be reverse biased by reversing SW.1. This should definitely show a high D-S resistance, >>1MΩ.

10.5.3 Testing diodes, zeners and LEDs

As two legged parts, zeners and diodes are easily tested in isolation by desoldering and extracting one leg alone. Testing requires a current-limited source (usually a dedicated bench supply) with enough voltage to turn the tested part hard on.

Diodes

Diode testing is like testing BJTs. In the *forward* direction, and assuming silicon (Si), there will be a low resistance sustaining 600mV (within +/−150mV) if actively tested. In the other, *reverse* direction, very high resistance, ie. >1MΩ – unless the diodes are germanium or Schottky types, in which case leakage to say 200kΩ is plausible. These types typically drop between 250 to 400mV. The drop rises with test current. With germanium (Ge) diodes, used for some better V-I limiters and in RF detectors, the forward drop falls considerably with even modest increases in temperature, including junction heating from surprisingly low currents. Thus this type must usually be tested at low currents, say <20mA.

Zeners

In audio power amplifiers, zeners likely to fail occasionally are those used as simple supply regulators (e.g. to get +15v for cascode and front-end circuits, from +60v, with only two components); and for setting current limits on MOSFET gates.

Zener testing requires a current limited source (e.g. a suitable bench or lab supply) with enough voltage to turn the tested part hard on. The required voltage ranges from about 2.4v, up to 100v or more. The output voltage metering will display the zener's clamping voltage at the set current. Most zeners found in conventional solid-state power amplifiers are rated at below 30v and below 10 watts, allowing a regular 0-30v lab supply to be used. Set the limit at about 10mA. Connect the zener and turn up the supply. If it clamps at about 0.8v, it is connected in reverse. Turn down the voltage before reversing, then increase it again. It should clamp at within 10% of the expected or identified rated voltage. If not, or if it clamps in either direction at much less than 0.4v, the zener is blown.

LEDs and Optos (Opto-isolator LEDs)

When driven forward ('on'), Opto LEDs drop about 1.6v at room temperature, when turned on quite hard with about 20mA. Some test meters with a proper diode test range will manage this. When polarity is reversed, resistance will be very high (> 1MΩ), but breakdown is not very high - and a meter with over 5v of compliance may cause this. This is harmless provided the current is limited, to say 20mA.

Panel LEDs sustain a variety of voltages. Red LEDs require between 1.8 and 2.2v to glow reasonably brightly. Orange, Yellow, Green and Blue LEDs require progressively higher voltages. As with the Opto-LED, breakdown can occur above 5v, so a reading that clamps when the test voltage is reversed is not a suspicious indication.

10.6 Oscilloscope traces

Figures 10.6-8 show common power amplifier pathologies, as they appear on a 'scope.

Figure 10.6 shows how 50Hz AC hum appears, at two timebase settings, on top of a 500Hz test tone. Check input wiring and grounding. Internally, check shields.

Figure 10.7 shows bursts of RF on a 1kHz tone with a p.r.f. (pulse repetition frequency) about five times higher. *RF can appear in many forms anywhere in a complex waveform, and the characteristic thickening or 'bubble' may only be visible at certain scope timebase and Y-scale settings, with certain loads, and certain programme.* Check OPN parts, and HF compensation, especially if new transistors have been fitted.

Figure 10.8 shows continuous, low level RF. If this gets any bigger, switch off. Else proceed as above.

If the waveform is in any way visually distorted (e.g. partly squashed, bent or asymmetric) switch off and recheck. If the +ve half only is afflicted, double check that section if it is a conventional output stage or else if it is a CE type (see Chapter 4), check the opposite, i.e. negative half; and *vice-versa*.

Figure 10.6

Appearance on a scope of a steady test signal modulated by hum, at two 'scope settings.

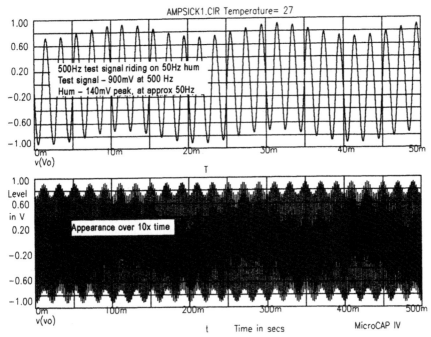

Figure 10.7

Appearance on a 'scope, of a steady test signal with bursts of RF, indicating instability.

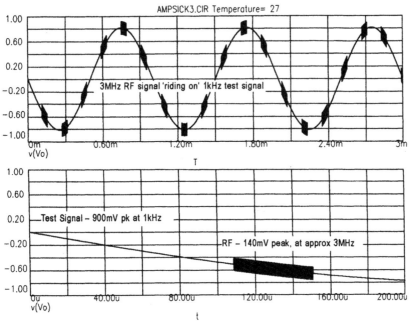

399

Figure 10.8

Appearance on a 'scope, of a steady state test signal, with 3MHz of continuous wave RF, caused by instability or RF pickup.

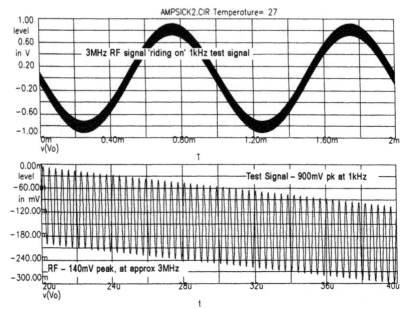

References

1   Towers, T.D, *Towers' International Transistor Selector*, Foulsham (UK).
2   Short, G.W, *DC matching of complementary pairs*, Wireless World, June 1973.
3   Towers, T.D, *International and other FET Selector*, Foulsham (UK).

Further reading

4   Daniel Metzger, *Electronic Components, Instruments, and Troubleshooting*, 2nd ed., Prentice-Hall, 1981. ISBN 0-13-250266-6.
5   Don & Carolyn Davis, *Sound System Engineering*, Sams, 1975 – see Appendix III, Recommended Wiring Practices.

Useful addresses for maintenance

A few of the internationally known makers of high performance, solid-state power amplifier makers are no longer in business in their own right. The official spares suppliers are:

HH Electronic: MAJ Electronics, Unit 1, Dawley Trading Estate, Stallings Lane, Kingswinford, West Midlands, DY6 7HU, UK.

Rauch Precision Engineering: Midland Sound & Light, 33 Duke's Close, Thurmaston, Leicester, LE4 8EY, UK.

Appendix 1

History

Some key events in the History of Power Amplifiers

1904, UK
The key to electronic amplification was first unlocked by Fleming's patent for the two element 'diode' valve, inspired by a meeting with Edison. The diode could not amplify but only one more step would be needed.

1906-7 US
Electronic audio amplification may have been first heard in the year Lee de Forest applied for a patent on the three element, triode tube he'd developed, on 25th October, 1906. By 1907, music had been informally broadcast by wireless for the first time.

1915-21, US
The first mass use of amplification to reproduce speech and subsequently music dated from the experimental efforts of Pridham and others at the Magnavox Co. beginning in 1915. By 1921, crowds of 125,000 were being covered at public meetings and music and opera had been broadcast down phone lines and reproduced by amplifiers in distant halls [1].

Figure A1.1

This Magnavox, made in the USA circa 1925, was probably the world's first mass produced audio power amplifier

1925-35
At the beginning of this period the everyday electro-dynamic 'cone' speaker *and also* electrostatic and horn speakers arrived in forms recognisable today. The early horn speakers' efficiency was still such that much less than one watt of amplifier power capability was needed to achieve sound levels that were adequate for reproduction of conversational speech in a home.

Also from 1925, audio power amplifier development was driven by the arrival of public radio broadcasting in the US and Europe, and also the first use of amplifiers to reproduce the new electrical music recordings on shellac discs, instead of relying on limited acoustic amplification. The first electric gramophone, the *Panatrope*, was launched by Brunswick in 1925.

In the USA, the development of 'talking' movies by Warner Bros., beginning in 1926, drove the development of amplifiers with substantially more power capability than the watt or so needed to satisfy radio listeners in homes.

In 1931, stereo was patented almost simultaneously in the UK by Blumlein working at the new E.M.I. (created from the merger of the Gramophone Company and Columbia Graphophone), and in the USA by Bell Labs. The first stereo PA system was publicly demonstrated in the USA on 27th April, 1933, when an orchestra was amplified in a distant venue.

1932, UK/US

Rodda and Bull of EMI labs conceive and patent the *beam* tetrode valve, to get around Mullard's patent on the pentode valve [2]. RCA shared the patent with EMI, and RCA's Otto Shade went on to make the brilliantly successful 6L6. The M-O Valve Co's variant, KT66 appeared in 1937. Both are still in production (1996) – even though their originating companies no longer exist.

1936, UK

Blumlein at EMI labs patented the *long tailed pair* circuit, a problem solved for the BBC as it prepared to launch the first ever public television service. This circuit later became and remains the bedrock of all modern, direct coupled transistor power amplifiers and other high performance gain blocks.

1937, UK

At EMI Labs, Alan Blumlein patents the ultralinear (or distributed load) output transformer connection.

1938-1946

By the late 1930s, the more up market amplifiers for 'gramophone reproduction' had reached the structure that is familiar to every Hi-Fi user up to this day - a power amplifier and possibly volume and tone controls packaged separately from individual loudspeaker cabinets.

In the US, typical cinema power amplifiers still used triode valves. Frequency response was commonly –20dB down at 15kHz [3]

Also in the US, guitar was played by blues musicians through (and modified by) the pickup input of radio set amplifiers, until the appearance of dedicated instrument amplifiers and the electric guitar, beginning with Leo Fender in 1940. This had a pronounced affect on audio amplifiers not just from then on, but also a quarter of a century later, when British and American electric blues players attracted the largest crowds in performance history.

1944, UK

In the midst of WW-II, Decca launch FFRR – Full Frequency Range Recordings. In the next decade this gives the impetus to drive-up the frequency response of audio amplifiers to at least 15kHz, while the best valve amplifiers will have hf rolloffs of 50kHz or above.

1947, UK

Ron Diggins, Public Address technician and radio shop owner in the ancient English port of Boston, begins operating the world's first mobile dance sound system, what would later be called 'discotheque' (a 1960s affectation). He built about a dozen of his all-in-one *art deco* consoles, called *The Diggola*. Each contained a 50 watt valve amplifier, and brought dance music at modest SPLs to halls of several hundred, at venues across the country [4].

Figure A1.2

Ron Diggins, inventor of the mobile discotheque, seen recently at one of his 50 year old Diggola consoles. Each contained a 50 watt valve amplifier. For today's dance music, solid-state amplifiers allow 100 to 3000 times the acoustic power to be developed in the space of a Diggola, to deliver rapture in the low bass regions.

1947, UK

D.T.N.Williamson (then at the MO Valve Co., UK; later at Ferranti) publishes a notable power amplifier design in *Wireless World* and establishes a *de facto* new standard for Hi-Fi.

1948, US

In June, operation of the first transistor at Bell Labs was announced, by a team lead by William Shockley.

1951, US

David Hafler reintroduces Blumlein's ultralinear connection [5].

1952, US

The *Darlington* transistor connection is patented at Bell Labs. In November, the J-FET is first proposed by Shockley. Workable, mass produced J-FETs are still 14 years away, though.

1952, US

Texas make the first silicon transistors. Again, commercial production is at least a decade away.

1952-62

Germanium transistors were all there was in this period. Power types were limited to low voltages mainly under 30v, and were very unforgiving of high temperatures, restricting affordable power to a few tens of watts at best.

1954, US

Interdigitation, precursor of 'Ring Emitter' BJTs, was first used by Fletcher, working for a small US maker, to enhance the speed of power transistors.

1954, US

The first transistor amplifiers appeared, for domestic use.

1956, US

While the first transistor computer was built, H.C.Lin [6] of RCA proposes the original *complementary* and *transformerless* push-pull amplifier circuit – a feat almost unmatchable by valves. Later, Lin's topology would become the basis of ordinary transistor amplifiers for the following 20 years.

1957, UK

The precision metal film resistor is patented. It will take 25 years before specific, refined makes become the norm for sonic quality in high performance analogue audio signal paths.

1958, UK

Pye announce first stereo LPs, almost simultaneously with Audio Fidelity in the US. Stereo rapidly becomes *de rigeur* in domestic Hi-Fi, and forever changes the way amplifiers are made.

1960, UK

Ferranti begin making the first commercially viable silicon transistors (BJTs) in Europe.

1961, UK

Dick Tobey & Jack Dinsdale describe a transformerless transistor audio amplifier in *Wireless World*, for DIY construction, based on Lin's 1956 proposals. Performance is sorely limited by Germanium transistors.

1963, UK

Mullard's BC 107 was one of the first of the more rugged and far faster silicon transistors (BJTs) to supersede germanium types.

1963, US

First J-FETs in production, but reliability is poor.

1964, US

Fairchild's μA702 was the first monolithic, IC op-amp.

1964, UK

Clive Sinclair launches the first working Class D amplifier module. Power: 10 watts/ch.

1964, US

Bipolar transistor action is modeled in depth by Gummel.

1965-75

In this period, the first affordable audio power amplifiers using transistors with above 25 watts output make their debut. The rapidly reducing price and widening performance envelope of the new silicon transistors makes this possible. By 1975, power capabilities of 500w into 8 ohms (with bridged channels) and over 1kW into 2 ohms are being achieved. Valve amplifiers are seen as increasingly anachronistic, except by musicians.

1965, US
Sharma publishes a transistor amplifier design capable of 200w but still employing transformer output coupling [7].

1966, UK
WEM (Watkins Electric Music) of London produces the first 100w$\Rightarrow$8Ω all silicon PA amplifiers using the latest RCA-transistors. Dozens premiered at the Windsor Jazz & Blues Festival, marking the beginning of large scale music PA, and the possibility of performing in front of crowds that would later top 250,000.

1967, US
Crown launch DC-300, the first 225w/4Ω/ch (stereo) all silicon power amp. Has patented class A-B/B biasing circuitry.

1968, US
Nestorovic of McIntosh patents a development of Blumlein's valve ultralinear connection with further distributed windings. Twenty years later, this technique would feature again in a domestic power amplifier, assisted by MOSFETs.

1970, UK
August 1970, WEM provide 5000 watts for the 2nd Isle of Wight Festival – more than had ever been used outdoors before [8].

1971, US
Quadrophonic discs appear but this attempt to sell twice as many amplifiers fails as makers squabble over formats.

1972, US
Phase Linear Corporation launch their Model 700 transistor amplifier for high headroom hi-fi, rated at 350w$\Rightarrow$8Ω/ch. Dozens are subsequently bought by the British Band Pink Floyd, for their '*Britannia Row*' PA system.

1973, UK
Midas introduce the first professional 4-band amplifier 'black block', for the actively-driven Midas PA system.

Figure A1.3

Two of Pink Floyd's famous and historic Phase Linear 700 *amplifiers, in their original 1974 flightcase, retired to domestic life in the London music room of John McHugh, with young offspring for scaling. These amplifiers were really made for US domestic Hi-Fi (hence the styling and power meters) but were pressed into use as the sonics were so good for the time. Other than powering Floyd's gigs worldwide from 1974-80, they were rented out to dozens of bands for tours in various territories over their 15 years in use at Britannia Row Productions. Note top air vents and the large rack 'u' space apparently wasted by the wiring panel, but really hiding the large transformers, that had to be mounted externally for roadworthiness.*
© John McHugh.

1973, Europe
Otala and Lohstroh publish a paper on a landmark power amplifier, with high slew limit, defined Z_0, low global NFB and a high biased 3 watt A/A-B transition.

1974, Japan
A new high power MOSFET for audio, invented by Professor Jun'ichi Nishizawa of the Electronic Telecoms Research lab at Tokuko University, is unveiled. The new part is being developed by Yamaha with backing from Japan's Technology Development Foundation. In Japan, Pioneer, Toshiba, JVC and Sony are also producing domestic amplifiers with power MOSFETs. [9].

1975, US
Crown patents changes in amplifier topology that are advantageous to bridging flexibility and protection circuitry sophistication.

1976, Japan
The first audio power amplifier design to employ MOSFETs as output transistors is launched by Yamaha [10].

1976, US
Nelson Pass patents a purist class-A biasing system [11].

1977, Japan
The first amplifier to exploit Class G is made by Hitachi for Hi-Fi [12].

1977, Japan
Hitachi launches lateral, audio power MOSFETs.

1977, UK
HH Launch S-500D, the first ultra-compact (2u) professional amplifier, rated at 500w/2.5Ω/ch.

1977, US
Crown's patented 'IOC' error indicator first appears on their industrial DC-300 II.

Nelson Pass publishes a design for a solid-state, single-ended amplifier in Audio magazine, precursor to Aleph 0 and Zen amplifiers.

1978, UK
HH Electronic launch first professional amplifier to employ Hitachi's lateral MOSFETs.

Figure A1.4

In 1978, HH (Cambridge, UK) was the first company to manufacture amplifiers for the world market, using Hitachi's dedicated lateral MOSFETs. © MAJ Electronics

1978, US
Nelson Pass patents *Stasis* for refined Class A amplification.

1978, Japan
Mr. Sano patents an improved Class A amplifier.

1979, US
Crown introduces first power amplifier with output protection controlled by an analogue-computer.

1981, US
Carver patents the M400A *'magnetic field amplifier'*.

1982, US
Super-Nova topology patented by Jim Strickland at Acoustat.

1983, US
QSC first uses Class H for high efficiency touring PA.

1984, US
Crown introduces ODEP, an improvement of their realtime analogue computer for monitoring power transistor stress.

1986, US
Crest launches 8001, the first audio power amplifier swinging 91v (1.2kW$\Rightarrow$4Ω) per channel in a 2u high case and exploiting class G.

1987, UK
Rauch Precision launch DVT-300s, the first audio power amplifier to employ a high efficiency resonant power supply, delivering 600w/ch in 2u size, class A-B.

1988, UK
Production ceases at M-O Valve company, founded 1911. A new chapter begins as equipment and skill is exported to China.

1989, UK
BSS Audio launch EPC-780, a unique combination of power density with high-end standards, delivering above 1kW/ch in 2u size, with unique class A/H, an asymmetric bridge and soft commutation.

1993, US
Crown launch MA-3600-VZ, using the adaptive power supply as previously advocated by Holman.

1994, UK
C-Audio launch XR-5001, the first audio amplifier capable of delivering over 100 volts rms/ch with audio-grade slew-rate and without bridging.

1994, US
Carver Research launches the *Lightstar* amplifier, claiming first radical departure from convention, into an adaptive psu with current-gate output device.

Nelson Pass publishes Zen, a single-ended MOSFET power amplifier [13].

1995, US

QSC launch *Powerlight*, a class G amplifier with intelligent HF switching power supply, delivering 2kW/2Ω/ch, in 2u. Has 'Dataport' remote control & monitoring connection.

Note: For classes G & H, Anglo-Japanese terminology applies. *This is the opposite of the designations used by many US makers.*

References

1 Green, I.W & J.P Maxfield, *Public address systems*, Transactions of AIEE, Feb 1923, vol.42.

2 *An Approach to Audio Amplifier Design*, GEC Co, UK, 1957. (Reprinted and available from Old Colony Sound Lab, USA or HFN Accessories Club, UK.)

3 McConnel, John.W, *Been there, done that !* (letter), Stereophile, Dec 1995.

4 Knott, Gordon, *The quiet revolutionary,* DJ Magazine (UK), No.138, May 1995.

5 Hafler, David and Herbert I. Keroes, *An ultra-linear amplifier*, Audio Engineering (US), Nov 1951.

6 Lin, H.C, *Quasi Complementary transistor amplifier*, Electronics, Sept 1956.

7 Sharma, Madan.M., *A 200 watt solid state power amplifier*, J.AES July '65.

8 Trynka, Paul (ed.), *Rock Hardware, chapter 10, The PA System*, by Ben Duncan, Outline Press (UK), Miller-Freeman (US), 1996. ISBN 1-871547-350

9 King, Gordon.J, *The Audio Handbook*, Newnes-Butterworths, 1975, p.103-108.

10 *Yamaha B1, Vertical FET power amplifier*, Electronics Today International, Feb 1977.

11 Pass, Nelson S, *Active bias circuit for operating push-pull amplifiers in class A mode*, US pat. 3995 228, Nov 1976.

12 Harris, Ron, *Class G*, Electronics Today International, Dec 1977.

13 Pass, Nelson, *The Pass Zen Amplifier*, The Audio Amateur, Q2, June 1994.

Further Reading

13 G.W.A.Dummer, *Electronic Inventions & Discoveries*, Pergamon, 1983.

14 Duncan, Ben, *A history of PA* in One, Two, Testing, June, July, Aug 1984.

15 Nameless, *Audio FET power transistors*, WW, July 1974.

Appendix 2

Makers' listings

A-Z, under principal use

This list is *not* intended nor able to define *high performance*. At each level, it is common for only arbitrary models made by some of the (e.g. lower budget or more industrial-oriented) makers to come within the category; or for the inclusion of the products of a specific maker to be hotly debated. *But businesses are living organisms.* It is therefore possible for any maker's quality to flip either way – to vastly better or worse, at any time. Unlikely in theory, but it has happened. *Caveat Emptor!*

Professional (Stage, Recording & Home Studio, & Broadcast)

Manufacturer's Name:	Principal Country of origin:	Classification by OPS active device:
Amcron - see Crown		
AB Int.	US	B
AMP	US	B
Ashley	US	M
ARX Systems	Australia	M
Audio Centurion	US	B
Australian Monitor Co	Australia	M
*Austin-Armstrong	UK	M
Avitec (badged C-Audio)	UK	M
BGW	US	B
Bi-amp	US	B,M
Boulder - *see domestic list.*		
Bryston	Canada	B
BSS Audio (Harman)	UK	M
Camco	Germany	B
Carlsbro	UK	B,M
Carver Professional	US	B

C-Audio Ltd (Harman)	UK	M,B
Cerwin-Vega	US	B
Chevin Research	UK	M
Citronic	UK	M
Cloud	UK	M
Crest	US	B
Crystal Precision (Dynamic Precision)	Norway	M
Crown	US	B
dB Technologies	Italy	M
* EAA	France	B
Electro-Voice	US	B
EHT	Australia	M
EV (badged Dynacord)	Germany	B
Fender	USA	B
FM Acoustics	Switzerland	B
Gallian-Kruger	US	B
Groove Tubes	US	V
Harrison	UK	M
IMC-Ross	US	B
HH (today owned by Laney)	UK	M,B
Hill Audio	UK	B
* HIT	UK	M
JBL (badged UREI, badged QSC)	US	B
KMD	US	B
Lab Gruppen	Sweden	B
Malcolm Hill Assocs. (now Chameleon Audio)	UK	B
MC² Audio	UK	B
* Millbank (1993-6 made Hill Audio's)	UK	B
Oceania	New Zealand	B
* Omniphonics	UK	B,M
* Otis Communications	UK	B
Pearce	US	B
Peavey (AMR)	US	B
* PSL	UK	M
QSC	US	B
QUAD Electroacoustics – *see domestic list.*		
Quested (badged MHA) – *see above*	UK	B
Randall	US	B
Ramsa (Panasonic)	Japan	B
Rank Strand (badged Hill)	UK	B
Rauch (Monitor Systems Technology)	UK	M
Renkus-Heinz	US	B
Rolec	UK	M
* SAE	US	B
Samick	Korea	B
Soundcraftsman	US	B

Soundtech	US	B
Stage Accompany	Netherlands	B
Stewart	US	B
Studer	Switzerland	B
Studiomaster (RSD)	UK	B,M
Sunn	US	B
TOA	Japan	B
Tubeworks	US	V
* Turbosound (pre Harman)	UK	B,M
* Turner	UK	B
Tytech	Korea	B
Yamaha	Japan	B,M
Yorkville (badged Tytech)	US	B

B = Bipolar transistor   M = MOS-FET   V = Valve/Tube.

*   indicates a maker that *either* no longer makes high performance amplifiers, but has
in the past; *or* one that is believed to no longer exist.

(Brackets) indicate either other, or older trading names, or where badged, the actual
maker.

Domestic   *K indicates part or whole kit form.*

Aaragon	UK	M
Aaragon (Mondial)	US	B
Accuphase	Japan	B
Acurus	US	B
Adcom	US	B
Airtight	Japan	V
Ameson-Cooper	UK	V
AP Components (ex-Audiokits) K	UK	B
Arcam	UK	B,M
Art Audio	UK	V
Audio Research	US	V,B
Audio Synthesis	UK	M
Audiolab	UK	B
Audionote	Japan	V
Aura	UK	M
B&W Nakamichi	UK/Japan	M,B
Balanced Audio Technology	US	V
Ben Duncan Research K	UK	M,B
Berning	US	V
Borbely Audio K	Germany	M
Boulder	US	B
Bryston – *see pro listing*	–	–.
Carver Corp.	US	V,B
Cary	US	V

Cello	US	B
Classé	Canada	B
Classic Audio	US	V
Conrad Johnson	US	V,B,M
Copland	Denmark	V,B
Creek	UK	B
Cyrus - see Mission	-	-
Denon	UK/Japan	B
DNM	UK	M
Edini	US	M
* Eidectic	Australia	M
Electrocompaniet	Sweden	B
Ensemble	Switzerland	M
Ensemble (Solen)	France	V,B
Exposure	UK	B
Fourier Components	US	V
Golden Tube Audio	US	V
Goldmund	Switzerland	M
Hafler	US	M
Harman-Kardon	US	B
Jadis	France	V
Jeff Rowland design group	US	B
Krell	US	B
LFD	UK	M
Linn	UK	B
Luxman	Japan	V
Manley (see also VTL)	US	V
Marantz	US	B
Mark Levinson	US	B
Mc'Cormack Audio	US	B
Mc'Intosh	US	V,B
Meridian	UK	B,M
Mesa Engineering	US	V
Michell	UK	M
Mission	UK	B
Muse	US	B,M
Musical Fidelity	UK	V,B,M
NAD	UK	B,M
Naim	UK	B
NYAL	US	V
Onix	UK	B
Orelle	UK	M,B
Parasound	US	B
Pass Labs	US	M
Perraux	New Zealand	B,M
Plinius	New Zealand	V

PS Audio	US	M
QUAD Electroacoustics	UK	B
Quicksilver	US	V
Rotel	Japan/UK	B
Rowland Research	US	B
Russ Andrews (RATA)	UK	M
Sonic Frontiers K	Canada	V
Sonographe	US	M
Sugden	UK	B
Sumo	US	B
Tandberg	Norway	B
Technics	Japan	B,M
Thorens	Switzerland	M
Threshold	US	B
Times One	US	M
Valve Amplification Co (VAC)	US	V
VTL (Vacuum Tube Logic)	US	V
Woodside	UK	V
YBA	France	M

A-Z of integrated loudspeaker-amplifier makers

This listing covers high performance audio power amplifiers built within the speaker cabinet or attached to it. It also includes power amplifiers which are discrete but expressly made and solely available with, (a) specific speaker system(s).

Manufacturer's Name:	Principal Country of origin:	Classification by OPS active device:
Domestic:		
ATC – *see Professional.*		
Audio Physic	Germany	B
B&W	UK	M
Meridian (Boothroyd-Stuart)	UK	B
Professional (Live sound & Recording Studio):		
ATC	UK	M
BGW - active subwoofers only	US	B
Coastal Acoustics (Harris-Grant)	UK	B,M
Genelec	Finland	B
Meyer	US	B
Quested	UK	M
* Red Acoustic	UK	B
Tannoy	UK	M

SUMMARY

Principal output device, by number of makers

Type	Domestic	Pro
BJT	47	53
MOSFET	30	25
Valve	27	2

Appendix 3

Active devices

The following listings cover active parts – over 140 transistors and over 60 valves – that are common or notable in modern high performance audio power amplifiers, and that are subject to wearout, or otherwise in practice require occasional replacement.

In the following semiconductor listings (A,B,C), low voltage (<50v), small signal (<50mA) transistors are excluded, since after burning in, few will fail during the lifespan of the owner, or before the remainder of the amplifier rusts and rots away !

The majority of parts are 'still current', ie. in manufacture at the time of writing. The fact that some are 15 and up to 25 years old, shows not necessarily near–perfection, but also the lack of further work by semiconductor makers in that area. A few 'classic' parts are included for performance comparisons – note in particular hfe (gain) and F_T (bandwidth) Vs. current (I_c).

In the subsequent valve listings (section D), input stage and rectifier valves are listed alongside 'large signal' types, since they all begin to wear just as soon as they are burnt in. Power tubes wear out even faster. Many of the more common types used in the past half century are listed. For older types, notably output triodes, consult makers or specialist valve literature suppliers, advertising in *Glass Audio* magazine (see *Publications* section of *References & Bibliography*).

A Bipolar transistors, silicon

1 Small & Medium Signal Types

*Below 2 Amperes collector current. *Includes some drivers*
In A–Z order. *Read on for pinouts.*

Model	Maker	Pol	Vceo (V)	Ic (A)	Pc (W)	hfe typ	F$_T$ MHz	Case No.
2N 3439	RCA	N	350	1	1	100	15	TO39
2N 3440	RCA	N	250	1	1	100	15	TO39
2N 4410	Motorola	N	80	0.25	0.15	180	60	TO92b
2N 5401	Motorola	P	–150	–0.6	0.5	120	200	TO92b
2N 5550	Motorola	N	140	0.6	0.3	155	100	TO92b
2N 5551	Motorola	N	160	–0.6	0.5	140	200	TO92b
2SA 872A	Hitachi	P	–120	–0.1	0.4	499	90	TO92b
2SA 968	Toshiba	P	–160	–1.5	25	155	50	TO220
2SA 970	Toshiba	P	–120	–0.1	0.3	200	50	TO92c
2SA 1016k	Sanyo	P	–120	–0.05	0.4	560	130	TO92a
2SA 1085	Hitachi	P	–120	–0.05	0.3	400	90	TO92c
2SA 1208	Sanyo	P	–160	–0.07	0.9	250	150	TO92a
2SA 1306B	Motorola	P	–200	–1.5	20	160	100	TO221
2SB 631k	Sanyo	P	–120	–1.0	1.0	190	110	TO126
2SB 716	Hitachi	P	–120	–0.05	0.75	500	350	TO92c
2SB 1144	Sanyo	P	–100	–1.5	1.5	250	100	TO126ML
2SC 2238	Toshiba	N	160	1.5	25	155	50	TO220
2SC 2240	Toshiba	N	120	0.1	0.3	200	50	TO92c
2SC 2362k	Sanyo	N	120	0.05	0.4	560	130	TO92a
2SC 2547	Hitachi	N	120	0.1	0.4	400	90	TO92c
2SC 2910	Sanyo	N	160	0.07	0.9	250	150	TO92a
2SC 3298B	Motorola	N	200	1.5	20	160	100	TO221
2SD 600k	Sanyo	N	120	1.0	1.0	190	130	TO126
2SD 756	Hitachi	N	120	0.05	0.75	700	150	TO92c
2SD 1684	Sanyo	N	100	1.5	1.5	250	120	TO126ML
BD 135	SGS–Tompson	N	45	1	8	100	250	TO126
BD 136	SGS–Thomson	P	–45	–1	8	100	75	TO126
BD 139	SGS–Thomson	N	80	1	8	100	250	TO126
BD 140	SGS–Thomson	P	–90	–1	8	100	75	TO126
BF 469	Philips	N	250	0.2	0.8	70	100	TO126
BF 470	Philips	P	–250	–0.2	0.8	70	100	TO126
BF 869	Philips	N	250	0.03	1.6	70	60	TO202
BF 870	Philips	P	–250	–0.03	1.6	70	60	TO202
MJE 340	Motorola	N	300	0.5	20	135	10	TO126
MJE 350	Motorola	P	–300	–0.5	20	50	4	TO126
MPS A42	Motorola	N	300	0.5	0.6	40	50	TO92b
MPS A43	Motorola	N	200	0.5	0.6	40	50	TO92b
MPS A92	Motorola	P	–300	–0.5	0.6	25	50	TO92b
MPS A93	Motorola	P	–200	–0.5	0.6	25	50	TO92b
MPS 8599	Motorola	P	–80	–0.5	0.6	150	150	TO92b
TIP 29c	Texas	N	100	1	30	120	3	TO220
TIP 30c	Texas	P	–100	–1	30	120	3	TO220
Other TIP– suffixes: A = 60v, B = 80, D = 120, E = 140 (V$_{ceo}$).								
ZTX 653	Zetex	P	–100	–2	1.5	200	140	E–line
ZTX 753	Zetex	N	100	2	1.5	200	140	E–line

2 – Large Signal (output & driver) types

In order of voltage (V_{ceo}), then alpha-numeric.
*Including Darlingtons. * Read on for pinouts.*

Model	Maker	Pol	Vceo (V)	Ic (A)	Pc (w)	hfe typ	F_T MHz	Case No.
BD 131	Philips	N	45	3	15	40	60	TO126
BD 132	Philips	P	–45	–3	15	40	60	TO126
MJ 802	Motorola	N	90	7.5	200	100	2	TO3
BD 539C	Texas	N	100	5	45	30	–	TO220
BD 540C	Texas	P	–100	–5	45	30	–	TO220
TIP 31c	Texas	N	100	3	40	60	3	TO220
TIP 32c	Texas	P	–100	–3	40	60	3	TO220
TIP 41c	Texas	N	100	6	65	50	3	TO220
TIP 42c	Texas	P	–100	–6	65	50	3	TO220
Other TIP– suffixes: B = 80, D = 120, E = 140, F = 160 (V_{ceo})								
40872	RCA	P	–100	–7	40	50	4	TO66
BD 711	SGS–Thom	N	100	12	75	25	3	TO220
BD 712	SGS–Thom	P	–100	–12	75	25	3	TO220
2SD 1046	Sanyo	N	120	8	80	130	8	TO3P
2SB 816	Sanyo	P	–120	–8	80	130	8	TO3P
MJE15028	Motorola	N	120	8	50	40	30	TO220
MJE15029	Motorola	P	–120	–8	50	40	30	TO220
2SA 1077	Fijitsu	P	–120	–10	60	130	30	TO220
2SC 2527	Fijitsu	N	120	10	60	130	40	TO220
BDY 56	Fairchild	N	120	15	115	45	10	TO3
BDY 77	SGS–Thom	N	120	16	150	70	0.8	TO3
MJ 11015	Motorola	P	–120	–30	200	2k	4	TO204
MJ 11016	Motorola	N	120	30	200	2k	4	TO204
2N 3441	RCA/Ha	N	140	3	25	75	0.2	TO66
2N 3442	RCA/Ha	N	140	10	117	45	0.5	TO3
MJ 15001	Motorola	N	140	15	200	70	2	TO204
MJ 15002	Motorola	P	–140	–15	200	70	2	TO204
2N 3773	RCA/Ha	N	140	30	150	37	0.7	TO204
MJ 15003	Motorola	N	140	20	250	70	2	TO204
MJ 15004	Motorola	P	–140	–20	250	70	2	TO204
MJE 15030	Motorola	N	150	8	2	60	30	TO220
MJE 15031	Motorola	P	–150	–8	2	60	30	TO220
2N 6259	RCA/Ha	N	150	16	250	32	0.2	TO3
2SC 2565	Toshiba	N	160	15	150	40	35?	TO221
2SD 424	Toshiba	N	160	15	150	65	3	TO204
2SB 554	Toshiba	P	–160	–15	150	90	3	TO3
2SC 3519	Sanken	N	160	15	130	50	40	MT–100
2SA 1386	Sanken	P	–160	–15	130	50	40	MT–100
EBX 3516	Exicon	N	160	25	125	100	3	TO247
EBX 3616	Exicon	P	–160	–25	125	100	3	TO247
TIP 35F	Texas	N	160	25	125	30	3	TO3P
TIP 36F	Texas	P	–160	–25	125	30	3	TO3P
2N 3583	RCA/Ha	N	175	1	35	140	10	TO66
2SA 1302	Toshiba	P	–200	–15	150	60	25	TO247
2SC 3281	Toshiba	P	–200	–15	150	60	25	TO247
2SC 3263	Sanken	N	230	15	130	40	35	MT–100

2SA 1924	Sanken	P	−230	−15	130	40	35	MT−100
2SC 3264	Sanken	N	230	17	200	40	35	MT−200
2SA 1295	Sanken	P	−230	−17	200	40	35	MT−200
MJ 15025	Motorola	N	230	8	250	35	5	TO204
MJ 15024	Motorola	P	−230	−8	250	35	5	TO204
MJ 15011	Motorola	N	250	10	200	50	−	TO204
MJ 15012	Motorola	P	−250	−10	200	50	−	TO204
MJ 15022	Motorola	N	250	15	175	150	3	TO204
MJ 15019	Motorola	P	−250	−15	175	150	3	TO204

Bipolar Transistor Tables (A1,2) – Notes:

N = npn; P = pnp; '−' = not specified by maker. ? = ambiguous maker's information.

hfe > 800 indicates a Darlington type. F_T is usually a minimum figure.

ML = 'mica less', i.e. insulated base.

The TO3 and T0−204 cases are identical. T0−204 is now the preferred designation.

The T03P, TO247e and MT−100 plastic cases are very similar to each other.

Cited maker is the best known and/or believed to be original source. Ha = Harris

Universal Table of Connections – BJT & MOS

Pinout Ref:	Pin No: 1	2	3	Common Types:

Plastic cases:

TO92a	C	B	E	Most BC
TO92b	E	B	C	Most 2N, MPS
E−line	E	B	C	Most E−line (ZTX)
TO92c	E	C	B	Many 2SA, 2SC, some BC, special E−line.

Caution ! Some types occur in differing leadout versions. Verify leadout first.

TO3P, TO 247, TO264,	G	S	D	All lateral MOSFETs.
TO3P, TO 247, TO264,	G	D	S	All vertical MOSFETs, i.e. DMOS, HEXFETs.

Metal cases:

TO3, TO204, *Lateral* MOSFETs			
	G	S	D
TO3, TO204, *Vertical* MOSFETs			
	G	D	S

C = Collector; B = Base; E = Emitter.
D = Drain; G = Gate; S = Source.

Plastic cases

All viewed from above, except where stated

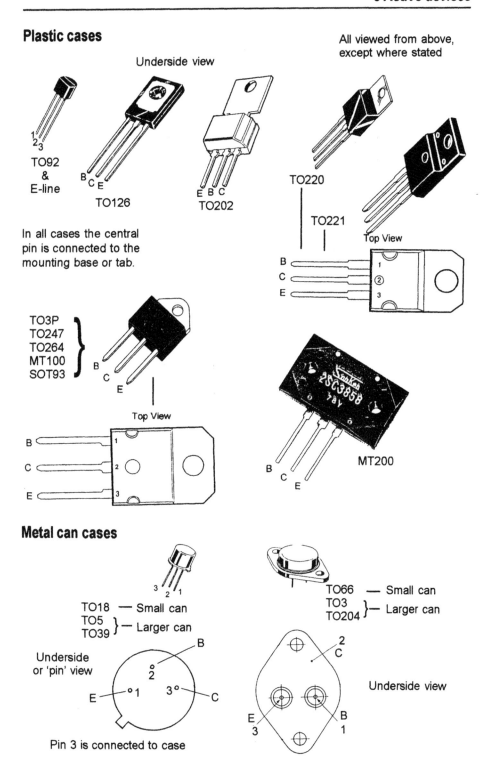

Underside view

TO92 & E-line

TO126

TO202

In all cases the central pin is connected to the mounting base or tab.

TO220

TO221

Top View

TO3P
TO247
TO264
MT100
SOT93

Top View

MT200

Metal can cases

TO18 — Small can
TO5 } — Larger can
TO39 }

Underside or 'pin' view

Pin 3 is connected to case

TO66 — Small can
TO3 } — Larger can
TO204 }

Underside view

B Lateral power MOSFETs

In order of voltage (V_{ds} max), then current handling (I_{ds}).

Model	Maker	Pol.	Vds volts	Ids A	Pmax (Watts)	Case Ref.
120v:						
2SK 133	Hitachi	N	120	7	100	TO3
2SJ 48	Hitachi	P	-120	-7	100	TO3
2SK 1056	Hitachi	N	120	7	100	TO3P
2SJ 160	Hitachi	P	-120	-7	100	TO3P
140v:						
2SK 134	Hitachi	N	140	7	100	TO3
2SJ 49	Hitachi	P	-140	-7	100	TO3
2SK 1057	Hitachi	N	140	7	100	TO3P
2SJ 161	Hitachi	P	-140	-7	100	TO3P
160v:						
2SK 135	Hitachi	N	160	7	100	TO3
2SJ 50	Hitachi	P	-160	-7	100	TO3
2SK 1058	Hitachi	N	160	7	100	TO3P
2SJ 162	Hitachi	P	-160	-7	100	TO3P
BUZ 900	Semelab	N	160	8	125	TO3
BUZ 900P	Semelab	N	160	8	125	TO247
ECF 10N16	Exicon	N	160	8	125	TO3
ECX 10N16	Exicon	N	160	8	125	TO247
BUZ 905	Semelab	P	-160	-8	125	TO3
BUZ 905P	Semelab	P	-160	-8	125	TO247
ECF 10P16	Exicon	P	-160	-8	125	TO3
ECX 10P16	Exicon	P	-160	-8	125	TO247
BUZ 900D	Semelab	N	160	16	250	TO3
ECF 20N16	Exicon	N	160	16	250	TO3
ECW 20N16	Exicon	N	160	16	250	TO264
BUZ 905D	Semelab	P	-160	-16	250	TO3
ECF 20P16	Exicon	P	-160	-16	250	TO3
ECW 20P16	Exicon	P	-160	-16	250	TO264
180v:						
2SK 175	Hitachi	N	180	8	100	TO3
2SJ 55	Hitachi	P	-180	-8	100	TO3
200v:						
2SK 176	Hitachi	N	200	8	100	TO3
2SJ 56	Hitachi	P	-200	-8	100	TO3
2SK400	Hitachi	N	200	8	100	TO3
BUZ 901	Semelab	N	200	8	125	TO3
BUZ 901P	Semelab	N	200	8	125	TO247
BUZ 906	Semelab	P	-200	-8	125	TO3
BUZ 906P	Semelab	P	-200	-8	125	TO247
ECF 10N20	Exicon	N	200	8	250	TO3
ECX 10N20	Exicon	N	200	8	250	TO247
ECF 10P20	Exicon	P	-200	-8	250	TO3
ECX 10P20	Exicon	P	-200	-8	250	TO247

BUZ 901D	Semelab	N	200	16	250	TO3
ECF 20N20	Exicon	N	200	16	250	TO3
ECW 20N20	Exicon	N	200	16	250	TO264
BUZ906D	Semelab	P	−200	−16	250	TO3
ECF 20P20	Exicon	P	−200	−16	250	TO3
ECW 20P20	Exicon	P	−200	−16	250	TO264
Above 230v:						
ECF10N24	Exicon	N	240	8	125	TO3
ECX10N24	Exicon	N	240	8	125	TO247
ECF10P24	Exicon	P	−240	−8	125	TO3
ECX10P24	Exicon	P	−240	−8	125	TO247

For pin connections, see *Universal Table of Connections* on previous page.

Items in *italics* are recently no longer made at the time of writing but were the sole hence universal types between 1977 and 1993. Other lateral MOSFETs may be substituted but may require small changes in compensation capacitance, protection threshold and biasing.

TO3 = metal can.

TO247 ≅ TO3P; TO247 is a plastic case, overmould version of TO3P.

TO264 ≅ TO3P; TO264 is a plastic case, *enlarged* & overmould version of TO3P.

C Power D–MOSFETs

* *In order of voltage (V_{ds} max), then current handling (I_{ds}).*
* *Types employed in audio OPS only.*

Model	Maker	Pol.	Vds volts	Ids A	Pmax (Watts)	Case Ref.
200v:						
IRF 540	IR/Samsung	N	100	28	125/150	TO220
IRF 9540	IR	P	100	19	125	TO220
2SK 413	Hitachi	N	140	8	100	TO3P
2SJ 118	Hitachi	P	140	8	100	TO3P
2SK 414	Hitachi	N	160	8	100	TO3P
2SJ 119	Hitachi	P	160	8	100	TO3P
IRF 640	IR	N	200	18	125	TO220

D Thermionic valves (electron tubes)

* in approximate order of ascending power handling.

European model No's:		USA model No's:		Internal
Common	SQ	Common	SQ	Structure
Preamp/Line level:				
ECC 81	M8162	12AT7(WA)6201		Double triode
& CV455, 6060	CV4024	B309		
ECC 82	M8136	12AU7(WA)	6189	Double triode
& 6067,		B329		
ECC 83	M8137	12AX7(WA)(A)(S)		Double triode
& 6057, 6L13, CV4004		B339	7025 (low noise)	
		12AY7	6072	Double triode
		12DW7	7247	Double triode
EF86		6BR8	6267	Pentode
Power Output: *(power ratings are approx. max and assume push–pull output)*				
		6V6	7408	15w, tetrode
EL84	N709	6BQ5		17w, pentode
&	6P15			
KT66		6L6	5881	25w, tetrode
			7027	28w, tetrode
EL34, KT77		6CA7	6550	100w, pentode
KT88			(6550)	100w, pentode
HT Rectification: *(all diode or double–diode)*				
EZ80		6V4		
EZ 81		6CA4	0A2, U709	
GZ 30		5Y3, 5Z4G	U50	
GZ31		6AL5	U52	
GZ 32		5U4	U51	
GZ 34		5AR5	U77,	
GZ37			U54	

SQ = Special Quality, ie. specially selected industrial type. If SQ is specifically called for, downwards substitution may give improper results.

Some model No's may be prefixed '–GT', which denotes 'Glass Tube'. In stage instrument amplifiers, G and GT types are not as rugged nor as suited to upside down running as the 'GC' type. Lifespan may be reduced by incorrect orientation!

Any model No's in brackets (XYZ) are near equivalents and require expert substitution and possible equipment adjustments.

Don't replace odd valves in an output set.

Appendix 4

Power amplifier terminology

Over 550 terms and abbreviations explained. Drawn from analogue power electronics, mechanical engineering; audio and electro-acoustics systems technology, music, the stage and recording production environments.

If the term or abbreviation sought is absent see the succeeding section, 'electronic units'.

abscissa Horizontal or X axis of a graph.

absolute value When quantity alone is the issue, and polarity (+ or −) is of no consequence. This makes no sense with money, but it has meaning with waveforms, where 'minus' is just a name for 'other side', not a debt. A full wave rectifier has the property of making all output voltage values one way, either positive or negative. This is absolute value conversion. When polarity information *is* expendable, conversion to absolute value in data processing or graph plotting can usefully increase dynamic range.

ac Alternating current. May be defined as a voltage source or current flow, that has both negative and positive excursions and when pure, that has an average value of zero. Ac with a single or dominant frequency is used for power transmission. Music signals are comprised of complex, highly 'impure' ac signals at more than one frequency.

ac coupled See *DC blocking*.

ac mains The public AC power line or supply.

ac millivoltmeter Used to measure audio signal levels, usually in conjunction with single tone excitation. The key specification is accurate rms reading of ac over a wide range of voltages and frequencies.

active device(s) Fundamental electronic components usually with 3 terminals or legs, able to amplify or otherwise control signals. See *BJT, J-FET, MOS-FET, valve*.

AES Audio Engineering Society.

AF Audio Frequency.

airfront The surface of a notional section of air, eg. one which begins with a cutaway at a cooling fan.

airpath The path of cooling airflow, usually forced.

am, a.m. Amplitude Modulated. Where one signal is 'multiplied' or 'ridden on' by another. The result of intermodulation.

amp (i) Colloquial abbreviation for 'power amplifier', 'guitar amplifier' or 'integrated pre+power amplifier'. (ii) Semi-official abbreviation for Ampere, not used much in this book, to avoid confusion.

amp rack Frame containing several (touring PA) power amplifiers.

ampere Unit of current or 'rate of charge movement'.

amplification Making bigger. Synonym for *gain*.

amplified diode Synonym for a common kind of V_{be} multiplier or 'VbeX'.

amplitude The magnitude or size of a signal. Conventionally seen as 'height' (the Y axis) on a graph or 'scope screen. Most signals are viewed in the dimension of voltage and may be expressed in 'V' or by a variety of variants, eg. mV, μV, dBu, dBR, or dBm. Sometimes signals are expressed in the orthogonal dimension, called current. Then the amplitude is measured in amperes. When the amplitude performs significant work, like driving a speaker, the amplitude may be expressed as the product of both dimensions, in watts, kilowatts or dBW. Note: 'Level' is similar but not an exact synonym.

amplitude distortion Generally harmonic or intermod distortion that is proportional to level, or level dependent, e.g. thermal distortion caused in speakers by voice-coil heating.

analogue, analog (US) (i) The class and part of all electronic circuits that are not digital. (ii) The representation or transformation of information through any continuously variable quantity operating in a manifold above 1 dimension, such as ink shape, voltage; phosphor brightness; or air vibration.

anodised, anodized An aluminum treatment yielding surface oxidation which is hard, a good high voltage electrical insulator, and a relatively good heat conductor.

ANSI American National Standards Institute.

anticlip Circuitry that *prevents* clipping.

antiphase Alias for the 'cold' or '−' side of a balanced signal; or the opposite of any prevailing absolute polarity condition.

antisurge A kind of fuse that *withstands* momentary 'surge' currents well above its rating.

asymmetric (i) Lopsided quality of music and other complex signals, where the sum of all positive signals is different from the sum of negative signals, over a given period, ie. the two do not sum towards a long term average of zero. (ii) Jargon for RF interference that is common mode. (iii) A circuit topology which is not symmetrical, either as drawn, or in terms of component physics.

audiophile A music (or sound) lover or enthusiast.

AUT, A.U.T, aut Amplifier Under Test. See *D.U.T.*

auto-null A circuit that uses feedback to automatically seek or tune to, a null or minimum. In the context of power amplifiers, it usually involves nulling or cancelling unwanted DC offset voltages resulting from imperfect part matching. Auto-nulling is also employed to seek and remove the stimulus' fundamental in relatively modern %THD+N analysers.

A$_{vol}$ Open-loop voltage gain: 'Amplification, voltage, open-loop'.

A-weighting (i) A filter used in testing. Emphasises defects at mid frequencies centered on 3.5kHz, where the ear is most sensitive, to help make measurements simulate better what is heard. The filter response is specified in the IEC 179 standard. (ii) The same filter used for SPL measurements, where it can improve corroboration between unwanted sound and measured levels, at least for modest sound levels.

BA Abbreviation of British Association, the name of small, fine, machine-screw thread series originated in Britain. Sizes commonly range from the large 0BA (approx. M6) down to 8BA (approx. M2.5).

back EMF A potentially damaging, counter-intuitive reverse-sourced voltage that occurs when the current flow in a magnetic device, e.g. inductor, motor or transformer is abruptly curtailed. May be many times higher than the preceeding signal voltage.

backline Musical instrument amplifiers. So named because when many are in use on a stage, they are arranged or stacked in a line or crescent across the back.

backtalk Signal leakage 'back down' the signal chain, from any later stage to any preceding stage.

balance (i) Equal but opposite signals and matched impedances to all other points, requirement for high CMR connections. (ii) tonal balance, akin to flat frequency response. (iii) level balance, between stereo L&R or between HF & LF sections.

bandwidth The range of frequencies over which a circuit, stage or component passes signals without more than a given *roll-off* attenuation, often –3dB or –1dB

bare-mounted Power semiconductors mounted without electrical insulation.

BBC British Broadcasting Corporation.

beat (i) Timing of sounds in music. (ii) The period intermodulation product or difference tone, created when two notes are sounded (acoustically) or frequencies mixed (electronically).

Belcher test An audio purity test designed by Dr. Alan Belcher while working at the BBC. Employs pseudo-random noise which simulates music program. The effect of an amplifier on the complex but known input components can be untangled afterwards.

belleville A compression washer used to reduce torque variations of bolted down electronic parts, e.g. with temperature cycling.

beta The raw current gain of a BJT. Strictly applicable only when the transistor is applied in the common emitter configuration.

bi-amped A two-way active speaker system or facility for such, employing separate amplifiers for each drive unit, i.e. 4 channels of amplifier are required for stereo bi-amping.

bias (i) A set of DC current and voltage conditions in a circuit that establish the device for operation. (ii) off centre. (iii) not impartial. (iv) the supersonic 'dithering' tone universally applied to enhance analogue tape fidelity while recording.

bi-filar (i) In resistors, a winding where the wire is folded back on itself, so the inductance is mostly canceled. (ii) In transformers, where windings are closely *mutualled*, to cancel leakage inductance.

B·I·L Formula expressing the force coupled from a speaker's voice coil, where B is flux density, I is current and L is length of wire immersed in the flux field.

bi-phase Two phase, usually referring to the opposing polarity (0,180°) drive, to push-pull output stages using transformer coupling. This approach was necessary when active devices came in only one polarity, ie. as with valves and early (pnp only) transistors.

bi-polar See *BJT*. Shortened name of the first widespread and most common kind of transistor. Can also mean 'bi-directional' (ie. able to go both ways equally), in context of monitoring circuitry.

BJT Bipolar Junction Transistor. See *bi-polar* and *FET*.

Bleeder A resistor arranged to bleed away stored charge at switch-off, for safety.

Blumlein integrator Synonym for *Miller integrator*.

Bode Plot Composite graphs of gain (in dB) and phase (usually in °), enabling margins of stability to be gauged from inspection.

bootstrapping Circuitry with behaviour analogous to lifting itself up by its own bootstraps. Usually refers to a capacitor in single supply rail versions of the Lin topology, which helps full output drive to be almost as large in one direction as the other. Can also refer to 'wobbled' supply rails, where half of double the rated supply voltage is available to the OPS at any instant.

BOS Bipolar (BJT) Output Stage

bottomed (i) When a loudspeaker cone hits the magnet pole piece, after being overdriven. This may be caused by a over-powered amp but more commonly arises when driven by high level (or excessively boosted) bass frequencies which attempt to exceed the drive-unit's mechanical excursion limits, Xmax. (ii) When a transistor (of any kind) is turned as hard on as it will go. See *Saturation (i)*.

box Concert PA, colloquialism for speaker cabinet.

breakdown (i) Electrical discharge or substantial current flow, in an insulator. (ii) Catastrophic transition in a semiconductor, to a permanently low resistance, with associated complete loss of control.

breaker Circuit breaker, i.e. resettable overcurrent protection against electrocution and/or fire.

bridge An ambiguity! Often refers to the bridge rectifier in PSUs. *No connection* with amplifier *output bridging* or *bridge mode*. Context usually makes the distinction.

bridge mode Where two outputs having the same absolute signal input and gain are in phase (or polarity) opposition, and where the speaker is connected across them. It amounts to floating balanced output drive.

bridging (i) See 'bridge mode'. (ii) When the input to a power amp has medium to high Z and ideally does not unbalance the line feeding it. (iii) Measuring or linking between two points in an electrical system, alias *jumping*.

brownout Sustained AC power pollution, where the PMR is depressed and harmonic content is especially high.

BS British Standard, a standard issued by BSI.

BSI British Standards Institute, equivalent of ANSI or DIN.

buffer A building-block part or circuit used to reduce signal-damaging interaction and improve relations between circuits or stages. Ideally draws only minuscule current from the source signal, has unity gain, and can drive awkward destination loads with little change in performance.

bulb (i) An incandescent lamp. (ii) The glass envelope surrounding a thermionic valve (vacuum tube).

burst power See *dynamic power rating*.

bush Plastic or other insulating, stepped collar used to align and principally isolate fixing screws from semiconductor bodies or heatsinks.

cascode A 2-device fundamental building block circuit. One device is connected in cascode, to shield the other from anode, collector or drain voltage swings (as applicable), hence reducing modulation effects and distortion. May be performed with the cardinal active devices (Valve, BJT, FET) and their combinations.

cathode (i) Negative terminal of a valve. (ii) Most negative end of a conducting diode or LED. (iii) *Positive* end of an active zener or 'Tranzorb' diode.

cathode follower A first order valve circuit, where the output is taken from the cathode. This gives unity voltage gain (no gain, no loss) but high current gain, so medium-to-low impedances (*ca.* 1k) can be driven without interposing an output transformer. Absolutely analogous to Emitter Follower (BJT) and Source Follower (FET).

cc Common collector. Alias for '(emitter) follower', a basic BJT circuit configuration.

CCIF (i) A test method for inter modulation distortion. (ii) An obscure Franco-European broadcasting standards organisation.

CCIR (i) A standard filter used for noise measurements. (ii) A Franco-European standards organisation.

CCW Counter clockwise. Anti clockwise.

CD Compact Disc, a digitally encoded recording.

CE, ce Common Emitter, a basic BJT circuit configuration, where the emitter terminal is at or nearest to ground. In its raw state without feedback, this configuration has medium voltage gain, and medium input and output impedances.

Cerafine A refined, novel dielectric composition for electrolytic capacitors, specially developed for audio by the capacitor maker Elna, in Japan.

CFB Current FeedBack. A loose description. See 4.10.1.

ch Channel, ie. a self-complete amplifier path.

channel separation Opposite of crosstalk. When one is high, the other is low. Usually expressed in −dB. A frequency or range must be given to be meaningful.

characteristic (curve) Behaviour of a semiconductor or other component, when relations between any two (or three) parameters are graphed together.

chip Vernacular for an IC (i.c., Integrated Circuit or monolithic circuit).

choke An inductor, usually one carrying mainly DC, with a smaller ripple (AC) component.

class A An output stage where the current delivered into the loudspeaker (or load) is never greater than twice the (high) bias current that otherwise passes through all output devices. May be single ended or push pull. Highly inefficient with music.

class B An output stage where except at crossover, the output current is provided solely by one of two conjugate 'push pull' halves at a time; and where nearly all output stage current flows through the load − and thus where most of the current only flows when it is needed to make music. Always push pull. Inefficient in places but a lot better than pure class A.

class A-B An output stage where the topology and output current are as class B above, but with a higher 'over' bias current that is between 0.01% and say 30% of the maximum output current. Bias current may be set over a wide range for improved linearity in the crossover region with complex tradeoffs, depending on active devices. According to bias current, A–B ranges from slightly less efficient than class B, to 'still more efficient than class A'.

class C Exists but solely used for RF (not audio) amplifiers.

class D An output stage using PWM ('digital') techniques.

class G An output stage with decked supplies. Higher supplies are connected by series transistors when signal peaks 'touch' successive lower rails. Greatly improved efficiency with music signals results.

class H Similar to 'G': an output stage with decked supplies. Progressively higher supplies are connected by parallel transistors to handle signal peaks which 'touch' the lower rails. Improved efficiency with music signals results. Pioneered by QSC.

clip, clipping Total overload, where signal peaks are flattened, the signal is eventually 'squared off' and clean sound with detail ceases. An abrupt change in transfer function caused by attempting to exceed the signal voltage swing limits in a conventional power amplifier employing negative feedback. Occurs also in amplifiers without feedback, but far less abruptly.

coil An inductor or choke.

colouration, coloration (US) General term for an irregular or afflicted frequency response, often also associated with aberrations in the time domain.

common emitter See *CE. Not* a Euro-mark.

comparator (i) A device or circuit that compares (usually) two signals or quantities, and outputs the difference. This type would be better termed a 'subtractor'. (ii) A circuit that outputs one or other logic states depending on the input values. This type is typically used to switch when signals are above or below a value.

compensation (i) Correction of phase shifts at high (radio) frequencies usually using a capacitor or resistor-capacitor pair. (ii) Other correction.

complementary Equal but opposite. In the context of amplifiers usually transistors (BJTs or MOSFETs) of opposite polarity (NPN or n-channel *versus* PNP or p-channel) but otherwise matched.

compression driver Drive unit optimised for horn loading.

conjugate The matching reciprocal part.

conjugate crossover Where the reactive elements have been cancelled by the inclusion of reciprocal reactances, yielding a resistive load.

consonant Musically pleasing.

constant current source See *current source.*

convection Cooling by natural air movement. When air is heated it expands. Therefore a given volume weighs less. Thus it rises. In a confined space, cooler air is in turn drawn in behind it. Applies no less to any gas or liquid.

cooling tunnel Heatsink formed into a tunnel down which a fan blows air.

CM Common Mode. A condition where equal and opposite transmissions, conductors, circuits or legs separate from ground have a common signal relative to ground super-imposed on them. This signal is nearly always environmental garbage.

CMG Common Mode Gain. The alternative perception of common mode rejection, when analysing circuit stages.

CMIR Common Mode Input Range. The range of common mode signals that will not cause damage or malfunction.

CMR Common Mode Rejection, ie. riddage of CM signals. Usually expressed as a ratio in decibels, e.g. −60dB is a 1000 fold reduction.

CMRR Common mode rejection Ratio, eg. −100dB, or 100,000:1, but unlike CMR, referred to the differential gain of the device. This can yield high but meaningless figures.

CMV Common Mode Voltage.

cold In balanced circuits, an *arbitrary* notation for one polarity, side or half, of the live signal. Opposite of hot.

CPC Circuit Protective Conductor. The 'earth' in AC mains wiring.

cps Cycles per Second – same as Hertz.

crest factor See *PMR*.

cross-conduction In output stages, where one side turns on substantial current while an uncontrolled current continues to flow in the other half, after it has nominally switched off. A BJT problem. Readily fatal.

crossover (i) Loudspeaker frequency divider & routing network. (ii) Point where music signal polarity flips twice every cycle, where the voltage passes through zero.

crossover distortion Distortion in non-class A output stages, caused by imperfect, non-monotonic (kinked) gain each time signal passes through zero volts.

crosstalk Leakage from one channel to the other in a 2 channel audio device or system, or between any other channels.

CSa Cross sectional area of a conductor, usually in mm^2.

CT, ct Centre tap – of a transformer; or the midpoint of resistors or reservoir capacitors.

current capability Amperes available from output. Values may be peak or rms, short or long term, and may vary with frequency and duty cycle.

current draw Current taken from the AC mains or other power supply.

current dumper The high current-handling part of the QUAD current dumping amplifier.

current dumping See *proprietary names* section – in appendix.

current feedback A feedback scheme where current is the sensed or controlled dimension at some juncture – rather than voltage.

current mirror Active circuit, element or IC which repeats (copies) a current in any required ratio, e.g. 1:1. Used for direct coupling, signal splitting and transformation.

current source Active circuit which ideally drives a chosen current (say 10mA) irrespective of the load it is driving. Ideally behaves like a near infinite voltage drawn through a near infinite resistance.

CW, cw ClockWise.

DAC Digital to Analogue Converter.

daisy chain(ing) A 'looping through' connection from a signal source into a number of power amplifiers, alias parallel or cumulative, so-called 'tandem' wiring.

damping The antithesis of resonance. The gradual dissipation of energy.

damping factor A notional or approximate figure of merit, based on the difference between the nominal impedance of a speaker, and the impedance of the source.

Darlington A second-order BJT circuit building block. Generally, the two BJTs are cascaded so their betas multiply. This yields very high current gains, from 2000x to above 250,000x. But HF stability and charge flushout, hence switching-off speed are problems.

dBA Weighted sound levels emphasising mid frequencies and based on approximate human hearing sensitivity around 70dB$_{SPL}$. See *A-weighting*. Largely ignores LF & HF frequencies. May be used to give some indication of effective audibility at moderate SPLs.

dBC Weighted sound level decibels. With music, dBC is almost a flat response test which measures most of what is really present.

dBm DeciBels referenced to a level of 0.775v, specifically into 600 ohms. Long obsolete but still often misused in place of dBu or dBv.

dBr DeciBels (dBs) referred to some defined or stated level, usually a voltage if unspecified.

dB$_{SPL}$ Decibels indicating (acoustic) Sound Pressure Level, with 0dB taken as the nominal threshold of human hearing.

dBu Decibels referred to a level of 0.775v, any impedance.

dBv The same as dBu. Note *small* v.

dBV Decibels referred to 1 volt, any impedance. Note big V.

dBvr/Ω dBs referred to full rms output voltage. '/Ω' is the load impedance, eg. /8 means 'into or when loaded with 8 ohms'. In this book, $\Rightarrow$ replaces /.

dBwr/Ω dBs referred into a given or the lowest rated impedance. '/Ω' is the load impedance, e.g. /3 means 'when loaded with 3 ohms'

dB$_{ZOL}$ dB referred to Zero (operating) Level. See ZOL.

DC, d.c., dc Direct current. Ideally, wholly invariant, fixed potential difference, defined as any having an average value over time that is non-zero. In reality, may have AC 'ripple' 'noise'; or other signals superimposed on it, or may be seen *as a component of* polluted AC.

DC balance When dc conditions for differential pairs of parts, or conjugate halves of a circuit, are equal but opposite, so the DC can be cancelled.

DC blocking A series capacitor guards the input to a stage.

DC coupled See *direct-coupled*.

DC/LF Low (audio) frequencies down to zero frequency (taken as DC).

DC offset DC voltage at an amplifier's output. Should be low or absent. Levels above 10mV ($^1/_{100}$th volt) can cause audible clicks or plops if speakers are plugged in when powered-up. DC above 1v will use up some of the excursion and power handling capabilities of hf and small lf drive-units, and above 5v, can cause burn-out.

DC protection Protection circuitry intended to protect the speaker and/or amplifier from damage due to excess DC voltages present at either input or output, due to defects in either the amplifier or the source(s) preceding it.

DC servo See *Servo*.

decade *Any* interval of 10, e.g. the difference between 10k and 1k; 10 and 100; or 0.1 and 0.01.

deciBel One tenth (deci) of a Bel. An aid to calculating levels in audio circuits, invented at Bell labs USA and named after Alexander Graham Bell, the Americanised, Victorian Scotsman who invented the telephone and was the first electro-acoustic engineer.

decoupled Preventing the passage of signals or coupling of parts of a circuit that may carry signals.

derate To use below maximum ratings.

derating The amount by which a component is used below its full ratings, usually to allow it to live far longer.

device Short for 'active device', synonym for 'a transistor'.

DFM Distortion Factor Meter.

differential Dual, two conductor or 'two phase' circuitry that is notionally balanced (in level, and impedance to all other points) but may not meet all the criteria. A more general term than *balanced*.

digilog Mixed circuitry which is both *digi*tal and ana*log*.

DIM Dynamic InterModulation, a relatively modern test developed to be sensitive to distortion that is produced during transient conditions typical of music program.

DIN *Deutsches Institute fur Normung*. German standards organisation equivalent to ANSI (USA) or BSI (UK).

direct coupled Circuitry that omits DC blocking capacitors between stages. Direct (also called 'DC') coupling has benefits but also setbacks.

discrete Composed of individual transistors, diodes, etc., cf. monolithic integrated circuits. With modern surface-mount parts, discrete assemblies need not be as bulky as with the larger parts used in the past.

dissipation Emission or disposal of waste energy as heat.

distortion General term for corruption, deviation from the original.

distro Professional touring abbreviation for signal (or mains power) *Distri*bution.

DMG Differential Mode Gain

DMM Digital Multimeter

DMOS A type of power MOSFET, optimised for switching but also applicable to linear amplification.

DNR Dy Namic Range. Sum of noise below $0dB_{zol}$ (zero operating level) + headroom above eg. $-80 +20dB = 100dB$ DNR.

DOM Digital Ohmmeter

drain One of the main terminals on a MOS-FET. It can be thought of as where the electron (or hole) flow is heading.

driver (i) A loudspeaker driver or drive-unit. (ii) A transistor in a penultimate stage.

(iii) The guts of a power amplifier excepting the final stage – and associated high current parts.

driver transistor See *driver, (ii)*.

DSP Digital Signal Processing.

DTSEC Differential To Single Ended Converter. Shorthand for any differential or balanced input receiving stage, that delivers an unbalanced output onto the equipment's subsequent, internal circuitry.

dual (i) A device having reciprocal behaviour, eg. an ideal capacitor is the dual of an ideal inductor. (ii) A semiconductor, usually MOS-FET, comprising two paralleled dice.

dummy load A load for testing, that simulates a speaker but produces no sound.

D.U.T, dut Device Under Test.

DVM Digital *V*olt Meter.

dynamic power rating Power delivery available for short periods, under 1 second, with implication that this is higher than the long term power delivery capability.

dynamic range, DNR The difference between the maximum output level (MOL or clip level) and the noise floor, all usually expressed in dB. It can be calculated from common hardware specs by adding SNR (signal-to-noise) and headroom figures, eg. If SNR = 110dB below zero level, and headroom = 2dB above zero level, DNR = 112dB.

Dynamic range for all music is ideally 160dB, covering SPLs from –15 to +145dB. Live analog audio in concert sound systems comes closest to this, with over 140dB range being theoretically possible, room and crowd noise allowing. 16 bit digital audio is notionally about 100dB. Pure 20 bits increases the DNR to 120dB. Once lost deep in noise, the true content of the dynamic range of a signal is irrecoverable – however much the surviving range content can be stretched by expansion.

dynamic speaker The most common, conventional type of drive unit (not enclosure), *cf.* an ESL or ribbon type.

Early effect Change in the signal handling parameters of a BJT, as the collector-emitter voltage swings over a wide range. Discovered early on, by Dr. Early. Esoteric amplifier topologies that have little or no internal voltage swing e.g. cascodes, are aiming to avoid distortion caused by the Early effect. Also prime cause of second breakdown.

earth (i) The natural soft surface of the planet, the ultimate reference for dissipating RF and power grid currents. Called 'Earth ground' in US. (ii) The green-yellow wire on power cables in most countries, eventually bonded to earth (i). (iii) Building frames and other large areas of metal or other conductors connected to earth, (i).

eddy currents Parasitic currents mutually induced into conductive materials that are neighbouring inductors or transformers, or other conductors passing high currents.

EHT Extra High Tension. Usually a voltage over 1kV. Normally used only in high power and esoteric valve amplifiers, or where cathode ray tubes are used, as in 'scopes, computer monitors, TV and other video screens. EHT can *kill. Observe precautions!*

EIA Electronic Industries Association (US).

elcap Electrolytic capacitor.

electro-acoustics Audio engineering where loudspeakers, headphones or microphones, and by implication, acoustics, are involved.

electrostatic coupling Induction of noise or signals into neighboring conductors or circuitry by electric field.

electrostatic shield Inside transformers, one or more *e.s.s.* or *i.w.s.* prevents capacitative leakage between primary and secondary or other or all windings. May also be used for electrical safety, diverting any flashover current to earth.

EMC Euro-legislation for Electro Magnetic Compatibility.

EMF Electro-Motive Force, the voltage behind a battery or other driving source.

EMI Electro Magnetic Interference. Electrical, magnetic, RF and/or static noise or spuriae released by human-made apparatus, or naturally by lightning and solar activity.

emitter follower A bipolar transistor connected so the signal emerges with a low source impedance, from the emitter, and where voltage gain is almost x1 (unity) but current is amplified, approximately by *beta*. Also called 'common collector'.

EP A proprietary heavy duty multi-channel (commonly 6 or 8 pin) speaker connector used to connect touring PA cabinets to amp racks.

EQ (i) 'EQualisation', alias *ad hoc* tonal manipulation. (ii) specific, usually fixed, frequency response tailoring.

E-Series A constant percentage logarithmic group. For example, the E3 series splits every decade into three nearly equal % parts: 1 to 2.2; 2.2 to 4.7 and 4.7 to 10.

ESL ElectroStatic Loudspeaker. Radically different to the common moving-coil kind, while a contemporary invention of the late 1920s.

ESR Effective Series Resistance. The resistance of an electrolytic capacitor. Varies with frequency,. therefore a non-linear resistance.

es shield see electrostatic shield

ETC Energy-Time curve. Plot of energy *Vs.* time.

EU European Union. Previously known as E.C. and E.E.C.

$f_{0.1}H$ where HF is −0.1dB down.

f_3H where HF is −3dB down.

f_3L where LF is −3dB down.

fan Electric cooling fan or other air blower.

FCC Federal Communications Commission, US agency responsible for RF interference matters.

FCW Fully Clockwise

feedback, negative Negative Feed Back (NFB) is the universal method of error-correction, found in the cosmos, in nature, in social systems and engineering. In all analogue electronics including amplifiers, it may be applied locally, globally or in some combination, nestedly. Equipment made without any or much NFB generally requires substantial matching and tuning effort and is more prone to deviate with use.

feedback, positive Destructive type of feedback that uncritically re-inforces the signal. The initial condition is magnified. The outcome is either saturation or lock-up; or oscillation or howling. But small amounts may be use to realise control over some analog and amplifier circuits, eg. to set specific output impedances, or to reduce %THD by increasing gain within an NFB loop.

feedforward Alternative kind of error correction, potentially superior to traditional negative feedback. The error is derived concurrently, sent forward, and subtracted from the output.

ferrite A concocted magnetic material suited to high frequency use, and also speaker magnets. Made from aniron (Fe) metal oxide powder.

FET Field Effect Transistor (J-FETs, MOSFETs). JFETs are frequently used in amplifier input stages, as the long tailed pair. MOSFETs are commonly output devices. In a minority of designs, they are used as drivers and ss devices.

fidelity Exact correspondence and faithfulness, in reproduction.

Fletcher-Munson Usually refers to the first widely published set of equal-loudness contours for human hearing, produced by the named at Bell Labs, USA. But these measurements were made over 60 years ago and are at best an approximation, as human hearing has a wide spread of at least 20dB.

floating Not referred to (nor needing referral to) ground/0v.

flown The suspension of loudspeakers and sometimes the associated power amplifiers, above the stage height (PA jargon), or wherever speakers are hung from wires, ropes or chains.

flushout A resistor placed between base and emitter of Bipolar power devices, used to speed-up the removal and dissipation of charge carriers (electrons or 'holes'), for faster or cleaner switching.

follower A common way of using transistors in final output stages. See 'emitter follower' and 'source follower'.

forward active mode When all semiconductors in a linear, usually direct coupled circuit, are adequately biased, eg. all silicon junctions are between above 450mV, up to 700mV (at 20°c).

forward bias As *forward active mode*.

FPBW Full Power BandWidth

frequency distortion Synonym for ragged (or non-uniform) frequency response.

frequency range Often used to denote where response limits have fallen by −10, −3 or −1dB. See *frequency response*.

frequency response How relative level or gain (usually in dB) varies with frequency (in Hz).

FTC Federal Trade Commission (USA).

FWR Full Wave Rectifier

gain Increase in signal magnitude (size, level). Increase may be principally manifest as current or voltage. Increase in either may also involve a power gain.

gain control A knob (or any incremental control in software) connected to a pot or VCA or other control element, usually attenuating the amplifier's preset, fixed gain.

geometry The 2 dimensional topography of a planar type of semiconductor.

germanium The element from which the first transistors were fabricated. Replaced by *silicon* semiconductors between 1960 and 1970. Germanium transistors have a low and gentle turn-on voltage *ca.* 0.25v at room temperature. Their setback is high leakage current (I_{cbo}) at even modestly high temperatures. They are thus particularly unsuited to power stages.

grid Control port of a valve. Literally a wire grid-mesh.

ground As earth, but more often an isolated or at least separated reference conductor for equipment's internal circuitry.

grounded bridge A circuit (patented by Crown in 1979) where two channels each have floating power supplies. Each output is a bridge type, and unusually, as a function of the original grounding scheme, each output can be mutually bridged.

group delay Different, but frequency-related, frequency-dependent delays in a signal, eg. bass portion of a sound arriving after treble.

ground lift A jumpering arrangement or switched component or combination of simple components, usually resistors, chokes, capacitors or diodes, used to isolate chassis/safety earth from signal 0v or ground.

ground loop A circular or multiple connection of ground conductors. The circuit formed acts as a shorted turn in the presence of power line magnetic fields (50 or 60Hz), hence high currents may flow, causing substantial 50/60Hz 'hum' voltage differences to appear along the wire in such a loop, even if the wire's resistance is low.

hard shorting A short circuit with a very low resistance and high current capacity, i.e. a solidly made short, *cf.* say a sliver of wire accidentally just touching a dirty live surface.

harmonic In amplifiers – a spurious tone that is an integer product of a lower frequency tone. The latter is called the fundamental in audio & electronics engineering – but musicians call the same 'first cause' their first harmonic, and all resulting harmonics are called overtones.

harmonic distortion The distortion caused by the generation of harmonics. Although always caused by 'bending' of the signal, and still occurring in all power amplifiers in measurable quantities at higher audio frequencies, variations of harmonics are infinite and perceived effects vary widely.

HCMV High Common-Mode Voltage.

headroom Level or 'acting space' above *zero level* before *clip*. Usually in dB and generally at least 15dB for professional equipment. May vary with gain or mode settings. Does not usually apply to power amplifiers or D/A convertors.

heat exchanger Synonym for a heatsink.

heatsink A dissipation aid. Generally a highly conducting metal connecting to the power semiconductors' bases, and having far greater thermal mass and dissipative capacity than the semiconductors.

hermaphrodite one gender connector that plugs into its own kind.

H-field (i) Magnetic field (ii) The magnetic part of an e-m wave.

high mains When AC mains voltage is above the norm.

HMP High Melting Point, solder that withstands high operating temperatures.

HNFB High Negative FeedBack. True of most transistor power amplifiers designed since the early 70s, excepting some esoteric types. Most valve amplifiers are only able to use moderate negative feedback.

holes The opposite of electrons, acting like virtual positrons, that carry current in pnp BJTs. Holes' speed of propagation is slower than that of electrons, hence pnp BJTs are intrinsically slower than npn of the same geometry.

hot The unearthy 'high' side of an unbalanced signal; or an arbitrary notation for one polarity, side or half, of a balanced signal.

HPF, h.p.f High Pass Filter, removes low frequencies.

HT High Tension. (i) The high voltage supplies (generally > +100v) required for anode and screen grids in valve/tube amplifiers. (ii) In transistor power amplifiers, the supply used by the output stage – even if this is only (say) 25v.

hybrid (i) A micro-circuit that contains both chips (dice) and discrete components (usually surface mount). (ii) Any circuit that involves different active devices or technologies, e.g. J-FET/BJT; IC op-amp/Valve.

Hz Hertz, unit of frequency, ie. rate in cycles per second.

IC Integrated circuit, 'chip' or monolithic circuit, made by building up on, and etching patterns onto, successive layers of metals, semiconductors, and insulators, beginning on a substrate of silicon.

IEC International Electrotechnical Commission. A worldwide standards body.

IEE Institute of Electrical Engineers (UK).

IEEE Institute of Electrical & Electronic Engineers (US).

IHF Institute of High Fidelity (US)

IMD, i.m.d Inter-Modulation Distortion or *Intermod* distortion. Artifacts that are not harmonically related to (ie. integers of) the stimulus signal's frequency. Such artifacts are highly discordant.

impedance Resistance to AC (including music) signals.

impedance match A precise requirement for RF and digital transfer, where cables have definite characteristic impedances. Generally, impedances are 50 or 75 ohms for coaxial lines; and 100 up to 600 ohms for bi-axial or spaced lines. There is no optimum condition like this for most analogue audio connections, excepting professional microphones (approx. 1200Ω) and the setting of output transformers, to get the maximum power and damping.

inductance A reluctance for current to cease flowing immediately, according to the strength of magnetic field caused by the current flow. Thus inductive impedance increases with frequency. Inductance in wires is increased by length, somewhat by thinnness, and spacing from the return conductors. When required, inductance is parcelled by coiling up a long wire.

infrasonic Frequencies below normal audibility. Often called 'sub-sonic'.

inharmonic Musical term for harmonic partials that don't have an integer relationships with the stimulus. See *IMD*.

input impedance The loading or impedance seen looking up the input terminals of an amplifier or other audio equipment.

inrush current Current surge in mains supply occurring when a large amplifier (generally rated above 100w) is initially switched on. Severity depends on timing against instantaneous mains voltage.

instability (i) RF oscillation, either fully continuous (CW), or periodic, or triggerable. (ii) Audio frequency, e.g. LF oscillation – 'motorboating' – was common in tube and especially battery powered R-C coupled amplifiers, up to 1970. Almost unknown in direct coupled, transistor amplifiers. (iii) any other triggerable ill event, random noises, and other effects of design defect, decaying parts, bad connections, or chemical contamination; or sonic variability.

interdigitation In power transistors (esp. BJT), the combining of a large current carrying capacity with speed, by splitting emitter &/or collector lands into many thinner sub-areas. These are commonly formed into maze-like patterns.

interstage crosstalk Signal leakage between stages of analogue electronic circuitry. More general than *backtalk*.

inverting A polarity flip or 'phase inversion' of 180°. In most cases inaudible by itself, but presence or absence could affect rf stability, if an amplifier's input and output cabling is *incorrectly* run close and parallel.

invertor (i) A circuit that inverts. (ii) A power supply circuit that provides AC mains from 12v DC or similar.

IOA Institute of Acoustics (UK)

I_q Quiescent current

ISO International Standards Organisation

ISVR Institute of Sound & Vibration Research (UK)

IWS, i.w.s. Inter Winding Shield – see *electrostatic shield*.

J.AES Journal of the Audio Engineering Society.

JFET Junction Field Effect Transistor. A small signal, low voltage rated species of active device that is fundamentally different to a BJT while achieving sometimes similar end-results.

JIS Japan Industries Standards.

jitter Digital timing inaccuracy. Major cause of poor sonic quality in digital audio. Also called 'phase jitter', and defined as 'short term non-cumulative variations of the significant instants of a digital signal from their correct positions in time'.

joule, J Unit of energy. 1 J per second expends 1 watt of power.

junction The guts of a power semiconductor. Typically an area of a few square millimetres (or less) where the heat originates from.

junction temperature The temperature at the junction. Usually nowhere else will be hotter.

kHz kilo-Hertz, thousands of Hertz

kilowatt, kW thousand(s of) watts.

knee Where a curve of how a signal behaves, makes a quite modest but distinct change in direction.

k-ohms or kΩ kilo-Ohms, thousands of Ohms

kWh kiloWatt–hour(s). Rate of electrical power consumption.

L_{eq} An SPL reading that is 'integrated' (averaged) typically over a period of several minutes. The period should be cited. It is akin to long term or 'slow' rms.

large signal A classification for circuit analysis. May be defined as a large-to-relatively large signal, usually causing swings within the top 8dB or >20% of the amplifier's output current or voltage range. For example, with a 500w into 8 ohm rated amplifier, peak output currents *above* 1.6A and likewise voltages above 20% of rated levels, ie. 12 volts, would be considered a *large signal* condition. Generally, the context is provoking non-linearity, protection, slew limiting, over-heating or other problems which do not occur at all below a certain threshold that is relatively close to the maximum.

LAT-MOS See *L-MOS*.

LDR Light Dependent Resistor.

leakage inductance The parasitic, residual inductance in a transformer winding, when all other windings are shorted. Degrades performance, eg. limits hf response of a valve amplifier's output transformer. Can be greatly reduced by careful design and construction.

Lemo Proprietary, miniature, rugged multipin connector used to daisy chain 3 or 4 bands of audio around professional amp racks.

levity Low in mass for size, i.e. light in weight.

LF Low Frequency.

limiter A compressor with a high to notionally 'infinite' dB_{in}/dB_{out} ratio, intended to prevent hard clipping even against over 10dB of overdrive.

Lin circuit The original transformerless, asymmetric transistor power amplifier topology (Lin, 1956) around which many amplifiers are still based.

line level The signal level at the input of an audio power amplifier – as opposed to usually higher voltages at amplifier outputs.

linearity Straightness of transfer (output/input) function. High linearity is desirable for accurate reproduction in amplitude. But beyond a point of relative perfection, other factors have greater effect on sonic quality. Also, the shape of any non-linearity can be of far more importance than its relative size.

listening level An optimum or comfortable level for listening. This may be system and room, as well as listener-dependent.

LMOS, L-MOS Lateral (channel) MOSFET. Early derivative of D-MOS, pioneered by Hitachi *ca.* 1977, for audio, linear ultrasonics and low RF.

LMP Low Melting Point, solder used for delicate repairs.

load Usually comprises a loudspeaker, or drive units; cabling and passive crossovers. Sometimes a loudspeaker is fed via transformers, eg. public address but also ESLs.

load line A (usually sloping) line that represents resistance on a graph of voltage and current excursion in a circuit, usually concerning operating conditions of an active device.

long-tailed pair, LTP The differential amplifier topology invented by Alan Blumlein in 1936, to transmit the original BBC TV signals down cables. The 'long tail' was the ultra high resistance connection to the cathode terminals, which are joined. The original valve circuit has long been translated for use with BJTs and FETs, and forms the input stage(s) of most op-amps and nearly every power amplifier using high negative feedback.

logstep A logarithmic step, i.e., a multiplier, eg. x1.12, rather than an arithmetic step, eg. keep adding 1.

L&R Left and Right channels of a stereo signal path.

LS, ls Loudspeaker

LT, lt Low Tension, alias low voltage, usually under 50v.

LTP Long Tailed Pair.

LV, lv Low Voltage. Definitions vary from below 1v to below 1500v, depending on context.

LVD Low Voltage Directive, a European electrical safety law.

M2, M3, M8 Sizes of medium-coarse metric machine screws and nuts.

magnetostriction Convulsion noise caused by change in size, made by magnetic cores and associated winding wires when magnetising current is (first) applied or changes abruptly.

matched pairs/sets Usually output valves or transistors, matched as complements, or for gain (H_{fe}) or (specific, benign) distortion spectra, all to enhance sonics.

mica Transparent insulator/conductor. A natural mineral, cleaves and related to slate. Superb electrical insulator but equally a relatively good heat conductor (as insulators go). Quite fragile and may be punctured or exhibit local hv breakdown if fractured.

microphony When vibration generates electrical noise or signals – which may be in the form of highly corrupted music. Used sparingly and appositely, microphony has added special qualities to sound in many recording. When in playback amplifiers, and thus affecting listening, it will blur details and add false euphony, depth, etc.

midband The middle of the human hearing range, where the ear is most sensitive, typically 500Hz to 5kHz, centering on 3kHz.

Miller capacitance Parasitic plus any explicit capacitance appearing between the high level and control terminals of an active device, hence C_{ga} for a triode valve; C_{bc} for a BJT; or C_{gd} for a FET. The effective capacitance is magnified by any voltage gain between the two terminals.

Miller integration A descriptive synonym for *Miller capacitance*, but may include effects of external capacitance.

mm² measure of a conductor's cross-sectional area, hence resistance per unit length and current rating.

mode conversion In balanced connections, when a common mode signal is irretrievably converted into a differential signal, e.g. due to bad grounding and/or cable capacitance imbalance.

modulation noise An error or noise signal that is only present concurrent with programme. It may therefore not be clearly audible for what it is, although its own detail-masking effects may be quite audible.

monotonic A variation in some parameter than continues, once begun, to either decrease or increase, and not waver or revert, e.g. a non-dipping slope on a graph.

MOS-FET Metal Oxide Semiconductor - Field Effect Transistor. 'MOSFET transistor' is a tautology! To stress transistor *cf.* valve, say 'MOS transistor' instead. There are several types of MOS FETs, constructed differently to each other and to J-FETs. See *FET*.

MSR Mark–Space Ratio. The durations of alternating conditions in squarewaves and also in complex but periodic signals.

MTBF Mean Time Before Failure. A *statistical* average lifespan of a component under defined conditions.

mu-metal An alloy providing powerful magnetic shielding for or against transformers and inductive parts. Expensive, and the shielding performance can be ruined by shock, bending or working.

music power A long disreputable means of hyping a the 'power rating' of an amplifier, which itself has only a faint connection with how loud it will eventually go. Still used in emerging markets. Generally, the power is derived by squaring peak rather than rms voltages, or by simply doubling or even quadrupling the *average* power figures (based on squaring true rms voltage).

musicality Abstract noun for equipment that is perceived as having 'good sound' and in particular, allowing through emotional undertones and rhythmic infection that are not transmitted by other equipment that may measure better – in the limited areas that are measured.

NFB Negative Feed-Back. See *Feedback, Negative*.

non-inverting When signal polarity is not overall flipped between the measurement points, usually input and output.

non-linear When the output of an amplifier (or other linear path device) is not linearly related to the input signal, ie. the in/out relationship is kinked and/or otherwise distorted.

non-linearity The abstract noun of 'non-linear'.

ntc See *thermistor*.

octave An interval of 2:1 or 1:2. And thus a measure of frequency that approximately follows human perception, in being ratiometric (or logarithmic). An octave upwards is a doubling; two octaves, a quadrupling of frequency. An octave down is a halving, two octaves a quartering.

offset voltage Stray DC voltage, ideally small, preferably in tens of millivolts or far less, present in internal circuitry and also at the output terminals of most direct-coupled, usually transistor amplifiers.

ohmage Colloquialism for resistance (in ohms).

on resistance (R_{on}) DC resistance between a FET's drain and source terminals, when turned on fully by gate-source voltage.

OPN OutPut Network, comprises several passive components, required to isolate power amplifiers with global NFB, from cabling and the load, above the audio range.

OPS OutPut Stage. Generally the high current, final stage(s) of a power amplifier.

OTL Output TransformerLess.

output impedance The 'source impedance' seen looking up the source's output terminals.

output transformer Used in most valve, and early transistor designs, to enable the valves (or transistors) to 'see' a suitable, medium impedance, while driving a speaker with a low impedance. In effect, used to exchange high voltage swings for lower ones, but at higher current.

out-of-phase (i) Analogous to a polarity reversal (when 180°). (ii) A phase shift or difference of any other amount, e.g. 13½°.

overall feedback Synonym for global feedback.

overload level Similar to 'clip level', also used where hard clipping does not occur, and defined as the level above which sonic quality passes acceptability due to discomfort, gross distortion or risk to the equipment.

overvoltage Excess voltage; beyond rated limits.

PA (i) Power Amplifier. (ii) Live music reinforcement system. (iii) Public Address system, for speech. (iv) Personal assistant

passive component(s) Resistors, capacitors, inductors, transformers; any part in an amplifier that isn't an active component or essentially mechanical.

parallel mode Two (or more) amplifier channels connected in parallel at both inputs and outputs. All gains must be closely matched. With two channels, impedances of up to half the rated per-channel-minimum may be driven. See *tandem*.

parasitic(s) Stray capacitance, inductance or resistance.

PC Personal Computer.

peaking out See *clipping*

peak limiting See *clipping*

pentode An amplifying valve with five active terminals, 3 of them grids.

PFC Power Factor Correction. Increase percentage utilisation of plant.

phase (i) Differences in relative or apparent timing of two signals, based on division of a cycle into 360 degrees (°), and with the earlier, stimulus or reference signal being called 0°. (ii) Colloquial misnomer for *polarity* (see).

phase shift Any deviation in phase, usually from 0°, the arbitrary reference. May be relative or absolute, and positive, negative, or complex.

phase splitter Not really splitting 'phase'. A circuit with a bi-polarity or push-pull outputs, havaing 180° phase difference between them.

pi-mode A variant of class A-B.

pk Peak

peak (i) Largest momentary value (in a period). (ii) Instantaneous momentary values *cf.* average, rms. (iii) An optimum point.

pig A large mains Variac, usually 100A or more, used to convert kilowatts of equipment mains voltage ratings without resorting to individual setting. Also used for large isolation transformers, used for safety.

pinfin A 'pin cushion' heatsink first made by Pantechnic (UK) in 1981.

pink noise Noise with equal energy per octave or decade. The sum of all music approaches this attribute. Pink noise is accordingly a good pseudo-music test signal, that's random but has a predictable behaviour on average, and is used as such. Making good pink noise is another matter.

piv, PIV Peak Inverse Voltage. The maximum rating of a diode.

PLASA Professional Light And Sound Association (UK).

PMR Peak to Mean Ratio, synonymous with *Crest Factor*. A characteristic signal voltage ratio. PMR = 3dB for a pure sinewave, about 10dB for compressed music, 15dB to 20dB for most music styles, and around 30dB for orchestral works.

P_0 Power output.

polarity The 'parity' of an audio signal. There are two states. Moving to the other one is called 'polarity inversion'. Polarity matters most when signals are mixed or when they return to the air via speakers as air is asymmetric.

polycarbonate A medium grade of capacitor dielectric.

polyester A medium grade of capacitor dielectric. Full name *polyethyleneterephalate*. Alias *Mylar*.

polypropylene A high grade of capacitor dielectric. Non-polar with very low loss.

polystyrene A high grade of capacitor dielectric. Non-polar with very low loss.

pot Potentiometer, variable gain or attenuator, or other sweepable control.

power bridging Same as 'bridging'

power density Power output capability expressed in terms of physical, spacial volume, eg. watts per cubic foot, or per rack unit.

power factor, PF In AC power usage, generally the ratio of actual power consumed to 'apparent power'. A typical power amplifier's DC supply has a notional PF of 0.5 to 0.7. When PF approaches a best case of 0.999, the load looks purely resistive, draws a sinusoidal current, and makes best use of given cabling, transformers and other AC distribution plant. See *PFC*.

p-p Peak-to-peak, the maximum value of a varying quantity, usually voltage or current.

ppm (i) Peak program meter, makes peaks visible by holding them briefly. (ii) Parts per million, e.g. $0.001\% = 10$ ppm.

pre-driver Transistor(s) before the driver(s). May be first stage of triple.

prf, p.r.f. Pulse Repitition Frequency, the rate at which a pulsed signal reccurs, rather than the frequency of the signal that makes up the pulse.

PROM Programmable Read Only Memory. An IC used to program a microprocessor, usually pluggable for ease of update.

program (US), programme (i) A music signal, usually composite, having typical dynamics and spectral content. (ii) A speaker rating based on a signal having these characteristics.

psophometric filters Filters that account for psychoacoustics, to aim to give a closer account of 'how it sounds', inevitably under quite specific conditions.

PSR Power Supply Rejection. Avoidance of coupling of aggressive non musical signals present on most unregulated power supply rails, into the signal path.

PSRR Power Supply Rejection *Ratio*. Rejection of signals on supply rails, expressed in dB or more archaically, as a crude ratio eg. $10,000:1 = -80$dB.

PSU, p.s.u. Power Supply (Unit). In audio power amplifiers, most often accepting AC mains input and giving DC output(s).

push-pull Where opposing transistors or valves alternately pass signal current, usually in an output stage.

PWM *P*ulse *W*idth *M*odulation. A 'digital amplification' technique, where signal amplitude is represented by the duty factor of a pulse train. Also used for switching psu control.

Q Sharpness of the approaches to a resonance. An important concept in EQ, and for EMI control, as at RF, even grounds and rails can have 'Q's causing malfunction.

QA Quality Assurance. Also 'Q-A'.

quiescent Standing, with connotation of inert, invariant.

quasi-complementary Usually refers to an output stage architecture, popular between 1966 and 1972, where complementary transistors are 'faked' by constructing a power pnp (say) from a power npn and a small signal pnp. This scheme requires care to attain the symmetry implied by the 'complementary' tag. This arrangement was superseded by the widespread introduction of (almost) complementary pairs of transistors, but is still favoured for engineering elegance and cheapness.

ramping, up/down Increasing or decreasing some quantity in a linear or smooth fashion.

RCA Radio Corporation of America. Late.

RC, R•C, R-C Resistor-Capacitor duo. If in series, may be called a Zobel. In either series or parallel, most often used in power amplifiers for stability compensation.

RCCB Residual Current Circuit Breaker. Protects against shock or fire.

RC coupling (Resistor-Capacitor) A synonym for transformerless yet DC blocked coupling, particularly between each stage of simple circuitry. Compare to *direct coupled.*

rectifier a diode dedicated to converting AC power to DC *cf.* small signal duties.

reflection When energy in an electrical system propagates bi-directionally; or is viewed in the counter direction.

regulation Control of some parameter by feedback loops, usually negative. In amplifiers, normally DC voltage and (DC) bias and stage current are or may be regulated.

regulator In amplifiers, usually an IC &/or discrete circuitry that provides a fixed, regulated voltage source, either as a supply rail or as a reference point.

reservoir Bulk DC supply storage, decoupling and smoothing capacitors.

residue Unwanted signals, i.e. noise and distortion left after negative feedback or some other error-reducing or cancelling mechanism has done its best.

resolution (i) A measure of the smallest possible increment of change that a given device is capable of. (ii) Sonic coherence, detailing or holographic action.

resonance Natural interchange of energy between associated capacitative and inductive elements leading to oscillation at a specific frequency.

resonant PSU A power supply where energy that would be wasted as heat is recycled through synchronous resonant energy storage. The result can be cleaner energy conversion with less RF noise

returns multi In concert PA, the multicore cable carrying the 3 or 4 bands from the crossover at the mix position, to the power amplifiers, in the stage wings.

RF Radio Frequency. Generally any frequency above 20kHz, and certainly above 150kHz, where the Long Wave AM broadcast band begins.

RFI Radio Frequency Interference, either unwanted breakthrough of radio transmissions, or EMI.

rf oscillation An amplifier excites and sustains its own signal at radio frequencies, often far above audio frequencies. Inaudible yet can harm the amplifier and speakers. Can be triggered by exceptional lead lengths, unsuitable speaker cable, open circuit cabling, and high RF field strengths. Rare in fixed systems with short leads.

rf protection Circuitry which protects an amplifier if RF oscillation or signals are detected. May also refer to RF filters and traps preventing ingress, and stabilising parts, that may prevent RF oscillation.

ringing A latent oscillation, on the verge of being continuous.

ripple (i) AC/DC conversion artifact, product of rectification, occurs on unregulated supplies. See *ripple current, ripple voltage*. (ii) Local, cyclic variations in frequency response, appearing wavy on a graph, caused by steep filtering, eg. for anti-aliasing in ADCs.

ripple current The pulsating current flowing in the DC reservoir capacitors of any conventional or switching AC-to-DC power supply. Causes heating in reservoir capacitors and is the main cause of their decay (if electrolytic), and eventual failure.

ripple voltage A sawtooth shaped signal that appears on unregulated DC supplies that are AC-derived. Fundamental frequency is twice the AC frequency if rectification is full-wave. Harmonic content is high. Cause of familiar raspy background hum in decaying or low performance amplifiers.

rms, r.m.s. Root mean square, shorthand for the mathematical operations performed to compute effective average ('mean') value of any complex AC signal that has the same heating effect as a DC voltage of that value, e.g. 100v DC = 100v AC *if* rms.

rolloff point Where HF and LF responses ultimate decay has set in by −3dB (70%) or any other specified amount.

rubbing coil A loudspeaker drive units' voice coil damaged by overheating or dislodgement, so it rubs, scrapes or chafes in the gap.

saturation (i) When power transistors are driven to or beyond 'hard on'. In this condition, if BJT, it may take a long time to switch them off. (ii) When a magnetic core in an inductor or transformer is pushed to it limits, above which its ability to magnify natural permeability drops abruptly. (iii) when a sound system gets no louder when driven harder, due to compression effects.

schematic Circuit drawing.

screen See *shield*.

second breakdown See *secondary breakdown*.

secondary breakdown Primary breakdown of any transistor is simple excess dissipation, eg. above 100 watts at *any* combination of current or voltage. Second breakdown is where the critical wattage reduces unexpectedly, due to thermal stress mechanisms with a potential for positive feedback.

self-bias In valve circuits usually, where the voltage drop across a cathode resistor, or contact potential, is used to develop a stage's biasing voltage.

sensitivity Amount of input voltage needed to achieve a given level at a power amplifier's output. *Low* sensitivity means a *large* input voltage is needed, and *vice versa*.

separation The isolation, a figure of merit, usually between 2 channels. The antonym of *Crosstalk*.

series feedback One of the possible connection schemes for global negative feedback. Commonly called an inverting stage.

series-resonant One of the two cardinal forms of LC or resonant circuit where the resonant elements form a series circuit.

servo Another name for corrective feedback. A DC servo enables a high degree of NFB to be applied at DC, with little or no additional feedback at audio or even subsonic (strictly infrasonic) frequencies. The DC servo's feedback action continuously works to reduce or 'autonull' DC offsets at the amplifier's output.

shield The conductive covering of a cable or enclosure. Drains away incident, induced RF currents, ideally protecting the audio signals travelling along the conductors inside. An equipotential surface, akin to a *Faraday cage*.

shoot-through In class D amplifiers, the equivalent of cross-conduction. Faster and more spectacularly destructive.

short A short circuit, a connection that has near zero resistance, relative to the source impedance of the current that does or will potentially flow through it.

shunt Components connected in parallel, ie. presenting an alternative path.

shunting Connecting components in parallel.

SID Slewing Induced Distortion. See 5.3.1.

Siemens Unit of conductance. Inverse or dual of Ohms.

silicon *cf.* older *germanium* transistors and semiconductors. Only applies to BJTs. All FETs are silicon.

silpad Insulating material for power transistors, that replace mica, made with silicon and chemically impregnated cloth.

single ended (i) An output stage (particularly) which is not push pull, nor symmetrical, nor employing complementary parts or halves. (ii) An unbalanced and non differential, circuit or input stage. (iii) A push-pull ('half-bridge') output stage operating from a single rail DC supply, thereby requiring an output DC blocking capacitor. (iv) A non-bridged power amplifier.

single pole (i) The basic rate of rolloff, –6dB/octave, developed by a single RC or RL pair. (ii) A single 'channel' of switch.

skin effect A error mechanism in conductors at high frequencies, where the signal is encouraged to see the path of least resistance in the surface of a wire. Thus the conductors' apparent resistance increases with frequency. The resulting impedance is $\sqrt{L}$, ie. 'the square root of an inductance'. This rises at +3dB/octave.

slave A low-budget PA amplifier with minimal facilities. Musician's parlance, dates from mid '60s.

slew limit The maximum rate of change of signal that an amplifier can process accurately, above which severe distortion sets in. Some large ultrasonic and RF signals are capable of exceeding the slew limit (or 'rate') of amplifiers with marginal slew rates and/or poor RF filtering. Usually measured in v/µS

slew rate, SR Same as slew limit. But can also refer to the actual rate of slewing in the stimulus (the signal) as opposed to the limits of the hardware.

sliding bias Dynamic bias for Class A, making it almost as efficient as Class A-B.

SLS Simulated LoudSpeaker

small signal A signal excursion small enough not to cause slewing or clipping or large-signal non-linearity, generally below 20% of rated output voltage.

SM Surface-Mount, miniature component(s).

solid state Any amplifier using BJTs, JFETs, MOSFETs or IGBTs (and any future semiconductors), instead of valves, which have since been called 'vacuum state'. Strictly, the two are not comparable as vacuum is not quite a state of matter.

source follower MOS-FET equivalent of an emitter follower. Has high input impedance, low output impedance, thus very high current gain. Voltage gain is about 0.9x.

speaker level as opposed to line level. Signal voltages peaking at from about 4v to over 230v rms.

Speakon Proprietary, purpose-designed loudspeaker cable connector family (Neutrik).

SMPS, s.m.p.s. Switch Mode Power Supply.

SMPTE Society of Motion Picture & Television Engineers.

SMPTE–IMD The original IMD test, originated by SMPTE.

s/n ratio, s.n. signal/noise. See *SNR*.

SNF Signal-to-Noise Figure – an average noise figure (an SNR in dB) across a specified band, eg. A or C weighted, or to CCIR – or some other standard.

SNR Signal to Noise Ratio, expressed in dB, usually below the rated power, ie. 0dBr. With the ease of noise analysis in 1/3rd octave bands, it may be taken at a spot frequency.

SOA Safe Operating Area. A graphical representation or map of the safe areas of current-voltage product, usually for the output transistors, and usually at a given temperature. See Chapter 5.

spike An abrupt, extreme increase in some quantity (usually voltage or current) reducing almost as quickly.

SPL Sound Pressure Level.

SR (i) Slew Rate. (ii) Series Resonant.

ss Small signal, usually referring to low current rated transistors

star earth, star ground A nodal point for more than two ground conductors.

steady-state Any test condition where the signal, while alternating, is periodic, i.e. exactly repeats, on and on.

stereo mode The normal condition of a 2 channel amplifier – each channel is independent and available. See bridge, mono, parallel and tandem modes.

stiff A low impedance – such that the voltage supply being referred to does not 'bend' even with heavy loading.

stray capacitance Unsolicited capacitance associated with other components or wiring. Synonym for parasitic capacitance.

stray field An unsolicited electrostatic or magnetic field. Typically a leakage after shielding and other preventative efforts.

subharmonic A harmonic that is an integer *fraction* of the stimulus frequency, eg.

1kHz is a subharmonic of 2kHz. Being unnatural in most acoustic instruments, such harmonics can be profoundly unnerving to the ear.

subjective Human perception unfettered by metallic instruments. Bio-instrumentation. Requires calibration.

sub-sonic See *infrasonic*.

substrate The 'land' that any transistor, IC or other solid-state device is built upon.

superimposition Enforced summation and mutual interference.

surge As with tides, a rise, usually abrupt and unplanned, and usually of voltage, or current, never 'power'.

sweep To monotonically increase (or decrease) the frequency, voltage or some other property of a stimulus, so the quantity ranges from a to b.

swing Maximum output excursion (usually in terms of voltage) of an amplifier. Alias 'voltage swing capability'. Determines the theoretical maximum power into any given impedance.

Sziklai (i) A 'complementary' power Darlington, where an PNP type is made by interconnecting a small pnp and a large npn. (ii) The Japanese inventor.

tails (i) Free short connecting cables or conductors. (ii) The short, desheathed ends of cables.

tandem mode American power amplifier makers' jargon for when two or more amplifier channels are driven equally by one source. If gain knobs are set the same, output levels will be the same also. See *Parallel* mode.

TDS, tds Time Delay Spectrometry.

tempco. / temperature coefficient The rate of change of some factor with temperature, usually expressed in ppm or % per °c.

termination (i) Whatever impedances the source and load see, respectively. (ii) Mainly at RF, a resistance or impedance that is attributed to a load and/or source for optimum performance, particularly in the time domain.

tetrafilar A method of winding output and other critical audio transformers, working in sections, for low leakage inductance and extended hf response.

tetrode Usually a valve with 4 active elements, i.e. with one more grid than a triode. But the beam tetrode has five elements, connected differently to a Pentode.. Both were invented in the mid 1930s in the US for audio power amplification.

THD+N, %THD+N 'Total Harmonic Distortion plus Noise'. The noise must be mentioned because it is significant or even dominant in many cases. Strictly, the THD is also (usually) a percentage (%).

thermalled out When an amplifier overheats and then switches off or down, to protect itself.

thermal runaway When heat dissipation in a transistor amplifier's output stage develops into a self stoking process, causing a junction temperature spiral that rapidly destroys one or more transistors. Endemic with inadequately heatsunk and protected

449

BJTs and can occur with switching MOSFETs. The problem does not afflict lateral MOSFETs nor valves

thermistor A semiconductor-based resistance, that varies with temperature, either in the same direction (positive tempco or ptc); or oppositely (negative or ntc).

thermionic A material which emits electrons (or ions) when heated.

THD Total Harmonic Distortion.

THF %THD Vs Frequency. A minimal, clear abbreviation designed for DOS computer files of %THD+N Vs. frequency measurements, allowing the remaining 5 characters to describe the DUT.

THX Tomlinson Holman eXperiment. A coherent methodology for cinema sound.

THY Total Harmonic distortion, dYnamic plot of. An unambiguous abbreviation designed for DOS type computer files of dynamic %THD+N measurements, allowing the remaining 5 characters to describe the D.U.T.

timbre Characteristic quality of a musical instrument, arising from the instrument's unique combination and pattern, of harmonics.

time out The end of a pre-determined duration or period.

tone burst A test signal used to examine decay spectra in the audio signal path.

tone generator Synonym for an audio signal generator.

tolerance Maximum (peak) variation within a population, usually of any apposite parameter of a manufactured part, e.g. V_{be} of a BJT. Commonly abbreviated to 'tol.'

topography The 3D physical geography & geology of a semiconductor or other monolithic part.

topology A generic amplifier circuit structure or collection of working structures.

torque Turning force or tightness, usually around 80cN•m for medium power semiconductors.

torque driver Used to set tightness of power transistor fixings.

toroid, toroidal Inductor or transformer windings around an annulus or 'donut' core. The first transformer, made by Michael Faraday in England over 160 years ago, was a toroid. Due to lack of a hard magnetic gap, efficiency is potentially higher than with ordinary E+I transformers.

trace The moving dot of light on a scope screen or a PCB copper track.

transconductance A voltage controlling a current; or vice versa.

transducer A device which is designed to convert one form of energy to another, here notably loudspeakers.

transfer function The in/out 'formula' of an amplifier or signal path.

transformerless Amplifiers without output transformers. Transformers were almost *de rigueur* with valves (and remain so) and were also used with transistor output stages until the mid-60's.

transient A sudden, steep, short lived change or burst in the level of voltage, current or some other quantity, in audio and electronics.

transition frequency F_T. A normalised measure of a BJT's hf response.

transistor An active amplifying device with 3 legs. See BJT, J-FET, MOSFET, IGBT.

tri-amp (i) A system using (an) active crossover(s) with three band outputs per channel. Individual power amplifier channels handle bass, mid and treble signals. (ii) A three channel amplifier, for tri-amp'ing.

tri-amplified See *tri-amp, (i)*.

trimcap A variable, *trimmable* capacitor. May be used to optimise HF/RF stability, or CMR, in power amplifiers.

trms true rms. Not a rigged average-responding reading.

triode The most basic valve that can amplify or control audio signals. So called because it has three active electrodes, analogous to the 3 terminals of BJTs and FETs.

triple A 3 piece BJT 'supertransistor'.

tube An '*electron* tube' (USA). Equivalent to a *thermionic valve* (UK). The first active device, and still in use for audio.

twisted pair An audio cable arranged to reject external noise fields particularly RF, and operating maximally in conjunction with balanced sources and destinations. The name describes the construction.

UL Underwriters' laboratories, setting safety standards for electrical components and electronic equipment. (US)

ultra linear A way of connecting tetrode, beam-tetrode and pentode valves to the output transformer, alias the 'partial triode' connection.

unbalanced (i) a single ended circuit, ie. having no symmetrical counterpart. (ii) A notionally balanced stage that is suffering some lopsided condition, preventing the benefits of balanced operation.

undamped When the reaction in a circuit or transducer to an impulsive stimulus, does not die down, or dies down slowly.

unity gain A gain of one – i.e. nil change in level.

unstabilised Generally refers to a DC (or less often AC) supply that is not regulated locally, and may therefore vary with demand and changes in the incoming supply.

VA Volt-Amperes (VxA), a measure of wattage that takes no account of power factor.

Va (i) Anode voltage. (ii) Main supply rail 'A' (in author's logical system).

valve Thermionic valve or vacuum tube, predecessor of the transistor.

Variac Variable AC transformer, originally a tradename of General Radio Co of USA also trading as Zenith in the UK. The transformer is auto-wired and quasi-toroidal. Its winding wires are bared and flatted along the top of the end-up core. The hot secondary is tapped off from a carbon wiper block, moved by a handwheel, or in large or servo-ed models, by motor. Usually they are arranged to sweep the AC mains from 0% to about 117%. Although they may be used for safer testing of equipment, a Variac is *not* an isolating transformer.

VAS Voltage Amplifying Stage

V$_{be}$X V$_{be}$ multiplier circuit. A sub-circuit used for biasing.

VCA Voltage Controlled Amplifier, with which gain can be controlled by electronic means, giving freedom from mechanics.

VDE *Verbund Deutscher Elektrotechniker*, association of German electrical engineers.

Veroboard A generic circuit board used for prototypes and ad-hoc fixes. Called perf. board in the USA. Invented and patented in the UK by Vero Electronics.

VFB, vfb Voltage FeedBack.

V–FET Conventional, vertical or switching MOSFET, named after its original 'V' groove shape. More strictly D-MOS.

V–I limiter Protection circuitry that prevents the Voltage-Current product (VxI) going outside its safe area. See SOA.

VHF Very High Frequency, above 30MHz - today merely quite high a high radio frequency. In audio terms, may be anything far above 20kHz.

Volt Unit of Electro-Motive Force (EMF) or 'pressure'.

V$_{os}$ Voltage, Off Set. See *offset voltage*.

Watt Unit of rate of work done or energy transfer over time, alias 'power'. Equal to One Joule per Second. After James Watt, British pioneer of steam power.

wobbling Where amplifier power rails are arranged to follow the (output) signal, in order to extend the maximum voltage swing without the higher cost or non-availability of parts rated accordingly.

wobbler A circuit in which the rails follow the signal.

wpc *W*atts *p*er *c*hannel. See also /ch. In this book, and as good general practice, 'wpc' implies average watts delivered at full power into the stated resistive load. If operation of other channels affects full output, then the lower output condition is cited. If no load is stated, then 8 ohms is assumed.

X$_{destruct}$ Maximum safe excursion (movement) of a loudspeaker drive-unit's cone. Usually less than 1/4" or 6mm on LF drivers, except in special units. Cone excursion for a given SPL increases abruptly at the driver's low end and becoming insignificant at higher frequencies.

XLR The 3 pin XLR is the universal professional connector for nearly all line level and some speaker connections. Invented by Cannon in US in the 60s. Improved by Neutrik in the 80s & 90s. 4 & 5 pin types are infrequently used for special or bespoke connections.

XOR A crossover network (active or passive).

Xtalk Crosstalk.

zero level The nominal average level of music signals or programme. Usually refers to voltage and typically 0dBu or +4dBu in professional applications. Typically –20 to 0dBu for 'consumer grade' equipment.

Z$_0$ Output impedance.

Zobel A series resistor & capacitor (R+C), one or more are placed across output of many amplifiers, of all kinds, to prevent RF oscillation under varying load conditions. Named after Dr.Zobel, of Bell Labs.

ZVS Zero Voltage Switching – usually of AC power.

0dBr Potentially an arbitrary reference level. But here, always the maximum output of a power amplifier. What 0dBr equals must always be stated clearly.

0v Zero volt connection, reference or 'ground', commonly pronounced '*Oh* Vee' in British English.

1u, 2u, 3u, 4u International standard equipment sizes, indicating different heights, where 1u = 1.75" or 44.45mm, so 2u = 89mm, 3u = 133mm approx., etc. Width is always 19" to ear tips. Depth is unspecified but typically 8" to 14" (200 to 350mm). An old *de facto* world standard based on historic British Post Office Telephones equipment practice.

/ch per channel, same as 'wpc'.

Electronic & physical dimensions, units & abbreviations

As used for specifications, technical discussions and circuit calculations

A,a	(i) Ampere, unit of current.
	(ii) Anode (of valve).
A	Signal gain or loss, in deciBels (dB).
A_v	Voltage gain.
A_{vol}	Open loop voltage gain (Amplification, voltage, open loop).
B, b	Base (of BJT); or
B	Magnetic flux density in Tesla (T) or Gauss.
B	Feedback factor.
β	Beta, H_{FE}
β	Beta, Transistor current gain, I_c/I_b, see H_{FE}, etc.
BW, Bw	Bandwidth.
c	centi, 1/100th.
c or C	Collector (terminal of BJT).
C_{ds}	Drain-source capacitance (of FET).
C_{dg}	Drain-gain capacitance (of FET).
C_{gs}	Gate-source capacitance (of FET).
C_{iss}	Short-circuited input (g-s) capacitance, common source.
C_{obo}	Output capacitance, open circuit, common-base.
C_{obs}	Output capacitance, short circuit, common-base.
C_{oss}	Short circuit output (d-s) capacitance, common source.
C_{rss}	Short-circuit reverse transfer capacitance, i/p to o/p with source shorted to gate, common source.
D,d	Drain (of FET).
dB	$20 \cdot \log_{10}$. +120dB = 1 million (re. voltage or current, not power). −120dB = 1 millionth. deci Bel (one tenth of one Bel, dimensionless unit named after

	Alexander Bell, hence the capital 'B').
E,e	Emitter (of BJT).
f	Frequency in Hertz (Hz).
F	Farad, unit of capacitance, after Michael Faraday.
f_T	*Transition* frequency, where h_{fe} is down −3dB relative to that at low (usually audio) frequencies.
G, g	Grid or gate; or
g1,g2	1st and 2nd grids, etc.; or
G	prefix for Giga, a thousand million ($1e^9$).
g_{fs}	Common source, small signal, forward transfer admittance ('real', resistive part of).
g_m	Mutual conductance − of any active device, in current per unit potential, eg. mA/v.
H	Magnetic field strength in Ampere/metre.
h_{FE}	Beta, raw, DC large signal current gain of BJT.
h_{fe}	Ac, small signal (incremental) current gain of BJT.
h_{oe}	Small signal, open circuit, output admittance, i.e. inverse of Zo, of BJT in common emitter mode.
Hz	Hertz, same as cycles per second (c.p.s.).
Ic	rms value of signal or ac current into collector of BJT.
I_{cbo}	*Base* leakage current in a reverse biased (hard off) BJT, measured with emitter unconnected. Beta times smaller than I_{ceo}
I_{ceo}	The leakage current that matters, E to C, with base open.
I_D	Drain current, dc.
$I_{D (on)}$	On-state drain current, may be maximum.
I_{DSS}	Drain current at zero gate voltage, i.e. leakage.
I_{GSS}	Gate current, ref. source, with drain-source shorted.
Im	'Imaginary' or non-resistive part of an impedance.
J	Joule, unit of energy. 1 watt = 1 Joule per sec.
k	Kilo, a thousand.
k	Cathode. Positive end if a zener or other reverse breakdown device, else (main) negative end of valve or semiconductor two or three terminal device.
K, °K	Degrees Kelvin, absolute temperature, = °c + 273.
m	milli, a thousandth.
m	metre, unit of length.
M, Meg	Mega, a million, or 1e6.
mV, mv	millivolts (units of 0.001v or $1e^{-3}$).
M	Mega, a million
n	nano, a thousandth millionth; or
n	A turns ratio in magnetics; or
n	A (any) number of.
p	pico, a million, millionth
P	Power in Watts.

P, Pri — Primary of a mains or audio transformer.

Q — Inverse of damping, the springiness of an oscillation.

R, r — Resistance, in ohms (W).

R — Repetitive (in device or equipment abuse ratings).

r_a — Anode resistance

rb — Base resistance (of BJT). May be internal parasitic or external in series with.

rc — Collector resistance (of BJT). May be internal parasitic or external in series with.

$r_{ds(on)}$ — Small signal, drain-source on-state resistance.

$r_{DS(on)}$ — Static drain-source on-state resistance.

r_e — Emitter resistance (of BJT). May be internal parasitic or external in series with.

Re — Real, the purely resistive portion of any impedance.

Rg — Generator resistance

R_{j-a} — Thermal resistance of semiconductor, junction (core) to ambient.

R_{j-mb} — Thermal resistance of semiconductor, junction (core) to mounting base (of part).

$R_{mb-sink}$ — Thermal resistance of semiconductor, mounting base (of part) to heatsink, may include resistance of the electrical insulating interface.

R_L — Load Resistance (or impedance).

R_{on} — On resistance, resistance of a MOSFET when turned fully on.

R_s, r_s — Source impedance

s (S) — Seconds (time or arc) or

S — Siemens (unit of conductivity, determinable by context) or

S, s — Source (of MOS-FET) or

S, s — Screen (grid of valve) or

S, s — Screen connection on a device.

T,t — Time (usually in seconds) or

T — Temperature. or

T — Tesla, unit of magnetic field strength.

T_A, T_{amb} — Ambient temperature.

T_j — Junction or (FET) channel temperature.

t_{off} — Turn off time.

t_{on} — Turn on time.

t_{stg} — Storage temperature (i.e. unpowered limits).

t_r — Rise time

t_{rr} — Reverse recovery time.

Tx — Transformer.

V, v — Electromotive Force or Potential Difference, in Volts.

V_{be} — Base-Emitter voltage of BJT, incremental small signal.

$V_{BE(sat)}$ — Base-emitter DC saturation voltage, of BJT, under specific conditions.

V_{ce} — Collector to emitter voltage of BJT, small signal, with base reverse biased.

V_{CEO}	Collector-emitter voltage of BJT, with base open.
$V_{CEO(sus)}$	Maximum c-e voltage with base open. Applies at relatively high values of collector current.
$V_{CER(sus)}$	Maximum c-e voltage with base connected to emitter via a cited, quite low resistance. Applies at relatively high values of collector current.
V_{DS}	Drain-source voltage, DC.
V_{GS}	Gate-Source voltage, DC.
$V_{GS(th)}$	Gate-source threshold, usually where a V-MOSFET's drain current is 250mA or some similarly low level.
V_{in}	Input voltage.
V_{out}, V_o	Output voltage.
$V/\mu S$	Volts per microsecond (rate of change of voltage).
W	Watt, unit of power.
Y_{fs}	Forward trans-admittance, of any FET.
Z	Impedance.
Z_{in}	Input impedance.
Z_o	Output impedance.
1e2	One hundred.
2e2	Two hundred.
1e3	One thousand.
1e6	One million.
1e-3	One thousandth.
1e-6	One millionth.
1e-12	Millionth of a millionth.
Δ	Pronounced 'Cap delta', incremental change.
μ	Micro, a millionth or $1e^{-6}$, or
μ	voltage amplification factor ('A' for valves), or
μ	Permeability (for magnetics).
μ_o	Permeability of free space.
ω	Omega, angular frequency – $2\pi f$, originally called 'pulsatance'.
Ω	'Cap Omega', commonly Ohm, unit of resistance.
$\circ$	Degree(s).
+	Positive, add, plus.
	Negative, subtract, minus.
$\cong$	Approximately equal to.
<	Less than.
>	Greater than.
$\Rightarrow$	Into.
$\leq$	Less than, or equal to.
$\geq$	Greater than, or equal to.
$-\infty$	Minus infinity, a nominal term used to indicate reduction of a signal to inaudibility.

Index

A

Abuse damage 377
AC mains
  3-phase 242
  Connectors 350
  Fuses and breakers 253–255, 349
  Impedance 347
  Interaction via 355
  Operable range testing 317
  Ring 352
  Spur 353
  Voltage drop 346
  Wiring recommendations 352–354
AC mains voltages
  Worldwide categories 342
Acoustic noise 328
Active cabinets 20
Active probing 394
Active systems
  Benefits 30
  Bi-wiring 31
Adverse loads
  Protection 200–210
  Table of 325
AES power rating 44
Air pollution 373
Airflow conventions 372
Amplifier
  Fitment 371
  Test, warm-up 267
  Wattage needed 333
Amplifier test preconditioning 266
Audio 2

B

Baker clamp diode 181
Balance of gains 275
Balanced input 59–63, 356-360
Bandwidth 272, 273
Bass qualities 15
Batteries 86, 235

Beta 89, 90, 172
Biasing 88, 91, 98–99, 121, 127, 134–136
  After repair 387, 388
Bipolar Junction Transistors: see *BJTs*
BJTs 171
  C_{ob} 190
  Compound 100
  Leakage 175
  Load Limits 194
  Selection 176
  Sharing 175
  SOA 173
  Voltage ratings 172
Breakthrough, signal 283
Bridge
  Full 115
  Modes 114
  Rectifier 234, 386
Bridged-bridge 117
Bridging 116, 336

C

Cascaded complementary pairs 93
Cascades 112, 113
Cascading 112, 113
Cascoding 103, 104, 112, 113
Channel separation 283
Class A 118
  Bias regulation 122
  Efficiency 119
Class A-B 127
  Efficiency 131
Class A/H 142
Class B 127
Class comparison 153
Class D 147
Class examples 154
Class G 138

Index

Printed and bound by CPI Group (UK) Ltd, Croydon, CR0 4YY

03/10/2024

01040433-0018